Lecture Notes in Computer Science 15482

Founding Editors

Gerhard Goos
Juris Hartmanis

The series Lecture Notes in Computer Science (LNCS), including its subseries Lecture Notes in Artificial Intelligence (LNAI) and Lecture Notes in Bioinformatics (LNBI), has established itself as a medium for the publication of new developments in computer science and information technology research, teaching, and education.

LNCS enjoys close cooperation with the computer science R & D community, the series counts many renowned academics among its volume editors and paper authors, and collaborates with prestigious societies. Its mission is to serve this international community by providing an invaluable service, mainly focused on the publication of conference and workshop proceedings and postproceedings. LNCS commenced publication in 1973.

Minsu Cho · Ivan Laptev · Du Tran · Angela Yao ·
Hong-Bin Zha
Editors

Computer Vision – ACCV 2024 Workshops

17th Asian Conference on Computer Vision
Hanoi, Vietnam, December 8–12, 2024
Revised Selected Papers, Part I

 Springer

Editors
Minsu Cho (ID)
Pohang University of Science
and Technology (POSTECH)
Pohang-si, Korea (Republic of)

Ivan Laptev
Mohamed bin Zayed University of Artificial
Intelligence
Abu Dhabi, United Arab Emirates

Du Tran
Google
Mountain View, CA, USA

Angela Yao (ID)
National University of Singapore
Singapore, Singapore

Hong-Bin Zha
Peking University
Beijing, China

ISSN 0302-9743   ISSN 1611-3349 (electronic)
Lecture Notes in Computer Science
ISBN 978-981-96-2640-3   ISBN 978-981-96-2641-0 (eBook)
https://doi.org/10.1007/978-981-96-2641-0

Foreword

Welcome to the 17th Asian Conference on Computer Vision (ACCV) 2024. This is the first time the conference is being hosted in Vietnam. The idea of bringing this prestigious event to Vietnam began in 2021, when two general and program chairs envisioned promoting the growth of computer vision in this region. By hosting the conference in a developing country, we aim to encourage other nations with less representation in the field to participate in this rapidly evolving discipline that is transforming lives globally. With Vietnam's affordable food and accommodation, we also hope that more participants will be able to attend and benefit from the conference.

ACCV 2024 is the first fully in-person ACCV event since the onset of the COVID-19 pandemic, with the last in-person edition held in 2018. This six-year gap meant that we couldn't rely on data from the past two conferences to accurately forecast submission and attendance numbers, making planning, budgeting, and logistics more challenging. As a result, we had to revise our estimates, plans, and budgets multiple times. We deeply appreciate the support of our Professional Conference Organizer, Nicole Finn, who enthusiastically (and courageously) agreed to help organize ACCV in a new region, outside the traditional North American and European venues she typically works with. Nicole worked tirelessly to navigate these changes, coordinating with hotels and convention centers to secure a venue that met both our needs and budget. We are also deeply grateful to our general chair, Richard Hartley, who took on the additional role of finance chair. Richard worked closely with Nicole, meticulously overseeing budget projections to maintain low registration fees while ensuring the financial sustainability of the conference. His dedication and sense of responsibility to the community, given his level of seniority, is a true testament to his commitment.

While ACCV 2024 aims for greater diversity and inclusion, our primary goal remains the exchange of scientific ideas while maintaining integrity and quality. This commitment is reflected in the program curation, led by our Program Chairs: Minsu Cho, Ivan Laptev, Du Tran, Angela Yao, and Hong-Bin Zha. They oversaw a rigorous process involving a thousand submissions and reviewers, ensuring that each paper received thorough and fair evaluation. Although smaller than conferences like CVPR, ACCV posed its own logistical challenges, including a tight timeline and budget constraints. Despite these difficulties, the Program Chairs were determined to hold an in-person Area Chair meeting to ensure high-quality decisions, even though this required additional effort and coordination—a clear reflection of their dedication. The area chairs who participated also valued the opportunity to connect and receive mentorship from more senior area chairs and the Program Chairs. As in-person meetings have become impractical for larger conferences, we hope that smaller conferences like ACCV can continue to offer this valuable in-person training for future area chairs.

This year's main conference is complemented by eleven workshops and two tutorials. We extend our thanks to the Workshop and Tutorial Chairs, Vineeth N. Balasubramanian, Li Liu, Anh Tran, and Kota Yamaguchi, who organized this crucial part of the program and coordinated with the ACML conference for a shared workshop day. This year, registered attendees for either ACCV or ACML can participate in workshops and tutorials across both conferences, fostering greater collaboration.

Our Publication Chairs, Yu-Lun Liu and Phong Nguyen, deserve special recognition for their outstanding work in assembling the camera-ready papers for the Springer volumes, ensuring their timely distribution. They courageously took on this challenging role without prior experience, fully aware of the limited time they would have to meet Springer's requirements once they received the camera-ready papers and final program. Their dedication and effort were crucial to the success of this process.

We are also grateful to our Local Chairs, Thi-Lan Le, Cuong Pham, and Minh-Triet Tran, who handled various local logistics, from working with vendors to organizing the area chair meeting and liaising with government officials for endorsements and approvals. VinAI, as a registered business entity in Vietnam, also played a crucial role in ensuring compliance with local regulations, coordinating with authorities, and handling endorsements. We thank the Vietnam's Ministry of Information and Communication for their support of this scientific event.

We are thankful to our Diversity and Inclusion Chairs, Supasorn Suwajanakorn and Miaomiao Liu, for their efforts in ensuring a diverse and inclusive selection process, even amidst budget uncertainties. Their work allowed us to support the participation of many who might not otherwise have attended. Financial support was made possible through sponsorships from several organizations: Asian Federation of Computer Vision, Sapien, Google, Springer, and the Australian Institute for Machine Learning. We also acknowledge the help from the Microsoft Research team, who provided the Conference Management Toolkit and technical support for ACCV 2024.

We hope you enjoy ACCV 2024 and take full advantage of the rich scientific program and networking opportunities this conference offers.

December 2024

Richard Hartley
C. V. Jawahar
Minh Hoai Nguyen
Dimitris Samaras
ACCV 2024 General Chairs

Preface

The 17th Asian Conference on Computer Vision (ACCV) 2024 was held in-person in Hanoi, Vietnam during December 8–12, 2024. The conference featured novel research contributions from almost all sub-areas of computer vision.

The ACCV 2024 main conference received 839 valid submissions (after desk rejections), which then entered the review stage. Fifty Area Chairs (ACs) and 1,020 reviewers made great efforts to ensure that every submission received thorough and high-quality reviews. Each submission received at least 3 reviews and was discussed by a panel of three expert ACs. Following previous editions of ACCV, this conference adopted a double-blind review process in which the identities of authors were not visible to the reviewers or area chairs, and the identities of the assigned reviewers and area chairs were not visible to the authors. The program chairs did not submit papers to the conference.

After receiving initial reviews, the authors had the option of submitting a rebuttal. In the post-rebuttal period reviewers reconsidered all paper materials including rebuttals and other reviews and provided their final recommendations. Area chairs were grouped into 17 conflict-free AC triplets. The area chairs led the discussions and made final recommendations based on reviews, author rebuttals, and discussions between reviewers. With the confirmation of three area chairs for each paper, 269 papers were accepted, making the acceptance rate of 32%. Among 269 accepted papers, 47 papers were selected for oral presentations.

ACCV 2024 also included eleven workshops and two tutorials. The ACCV 2024 workshops and tutorials were held on two days of December 8 and 9, 2024, with the first day (December 8th) being shared with ACML 2024 workshops. The proceedings of ACCV 2024 are open access at the Computer Vision Foundation website, by courtesy of Springer. The quality of the papers presented at ACCV 2024 demonstrates the research excellence of the international computer vision communities.

This year, ACCV 2024 also featured three keynote speeches given by Professor Michal Irani (Weizmann Institute of Science, Israel), Dr. Gérard Medioni (Amazon, USA), and Professor Deepak Pathak (Carnegie Mellon University, USA).

We would like to thank all the organizers, keynote speakers, area chairs, reviewers, and authors who made great contributions to ensure a successful ACCV 2024. Last but not least, we would also like to thank the attendees of ACCV 2024. Their presence showed strong commitment and appreciation towards this conference.

December 2024

Minsu Cho
Ivan Laptev
Du Tran
Angela Yao
Hong-Bin Zha
ACCV 2024 Program Chairs

Organization

General Chairs

Richard Hartley	Australian National University, Australia
C. V. Jawahar	IIIT Hyderabad, India
Minh Hoai Nguyen	VinAI and The University of Adelaide, Australia
Dimitris Samaras	Stony Brook University, USA

Program Chairs

Minsu Cho	Pohang University of Science and Technology, South Korea
Ivan Laptev	Mohamed bin Zayed University of Artificial Intelligence, United Arab Emirates
Du Tran	Google, USA
Angela Yao	National University of Singapore, Singapore
Hongbin Zha	Peking University, China

Workshop Chairs

Anh Tran	VinAI, Vietnam
Vineeth N. Balasubramanian	IIIT Hyderabad, India

Tutorial Chairs

Kota Yamaguchi	CyberAgent, Japan
Liu Liu	National University of Defense Technology, China

Publication Chairs

Yu-Lun Liu	National Yang Ming Chiao Tung University, Taiwan (R.O.C.)
Phong Nguyen	VinAI, Vietnam

Social Media Chair

Victor Escorcia   Samsung AI Center Cambridge, UK

Diversity and Inclusion Chairs

Miaomiao Liu   Australian National University, Australia
Supasorn Suwajanakorn   VISTEC

Technical Chairs

Seungwook Kim   Pohang University of Science and Technology,
                                         South Korea
Dayoung Gong   Pohang University of Science and Technology,
                                         South Korea

Local Chairs

Cuong Pham   PTIT and VinAI Research, Vietnam
Thi-Lan Le   Hanoi University of Science and Technology,
                                         Vietnam
Minh-Triet Tran   Ho Chi Minh City University of Science, Vietnam

Area Chairs

Abhinav Shrivastava   University of Maryland, USA
Anh Tran   VinAI, Vietnam
Asako Kanezaki   Tokyo Institute of Technology, Japan
Bohyung Han   Seoul National University, South Korea
Boxin Shi   Peking University, China
Chen Chen   University of Central Florida, USA
Fahad Shahbaz Khan   MBZUAI, United Arab Emirates
Gianni Franchi   ENSTA Paris, France
Go Irie   Tokyo University of Science, Japan
Guofeng Zhang   Zhejiang University, China

Hazel Doughty	Leiden University, Netherlands
Heng Wang	TikTok, USA
Hieu Pham	VinUniversity (VinUni), Vietnam
Hyung Jin Chang	University of Birmingham, UK
Jian Sun	Xi'an Jiaotong University, China
Jiwen Lu	Tsinghua University, China
Junsong Yuan	State University of New York at Buffalo, USA
Karteek Alahari	Inria, France
Khoi Nguyen	VinAI Research, Vietnam
Leonid Sigal	University of British Columbia, Canada
Makarand Tapaswi	IIIT Hyderabad, Wadhwani AI, India
Manmohan Chandraker	UC San Diego, USA
Ming-Hsuan Yang	University of California at Merced, USA
Ngan Le	University of Arkansas, USA
Rei Kawakami	Tokyo Institute of Technology, Japan
Saeed Anwar	Australian National University, Australia
Salman Khan	MBZUAI, United Arab Emirates
Seon Joo Kim	Yonsei University, South Korea
Shiguang Shan	Institute of Computing Technology, Chinese Academy of Sciences, China
Shijian Lu	Nanyang Technological University, Singapore
Shizhe Chen	Inria, France
Simon Lucey	University of Adelaide, Australia
Suha Kwak	POSTECH, South Korea
Tae-Hyun Oh	POSTECH, South Korea
Tae-Kyun Kim	KAIST/Imperial College London, South Korea
Tat-Jun Chin	The University of Adelaide, Australia
Triet Tran	Ho Chi Minh University of Science, VNU, Vietnam
Venkatesh Babu Radhakrishnan	Indian Institute of Science, India
Vicky Kalogeiton	Ecole Polytechnique, IP Paris, France
Vincent Lepetit	Ecole des Ponts ParisTech, France
Wen-Sheng Chu	Google, USA
Xiang Bai	Huazhong University of Science and Technology, China
Xiaodan Liang	MBZUAI, United Arab Emirates
Xiaojun Chang	Mohamed bin Zayed University of Artificial Intelligence, United Arab Emirates
Yasuyuki Matsushita	Osaka University, Japan
Yoichi Sato	University of Tokyo, Japan
Young Min Kim	Seoul National University, South Korea
Yusuke Matsui	The University of Tokyo, Japan

Zhanyu Ma   Beijing University of Posts and
                                   Telecommunications, China
Zuzana Kukelova   Center for Machine Perception, CTU in Prague,
                                   Czech Republic

Reviewers

Abbas Anwar   Aniket Roy
Abdallah Dib   Anil Batra
Abdul Jabbar Siddiqui   Anoop Cherian
Abhinav Kumar   Anshul Shah
Abulikemu Abuduweili   Anthony Hu
Aditya Arun   Anyi Rao
Aditya Sahdev   Ao Luo
Aditya Singh   Aoran Xiao
Adriano Fragomeni   Arjan Kuijper
Adrien Lafage   Asm Iftekhar
Ahmed Abdelkader   Atsushi Hashimoto
Ahyun Seo   Avideep Mukherjee
Aishik Konwer   Avinash Sharma
Akash Awasthi   B. V. K. Vijaya Kumar
Akin Caliskan   Bang Liu
Akisato Kimura   Baoteng Li
Akos Godo   Benjamin Killeen
Akshay Kulkarni   Berthy Feng
Akshita Gupta   Bharadwaj Ravichandran
Alberto Marchisio   Bin Chen
Alex Jinpeng Wang   Bin Chen
Alexander Binder   Bingfeng Zhang
Ali Athar   Binod Bhattarai
Ali Zia   Bo Li
Alper Yilmaz   Bo Liu
Ameya Joshi   Bo Miao
Amir Atapour-Abarghouei   Bo Sun
Amitangshu Mukherjee   Bo Wang
Anders Dahl   Bo Yang
Andong Tan   Bo Zhang
Andrew Gilbert   Boeun Kim
Andrew Beng Jin Teoh   Bogdan Raducanu
Aneeshan Sain   Bohong Chen
Anh Tran   Bojian Wu
Anh Vu Nguyen   Boseung Jeong
Anh-Dzung Doan   Boshen Xu
Anh-Quan Cao   Bowen Cai

Bowen Wen
Boyu Yang
Brian Clipp
Bruno Korbar
Buzhen Huang
Byung-Kwan Lee
Cameron Gordon
Carlos Rodriguez-Pardo
C. Kumar Mummadi
C. Hewa Koneputugodage
Chandra Kambhamettu
Chang Liu
Changmin Lee
Changtao Miao
Chanyong Jung
Chao Qu
Chao Wen
Chaojian Li
Chaoqin Huang
Chaowei Fang
Chau Pham
Che Sun
Chee Kheng Chng
Chen Feng
Cheng Chen
Cheng Long
Cheng Luo
Cheng Perng Phoo
Chengxin Liu
Chengxu Liu
Cheng-Yen Yang
Chengyuan Zhuang
Chengzhi Mao
Chengzhou Tang
Chenhongyi Yang
Chenwei Tang
Chien-Yao Wang
Chi-Han Peng
Ching Lam Choi
Chongjian Ge
Chongruo Wu
Cho-Ying Wu
Christos Kyrkou
Chuanguang Yang
Chuhua Xian

Chul Lee
Chun-Hsiao Yeh
Chunlei Peng
Chunxia Xiao
Chuong Huynh
Cong Wu
Congli Wang
Congpei Qiu
Daan de Geus
Daekyu Kwon
Da-Han Wang
Dahyun Kang
Daisuke Miyazaki
Dan Zeng
Danfeng Hong
Danpeng Chen
Darshan Singh S.
Dasong Li
David Chan
David Hart
Da-Wei Zhou
Decheng Liu
Deepika Bablani
Denis Baručić
Denys Rozumnyi
Deyi Ji
Di Yuan
Diego Thomas
Difei Gao
Dimitrios Sakkos
Ding-Jie Chen
Dingyi Yang
Dingyuan Zhang
Divy Kala
Divya Choudhary
Dong Wang
Dongdong Wang
Donggon Jang
Donghao Zhou
Donghwan Kim
Donghyeon Kwon
Dong-Jin Kim
Dongkai Wang
Dongkeun Kim
Dongliang Cao

Dongliang Chang
Dongqing Zou
Dongyoung Kim
Dongze Lian
Driton Salihu
Duc Vu
Duc Anh Nguyen
Duc Minh Vo
Duc-Tien Dang-Nguyen
Duolikun Danier
Duy Le
Duy Minh Ho Nguyen
Eldar Insafutdinov
En Yu
Erickson Nascimento
Eshika Khandelwal
Ethan Elms
Ethan Tseng
Fabio Pizzati
Fahad Shamshad
Faisal Qureshi
Fan Lu
Fan Yang
Fan Yang
Fangjinhua Wang
Fangyi Chen
Federico Stella
Fei Pan
Fei Xie
Fei Xue
Feiran Li
Fenggen Yu
Fengting Yang
Filippo Maggioli
Fiora Pirri
Florian Kleber
Fu-En Yang
Fu-Jen Chu
Fu-Jen Tsai
Fumihiko Sakaue
Fumio Okura
Furkan Kınlı
Gaku Nakano
Gangming Zhao
Gangwei Xu

Gianfranco Doretto
Gihyun Kwon
Giulio Rossolini
Gonçalo Dias Pais
Gongjie Zhang
Gu Wang
Guanglei Yang
Guangming Lu
Guangyu Sun
Guangzhi Wang
Guile Wu
Guo Chen
Guoqing Wang
Guorong Li
Gustavo Perez
Gwangtak Bae
Gyeongsik Moon
Gyuseong Lee
Hai D. Pham
Hai X. Pham
Haiming Xu
Haithem Turki
Haiyang Jiang
Han Hu
Han Qiu
Hanbyel Cho
Hangil Park
Hanjiang Hu
Hanjung Kim
Hanqing Sun
Hanxiao Jiang
Hao Wang
Hao Wang
Hao Zhao
Haobo Yuan
Haokun Lin
Haotong Lin
Haowei Tai
Haoyue Bai
Harry Cheng
Hashmat Shadab Malik
Heeseung Yun
Helder Araujo
Helena Maia
Heng Guo

Heng Li
Hengli Wang
Heran Yang
Heydi Mendez-Vazquez
Hideaki Uchiyama
Hieu Le
Himangi Mittal
Hiroaki Santo
Hirokatsu Kataoka
Hoàng-Ân Lê
Hoang-Quan Nguyen
Hojun Jang
Hong Wang
Hongchen Luo
Hong-Han Shuai
Hongje Seong
Hongji Guo
Hongjia Zhai
Hongyi Fan
Hsin-Ping Huang
Hsuan-I Ho
Hsu-Kuang Chiu
Hu Wang
Huafeng Wang
Huaiwen Zhang
Huan Wang
Huan Zheng
Huan Zheng
Huangying Zhan
Huanrui Yang
Huayi Zhou
Huei-Fang Yang
Hui Lin
Huijing Zhan
Huimin Ma
Huiming Sun
Huiyuan Yang
Hung Nguyen
Hung Tran
Hyeokjun Kweon
Hyung-gun Chi
Iago Suárez
Inho Kim
Inhwan Bae
Iuliia Kotseruba

Jaegul Choo
Jaeho Lee
Jaeseong Lee
Jaewon Lee
Jaime Cardoso
Janghoon Choi
Jean-Philippe Tarel
Jeonghun Baek
Jhih-Ciang Wu
Ji Hou
Ji Liu
Jia Wan
Jiachen Li
Jiachen Sun
Jiacheng Li
Jiafan Zhuang
Jiahao Nie
Jiahong Ouyang
Jiajun Tang
Jiali Duan
Jiaman Li
Jiaming Zhang
Jiande Sun
Jiang Liu
Jiange Yang
Jianglong Ye
Jiangpeng He
Jianing Li
Jianing Xi
Jianjia Wang
Jianqiao Zheng
Jianqiu Chen
Jianyang Gu
Jianzhong He
Jiaqi Li
Jiashuo Yu
Jia-Wang Bian
Jiawei He
Jiayi Ji
Jiaze Wang
Jiazhen Wang
Jichang Li
Jicheol Park
Jie Guo
Jie Hong

Jie Min
Jie Tang
Jie Yang
Jie Zhang
Jie Zhao
Jierui Lin
Jierun Chen
Jieyu Li
Jihyun Lee
Jin Fang
Jingchun Cheng
Jinghao Shi
Jinghua Hou
Jinghuan Shang
Jingjing Deng
Jingjing Xiong
Jingyi Zhang
Jinho Jeong
Jinhong Deng
Jinjian Wu
Jinsu Yoo
Jinsung Lee
Jinwoo Kim
Jinyoung Choi
Jinyu Cai
Jiyuan Liu
Jonathan Donnelly
Jongbin Ryu
Jongmin Lee
Jongwon Choi
JoonKyu Park
Joya Chen
Juan C. Sanmiguel
Jue Wang
Julian Tanke
Julie Mordacq
Junbin Xiao
Junbo Zhang
Junfei Yi
Jungchan Cho
Junha Lee
Junhan Chen
Junhao Dong
Junhyeong Cho
Junhyug Noh

Junhyun Lee
Junke Wang
Junlin Hu
Junsong Fan
Junwei Liang
Kai Chen
Kai Katsumata
Kai Wang
Kai Zhu
Kaihong Wang
Kailun Yang
Kamal Nasrollahi
K. Vaishnavi Gandikota
Kashu Yamazaki
Kavisha Vidanapathirana
Kazuhiro Hotta
Ke Xu
Keita Takahashi
Keke Tang
Keyan Wang
Khoi Pham
Kibok Lee
Kim Jun-Seong
Konstantinos Alexandridis
Konstantinos M. Dafnis
Koutilya Pnvr
Krishna Kanth Nakka
Kuan-Chih Huang
Kumar Ashutosh
Kun Fang
Kun Li
Kun Su
Kun Xia
Kun Xiang
Kun Zhou
Kunming Luo
Kwang Moo Yi
Kyong Hwan Jin
Kyoungkook Kang
Lang Nie
Latha Pemula
Lei Jin
Lei Tan
Lei Wang
Lei Zhu

L. Sampaio Ferraz Ribeiro
Leonardo Iurada
Leonardo Nunes
Li Ding
Li Niu
Li Song
Liang An
Liang Chen
Liang-Jian Deng
Liangke Gui
Lidong Yu
Ligong Han
Lile Cai
Lin Liu
Lin Geng Foo
Lingdong Kong
Linghao Chen
Linlin Shen
Liwei Yang
Li-Wei Kang
Lixiang Ru
Long Ma
Long Pham
Longfei Han
Longkun Zou
Lujun Li
Lumin Xu
Luwei Yang
Lv Tang
M. Yashwanth
Mahmoud Afifi
Manogna Sreenivas
Maoxun Yuan
Marc A. Kastner
Marco Piccirilli
M.-Luliana Georgescu
Marios Loizou
Martin Eisemann
Martin Mundt
Martin Weinmann
Matej Grcić
Matteo Dunnhofer
Matthew Beveridge
Matthew Gwilliam
Max Ehrlich

Maxwell Collins
Meghshyam Prasad
Mei Wang
Meirui Jiang
Meng Liu
Miaohui Wang
Miaomiao Liu
Michael Greenspan
Michael Wray
Mikhail Kennerley
Min Je Kim
Minesh Mathew
Mingde Yao
Mingfei Han
Ming-Feng Li
Mingfu Liang
Mingfu Xue
Minggui Teng
Minghui Hu
MingKun Yang
Ming-Kun Xie
Mingon Kang
Mingyuan Liu
Minh Luu
Minh Tran
Minkwan Kim
Mizuki Kojima
Mohammadreza Babaee
Mohammed Mahmoud
Mohit Goyal
Momin Abbas
Moon Ye-Bin
Mosam Dabhi
Mouin Ben Ammar
Mouxing Yang
Muhammad Maaz
Mutian Xu
Myungsub Choi
Nacim Belkhir
Nagabhushan Somraj
Nakamasa Inoue
Nakul Agarwal
Namyup Kim
Nanqing Dong
Nanyang Ye

Nayeong Kim
Necati Cihan Camgoz
Nhat Chung
Ni Zhang
Nico Messikommer
Nicola Garau
Niki Foteinopoulou
Nikita Durasov
Nikolaos Gkanatsios
Nikolaos Zioulis
Ningli Xu
Nirat Saini
Noranart Vesdapunt
Oh Hyun-Bin
Olivier Laurent
Omkar Thawakar
P. J. Narayanan
Pablo Garrido
P. Shivakumara
Paola Cascante-Bonilla
Parikshit Sakurikar
Paritosh Parmar
P. Kumar Anasosalu Vasu
Pedro Castro
Peiqi Duan
Peng Dai
Peng Wu
Peng-Hao Hsu
Pengliang Ji
Pengpeng Li
Pengpeng Zeng
Peter Kulits
Petra Bevandić
Pha Nguyen
Pingping Zhang
Pratik Vaishnavi
Pravin Nagar
Priyam Dey
Pulkit Kumar
Qi Bi
Qi Yu
Qiang Nie
Qiangmin Chen
Qiaole Dong
Qichen Fu

Qihao Liu
Qing Yu
Qing Zhang
Qingji Guan
Qingsong Zhao
Qingtian Zhu
Quan Dao
Quan Tang
Quang Nguyen
Quang Nguyen
Rahul Sajnani
Rakib Hyder
Ran Xu
Ratnesh Kumar
Rémi Kazmierczak
Renato Martins
Renjiao Yi
Renjie Wan
Reyer Zwiggelaar
Ricardo Garcia Pinel
Ridouane Ghermi
Robert Sablatnig
Robin Courant
Rohit Bharadwaj
Rohit Jena
Rohit Keshari
Rohit Kundu
Ronglai Zuo
Ronny Haensch
Rui Li
Rui Qian
Rui Xu
Rui Zhao
Rui Zhu
Rui Zhu
Ruikang Xu
Ruili Feng
Ruilong Li
Ruiqi Zhao
Ruixuan Yu
Ruizhi Shao
Runpei Dong
Runtian Zhai
Runze Li
Ruoshi Liu

Ruoteng Li
Ryo Furukawa
Ryo Hachiuma
Ryo Kawahara
Ryosuke Furuta
Ryota Yoshihashi
Ryuhei Hamaguchi
Salil Tambe
Salma Abdel Magid
Sampath Chanda
Sandesh Kamath
Sang Min Kim
Saptarshi Sinha
Sara Elkerdawy
Sara Pieri
Sateesh Kumar
Scott McCloskey
Sehyun Hwang
Sejong Yang
Seogkyu Jeon
Seonguk Seo
Seung Hyun Lee
Seungheon Kim
Seungho Lee
Seungjun Nah
Seungryul Baek
Seungwook Kim
Shady Abu-Hussein
Shailaja Keyur Sampat
Shaobing Gao
Shaolin Su
Shaoxiong Zhang
Shao-Yuan Lo
Shengeng Tang
Shengsheng Qian
Shenqi Lai
Sherry Chen
Shikun Li
Shin-Fang Chng
Shin'ichi Satoh
Shintaro Yamamoto
Shiqi Tian
Shiqiang Ma
Shivanand Venkanna Sheshappanavar
Shiyu Li

Shiyu Zhao
Shiyue Zhang
Shuaicheng Liu
Shuangrui Ding
Shuhong Zheng
Shuo Cheng
Shuvendu Roy
Shuwei Huo
Shuxiao Ding
Shuzhe Wu
Sicheng Zhao
Sihui Luo
Simon Reiß
Simon Woo
Sinisa Segvic
Sixun Dong
Siyuan Yang
Sonia Raychaudhuri
Soumya Banerjee
Srijan Das
Srinivas Rana
Sua Choi
Sucheng Ren
Sukrit Shankar
Sunny Bhati
S. Narasimhaswamy
Syed Talal Wasim
Sze Jue Yang
Taeyun Woo
Taihong Xiao
Taiki Sekii
Taisong Jin
Takahiro Kushida
Takahiro Okabe
Takashi Shibata
Takayuki Okatani
Takeshi Saitoh
Takuma Yagi
Takumi Karasawa
Tam Nguyen
Tao Hu
Tao Sun
Tao Wang
Tao Wu
Tao Wu

Tao Yu
Tao Zhang
Tao Zhou
Taoyue Wang
Tarun Kalluri
Tat-Jun Chin
Tengfei Liu
Thanh Le
Thao Nguyen
Thinh Phan
Thi-Thu-Huong Le
Thomas Leimkuehler
Thomas Westfechtel
Tiange Xiang
Tianpei Gu
Tianrui Chai
Tianyun Zhang
Tian-Zhu Xiang
Tiesong Zhao
Tingting Xie
Toby Breckon
Tomas F Yago Vicente
Tongda Xu
Tongyu Yang
Tooba Imtiaz
Toshihiko Yamasaki
Trong Thang Pham
Trong-Thuan Nguyen
Trung Pham
Trung-Nghia Le
Truong Vu
Tsai-Shien Chen
Tu Van Ninh
Tuan-Anh Vu
Tu-Khiem Le
Tung Do
Tung Kieu
Uy Tran
Van Nguyen Nguyen
Vasileios Mezaris
Vatsal Agarwal
Viet Nguyen
Vincent Cartillier
Vincent Gaudilliere
Vincent Tao Hu

Vinh-Tiep Nguyen
Viraj Shah
Vu Truong
Wataru Shimoda
Wei Liao
Wei Mao
Wei Wan
Weicheng Zhu
Weidong Cai
Weifeng Liu
Weihao Li
Weihao Xia
Weijie Lyu
Weijun Mai
Weilian Song
Weiwei Cai
Weiwei Xu
Weixiao Liu
Weixuan Tang
Wei-Yi Chang
Wenbin Li
Wenhan Yang
Wenhao Wang
Wenhui Zhou
Wenjia Wang
Wenke Huang
Wenqian Wang
Williem Williem
Wolfgang Fuhl
Woo Jae Kim
Woobin Im
Wooseok Lee
Xi Cheng
Xi Wang
Xi Yu
Xiang Chen
Xiang Gu
Xiang Wen
Xianghui Xie
Xiangwei Kong
Xiangyang Li
Xiangyu Xu
Xiankai Lu
Xiao Zhang
Xiaoguang Li

Xiaohan Chen
Xiaohua Huang
Xiaole Tang
Xiaopeng Ji
Xiaosong Jia
Xiaotao Hu
Xiaotong Luo
Xiaoyun Yuan
Xiaoyun Zhang
Xibin Song
Xijun Wang
Xin Feng
Xin Li
Xin Li
Xin Liao
Xin Wei
Xin Yang
Xin Yuan
Xin Zhou
Xindi Wu
Xingjiao Wu
xingkui Zhu
Xingtong Liu
Xingyu Liu
Xinhang Liu
Xinxin Zhu
Xiu Su
Xiujun Li
Xiujun Shu
Xiuwei Chen
Xu Cao
Xu Cao
Xu Yao
Xu Zhao
Xuan Ju
Xudong Liu
Xuepeng Shi
Xueqian Li
Xueting Liu
Xueying Jiang
Xugong Qin
Xuhua Huang
Xuqian Ren
Xuxin Lin
Yajie Wang

Yajing Zheng
Yan Wang
Yan Xia
Yan Yang
Yanan Li
Yanan Zhang
Yang Yang
Yangguang Zhu
Yanjing Li
Yannan Pu
Yanqing Shen
Yanru Xiao
Yansong Tang
Yao-Chih Lee
Yaojie Liu
Yaosi Hu
Yaqing Ding
Yash Bhalgat
Yash Mukund Kant
Yasunori Ishii
Yawen Zeng
Yazeed Alharbi
Ye Du
Ye Liu
Yeying Jin
Yi Chang
Yi Zhang
Yichang Shih
Yichao Cao
Yicong Li
Yifan Xing
Yifei Huang
Yihao Huang
Yihua Cheng
Yijie Lin
Yijie Zhong
Yilin Wen
Yingcheng Liu
Yingliang Zhang
Yinglin Zheng
Yingqian Wang
Yining Jiao
Yinjie Lei
Yinuo Jing
Yinyu Nie

Yipeng Qin
Yiqi Lin
Yiqi Zhong
Yiqing Shen
Yiqun Lin
Yiqun Wang
Yiran Guan
Yiran Xu
Yisi Luo
Yixuan Ren
Yizhak Ben-Shabat
Yizhen Lao
Yizhou Wang
Yizhou Wang
Yongtuo Liu
Yoonwoo Jeong
You Xie
YoungBin Kim
Youngseok Yoon
Youwei Lyu
Yu Liu
Yu Yin
Yu Zhang
Yuan Shen
Yuan Tian
Yuanhao Zhai
Yuchen Pei
Yucheng Zhao
Yuchong Sun
Yucong Shen
Yue Fan
Yue Xu
Yuecong Min
Yuedong Chen
Yueying Gao
Yueying Kao
Yufei Xie
Yuhe Jin
Yujia Chen
Yu-Jie Yuan
Yujun Cai
Yujun Tong
Yukang Cao
Yuki Fujimura
Yukun Huang

Yuliang Liu
Yu-Lun Liu
Yuming Gu
Yun Liu
Yun Xing
Yunfan Li
Yunhao Zou
Yunhua Zhang
Yunhui Guo
Yuning Cui
Yunqiu Xu
Yuqi Yang
Yuru Pei
Yusuke Hirota
Yusuke Kurose
Yusuke Sekikawa
Yusuke Sugano
Yuta Nakashima
Zakaria Laskar
Ze Yang
Zeeshan Khan
Zehua Sheng
Zerui Chen
Zeyu Xiao
Zezeng Li
Zhang Chen
Zhanzhan Cheng
Zhaodong Sun
Zhao-Min Chen
Zhaopei Huang
Zhaowen Li
Zhaoxin Fan
Zhaoyi An
Zhaoyu Chen
Zhen Chen
Zhen Chen
Zhen Liu
Zheng Chen
Zheng Chong
Zheng Qin
Zhengyi Luo
Zhennan Wang
Zhenwei Shi
Zhenyu Zhang
Zhenyue Qin

Zhepeng Wang
Zhewei Huang
Zhexiong Wan
Zhi Chen
Zhi Gao
Zhicheng Sun
Zhigang Tu
Zhihua Liu
Zhijian Huang
Zhijie Deng
Zhijie Wang
Zhikang Wang
Zhiliang Wu
Zhiming Zou
Zhineng Chen
Zhipeng Fan
Zhiqi Huang
Zhiqi Kang
Zhi-Qi Cheng
Zhixuan Yu
Zhiyong Wang
Zhiyuan Mao
Zhong Li

Zhongqun Zhang
Zhongyun Hu
Zhuangzhuang Chen
Zhuo Su
Zhuoran Yu
Ziang Cao
Zicheng Zhang
Zichun Zhong
Zikai Song
Zikui Cai
Zikun Zhou
Ziqi Huang
Ziqi Zhang
Ziqiang Li
Ziteng Cui
Ziteng Gao
Zitong Yu
Zixiang Zhao
Zixuan Jiang
Ziyao Zeng
Zi-Yi Dou
Zizheng Yan
Zutao Jiang

Contents – Part I

GAISynMeD

LAMM

LAVA

Contents – Part II

RichMediaGAI

WiCV

AWSS

BgSub: A Background Subtraction Model for Effective Moving Object Detection

Islam Osman$^{(\boxtimes)}$ and Mohamed S. Shehata

University of British Columbia, 3333 University Way, Kelowna, BC, Canada
islam.osman@ubc.ca

Abstract. Moving object detection is a core task in computer vision. However, existing deep learning-based moving object detection methods require a large number of labeled frames to achieve good generalization and performance. In moving object detection tasks, there is no such a large-scale labeled dataset because the labeling process requires a lot of time and effort. In this paper, we compiled a large-scale dataset by 1) combining existing moving object detection datasets. 2) using an in-painting deep learning model to transform datasets from video object segmentation task to moving object detection tasks. 3) generating synthetic datasets by combining random backgrounds with random foreground objects. Additionally, we propose a novel deep-learning model that performs background subtraction on the object level. This model is trained on the compiled dataset and shows superior performance in the moving object detection task. The model is evaluated using the CDNet dataset and results are compared with current state-of-the-art models. The results show that our model outperforms the best-reported state-of-the-art model by 1.6%.

Keywords: Moving object detection · Background subtraction · Deep learning

1 Introduction

Moving object detection (MOD) is a critical task in various computer vision applications, including object tracking, autonomous driving, and surveillance [1,19,39]. As a result, it has garnered significant attention from researchers in recent years. The task of moving object detection is to output a binary mask where the moving objects are represented as white pixels and the background is represented as black pixels. There are many challenges in this task such as dynamic background (e.g., trees and sea movement), shadows, illumination changes, and camera motion.

Techniques for moving object detection are generally classified into three categories: 1) unsupervised learning methods, 2) semi-supervised learning methods, and 3) supervised learning methods. Unsupervised learning methods use holistic approaches to differentiate moving objects from the background without

M. Cho et al. (Eds.): ACCV 2024 Workshops, LNCS 15482, pp. 3–16, 2025.
https://doi.org/10.1007/978-981-96-2641-0_1

the need for labeled data [8,44]. These methods often require careful tuning of hyperparameters to perform effectively across different datasets, making them task-specific. However, they typically do not perform as well as supervised methods in terms of accuracy. Semi-supervised learning methods, the second category, aim to achieve high performance by using a combination of labeled and unlabeled data [17,18]. While these methods can reduce the amount of labeled data needed, they usually require substantial training time to converge and may still face challenges in generalization. Finally, supervised learning methods [9,28] tend to deliver the highest segmentation accuracy among the three categories but at the cost of requiring large amounts of labeled data. Since there is no large-scale labeled dataset in moving object detection tasks, these models have limited performance. We define a large-scale labeled dataset as a dataset that has hundreds of thousands of labeled images such as ImageNet [14] in image classification, and COCO [27] in object detection.

In this paper, we proposed a novel deep learning model for moving object detection called BgSub as an abbreviation for Background Subtraction. This model is a mixture of a ViT transformer and a convolutional neural network. To overcome the problem of the limited labeled dataset, we compiled a large-scale dataset of more than $300,000$ labeled images. This dataset has three parts. 1) Real dataset: a combination of multiple moving object detection datasets. 2) Transformed dataset: transforming datasets of video object segmentation to moving object detection. 3) Synthetic dataset: a dataset generated by selecting from a random background and combining it with random objects at random locations. Finally, we trained the proposed model on the compiled dataset and evaluated the model using a benchmark dataset for moving object detection called CDNet [46].

The paper is organized as follows: Section 2 provides a literature review. Section 3 provides a detailed explanation of the proposed work. Section 4 depicts the results of the experiments that were conducted. Finally, Sect. 5 concludes and summarizes the paper and discusses the future directions.

2   Related Work

Moving object detection has evolved significantly, with methods generally categorized into statistical and CNN-based approaches. Statistical methods build a background model using intensity values and motion vectors (e.g., optical flow) to detect moving objects as pixels that deviate from the model. These include techniques like ego-motion compensation [3], feature-based methods [38], motion-based approaches [4], image-cue-based models [40], and subspace-based approaches [15,16]. While effective in certain scenarios, these methods often struggle with high computational costs and are prone to performance degradation in environments with dynamic backgrounds or significant camera movement [29,37].

CNN-based methods have shown improved accuracy in detecting moving objects, especially in complex environments with changing lighting or background motion [46]. Early models like ConvNets [10] were trained on datasets like CDNet2014 to perform simple subtraction between current frames and static background images. Over time, more sophisticated architectures emerged. For instance, Babaee et al. [5] incorporated pre-processing steps to adapt to dynamic backgrounds and post-processing like spatial-median filtering [41] to accurately detect camouflaged objects. End-to-end models, such as Cascaded CNN [45], introduced multi-resolution networks to detect objects at varying scales while ensuring spatial coherence. Advanced models, such as Johnander et al. [21], Zhang et al. [48], and FgSegNet [24,25], leveraged multi-scale feature extraction, optical flow, and temporal information to refine moving object detection. Furthermore, motion-guided attention models [23] integrated appearance and motion saliency. MODY-Net [35] uses an ensemble of multi-scale outputs to minimize the model uncertainty. Hence, reduces the false-positives. REFNet-TBPI [32] uses a multi-scale fusion network to capture objects' regions and boundaries. Hence, the model detects the object's fine details. Additionally, the model is trained using a continual learning technique called task-based parameter isolation (TBPI) to prevent forgetting across video sequences which improves the overall performance. TransBlast [31] is an attempt to use transformers in the MOD task with an augment loss function that maximizes the separation of foreground and background using a subspace learning module. This model has good generalization. However, its performance was limited because transformers require a massive amount of labeled data to achieve good performance. Hence, in this paper, we compiled a large-scale dataset to train a transformer-based model mixed with CNN layers to produce accurate masks achieving state-of-the-art performance in the MOD task.

3 Proposed Work

In this paper, we propose a novel deep-learning model for segmenting moving objects from frame sequences given the background frame. Hence, the model learns to perform background subtraction on high-level features instead of the traditional pixel-to-pixel background subtraction. This model is a combination of vision transformer (ViT) and convolutional neural networks (CNN). The transformer is used to extract high-level features from the current frame and background frame. On the other hand, CNN is used to learn background subtraction of the high-level features extracted by the ViT and produce the output mask of the moving objects. This model is trained using a large-scale dataset that we compiled from various of different sources. In this section, we show the model architecture, the process of compiling the large-scale dataset, and the training procedure.

3.1 BgSub Architecture

BgSub consists of four major components as shown in Fig. 1, 1) Frame encoder, 2) Feature pyramid module, 3) Multi-scale fusion, and 4) Segmentation head.

Frame encoder is used to extract feature maps from the current frame and the background frame. The encoder is based on a DinoV2 [30] transformer,

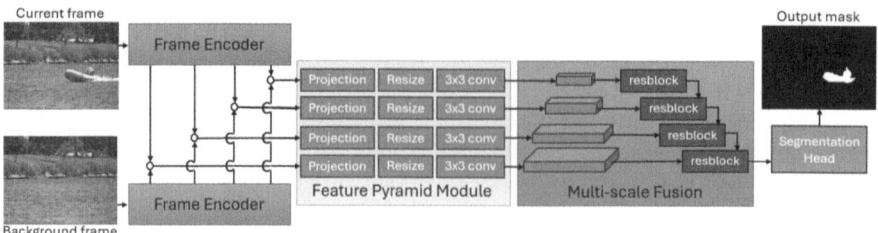

Fig. 1. The architecture of BgSub.

which is a form of vision transformer trained using self-supervised learning on ImageNet. The model consists of 12 transformer blocks with embedding sizes that vary based on the model size (e.g., ViT-s embedding size is 384, while ViT-b embedding size is 768). The patch size is 16 (i.e., the input image is split into non-overlapping 16×16 patches). The input frame size is 224×224. We use the output of the last 4 blocks of the encoder to generate the feature pyramid. The output size of each block is $C \times H/16 \times W/16$, where C is the embedding size, H is the frame height, and W is the frame width.

Feature pyramid module (FPM) is used to extract multi-scale features from the frame encoder. Feature pyramid has 4 parallel paths. Each path consists of 4 sequential layers. The first layer is a concatenation layer to combines both features of the current frame and the background frame. The second layer is a projection layer (i.e., 1×1 convolutional layer) to produce different numbers of channels for different scales, such that small scales have a larger number of features than large scales to keep the inference time of the model reasonable (as the time needed to performing convolution operation on a large scale is longer than that on a small scale). The third layer is a resizing layer to get different scales. The last layer is a 3×3 convolutional layer to produce the feature map of each scale. The output is a feature pyramid of 4 scales, where each level of the pyramid has a scale of $2\times$ the next level. The lowest level has a feature map of size $C_1 \times H \times W$, while the top level has a feature map of size $C_4 \times H/8 \times W/8$

Multi-scale fusion is used to merge the feature pyramid levels into a single feature map. The fusion of pyramid levels is performed sequentially. The pyramid level i level is merged with level $i-1$ through a residual block (i.e., three convolutional layers with a skip connection from the first layer to the last layer). Hence, the output of the residual block at level 1 (i.e., lowest level) has information on all feature pyramid levels.

Segmentation head is used to produce the final output, which is a binary mask that represents the moving objects in white pixels and the background in black pixels. The segmentation head uses the output feature map of the multi-scale fusion process using three convolutional layers, two of which are 3×3 convolutional and the last one is a 1×1 convolutional layer followed by a sigmoid activation function.

3.2   MOD Dataset

The compiled moving object detection dataset is a large-scale dataset of more than 300,000 labeled images. This dataset consists of three types of datasets. 1) Real datasets: a combination of existing moving object detection datasets. This dataset includes AAU [6], LASIESTA [13], SBM [12], CDNet [46], and urbantracker [20]. These datasets have frame sequences with their corresponding segmentation masks. We select the background frame for each frame sequence as the frame that has the least amount of white pixels in its segmentation mask (i.e., the frame with almost no moving objects in it). Sample frames of this dataset are shown in Fig. 2. 2) Transformed datasets: video object segmentation datasets focus on tracking and segmenting a specific object along a video sequence. The problem is that the object always exists in all frames. Hence, we can not select the background frame for any of the frame sequences. To overcome this issue, we used an in-painting model called ProPainter [49]. This model can remove an object from an image and replace it with a predicted background. For each frame sequence, we use the mask of the first frame and feed it to the ProPainter to generate the background frame. The datasets used are DAVIS16 [36] and YT-VOS18 [47]. An example of background generation is shown in Fig. 3. 3) Synthetic dataset: we randomly select a background frame from all frame sequences, and then we add foreground objects from a random number of randomly selected frame sequences. An example of the generation is shown in Fig. 4. The ratios of all datasets are shown in Fig. 5.

Fig. 2. Sample frames from existing MOD datasets.

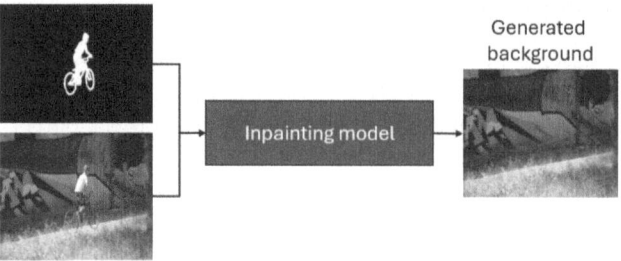

Fig. 3. Generating background frame to transform VOS dataset to MOD dataset.

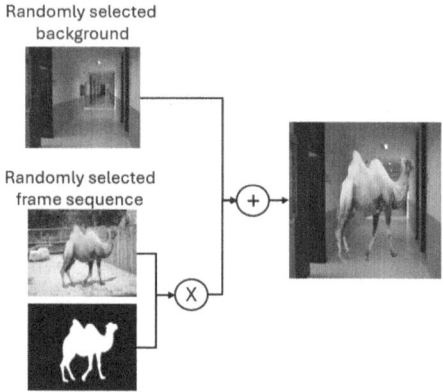

Fig. 4. An example of generating synthetic frame.

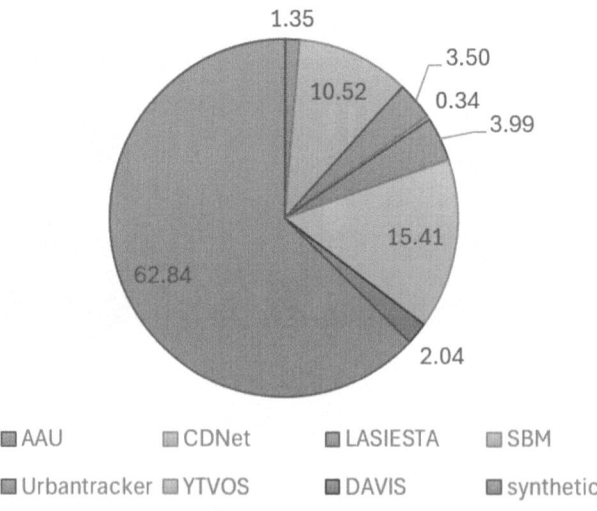

Fig. 5. The ratio of each dataset in the compiled dataset.

3.3   Implementation Details

Training is done for 120 epochs on the compiled dataset. The learning rate is $1e-5$, the batch size is 64, and the optimizer used is AdamW with weight decay of 0.01 and beta values 0.9 and 0.999. The data augmentation used are color jitter, random horizontal flips, and random affine (i.e., random rotation up to 15 degrees and random shear up to 10). The same augmentations are applied for both the current frame and background but with different values, the reason behind this is to make sure that the background frame does not look the same as the background in the current frame. Hence, preventing the model from learning to perform pixel-to-pixel subtraction. The same augmentations with the same values are applied to the ground truth mask except for the color jitter. The loss function used is a combination of focal loss [26] and IoU loss. The focal loss is calculated as follows:

$$\ell_{focal}(p) = \begin{cases} -\alpha(1-p)^\gamma \log(p) & \text{if } y = 1 \\ -(1-\alpha)p^\gamma \log(1-p) & \text{otherwise.} \end{cases} \tag{1}$$

where α is a weighting factor used to balance the importance of background and foreground pixels, p is the model estimated probability for the ground truth class, and γ is the focusing parameter which reduces the relative loss for well-classified examples, putting more focus on hard (i.e., misclassified examples). We calculated the ratio r of foreground to background pixels in all masks and set the value of α to $1-r$ to pay more attention to foreground pixels. The value of γ is typically chosen based on empirical testing resulting in $\gamma = 2$. On the other hand, the IoU (i.e., intersection over union) loss is calculated as follows:

$$\ell_{iou} = 1 - \frac{|A \bigcap B|}{|A \bigcup B|} \tag{2}$$

where A is the ground truth and B is the predicted regions. The reason why it is $1 - iou$ is that we want to maximize the iou while the loss function is usually minimized during training. The final loss function is:

$$\mathcal{L} = \ell_{focal} + \ell_{iou} \tag{3}$$

4   Experiments and Results

In this section, we evaluate the proposed BgSub using a moving object detection benchmark datasets called ChangeDetection.Net (CDNet) [46]. The results are compared against state-of-the-art moving object detection models. In this experiment, our model is trained using our compiled dataset which contains CDNet inside it. However, only 600 frames per frame sequence are in the compiled dataset. The rest of the frames are left for testing. Similarly, all other models are trained using the same 600 frames from each fame sequence in CDNet. The evaluation metrics used are recall (Re), precision (Pr), and F-score ($\mathcal{F}$).

Table 1. Results of BgSub against state-of-the-art moving object detection methods on CDNet.

Method	Re	Pr	$\mathcal{F}$
sEnDec [2]	0.903	0.798	0.842
FgSegNetV2 [24]	0.893	0.741	0.801
CascadeCNN [45]	0.821	0.771	0.786
IUTIS-5 [7]	0.789	0.808	0.771
SemanticBGS [11]	0.789	0.830	0.789
BSUV-Net [43]	0.820	0.811	0.786
BSUV-Net-SBGS [43]	0.817	0.831	0.798
DeepBS [5]	0.754	0.833	0.745
SuBSENSE [41]	0.812	0.751	0.741
WisenetMD [22]	0.817	0.766	0.753
PAWCS [42]	0.771	0.785	0.740
MODSiam [34]	0.823	0.692	0.714
TransBlast [31]	0.867	0.811	0.831
FeSh-Net [33]	0.889	0.813	0.851
MODY-Net [35]	0.903	0.822	0.865
REFNet [32]	0.961	0.827	0.881
REFNet-TBPI [32]	0.969	0.850	0.901
BgSub*	0.961	0.849	0.903
BgSub	**0.971**	**0.863**	**0.917**

Table 2. Detailed results of BgSub on the 11 different challenges in CDNet.

Method	# of videos	$\mathcal{F}$
Bad weather	4	0.950
Baseline	4	0.931
Camera jitter	4	0.941
Dynamic background	6	0.957
Intermittent object motion	6	0.926
Low frame rate	4	0.945
Night videos	6	0.887
PTZ	4	0.784
Shadow	6	0.945
Thermal	5	0.919
Turbulence	4	0.903

Table 1 shows the results of the proposed BgSub against state-of-the-art moving object detection methods using the CDNet dataset. We reported the results of 2 versions of BgSub. The first one is BgSub trained using CDNet only, referred to as BgSub* in the table. The performance of BgSub is 0.2% better than the top-performing model REFNet-TBPI. However, the second version of the model is trained using our compiled dataset. This version of BgSub outperforms other models by 1.6% in the F-score, which is 1.4% better than training the same model on CDNet only. These results highlight the effectiveness of training the model on a large-scale MOD dataset. The detailed performance on different challenges of the CDNet is shown in Table 2. As shown in the table, the performance of BgSub is superior on all challenges except for night videos, PTZ, and turbulence. The problem with night videos is that it is hard to locate the objects even with the naked eye. However, the model is still able to detect 88.7% of the moving object pixels. For the PTZ challenge, the performance of BgSub is limited due to the fact that the background scene changes dramatically in this challenge from one frame to another. Since the model is based on background subtraction when the background of the current frame is dramatically different from the background frame the model produces false positives due to uncertainty. As the model is not sure whether the new objects in the current frame is a moving object or a static object. Finally, the moving objects in the turbulence challenge are very small in size and the model is sometimes unable to detect tiny objects (Fig. 7).

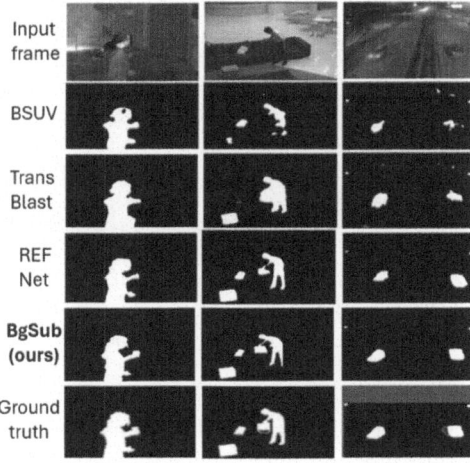

Fig. 6. Sample visual results of BgSub against other models.

4.1   Visual Results

Figure 6 shows sample results from the proposed model BgSub against other MOD models on sample videos from the CDNet dataset. As shown in the figure

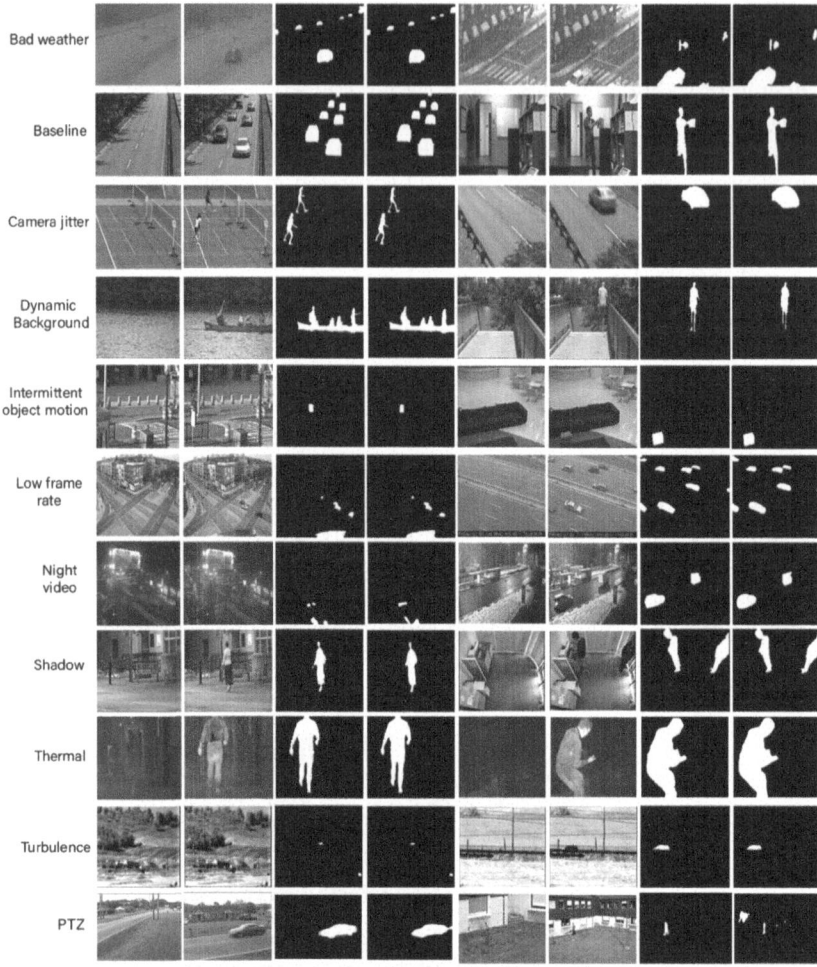

Fig. 7. Visual results of BgSub for each challenge in CDNet. The 1st and 5th columns are background frames, 2nd and 6th columns are current frames, 3rd and 7th columns are ground truth, 4th and 8th columns are BgSub output.

the output of BgSub is almost the same as the ground truth, unlike other models that have some false positives and some false negatives. Two sample videos from each challenge in CDNet are shown in Fig. 7. This figure highlights the effectiveness of BgSub in detecting moving objects under different challenges with superior performance. Even in the dynamic background challenge the model understands that the sea and trees are background objects and did not detect them as moving objects. The only downside of the proposed model is the PTZ challenge. This is due to the fact that the background massively changes from one frame to another. Hence, some new objects introduced to the scene may be detected as moving objects due to the model's uncertainty. Finally, to test

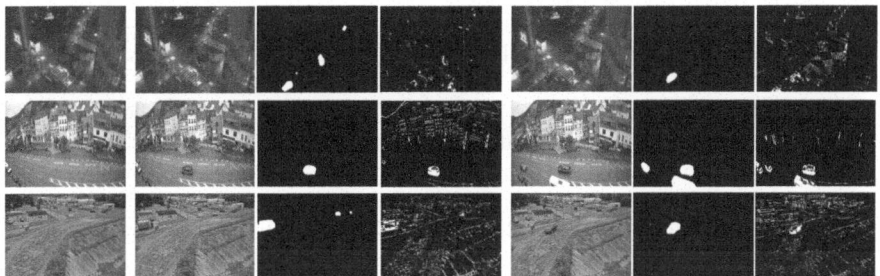

Fig. 8. Visual results of BgSub on real-world frame sequences captured from public IP cameras. The 1st column is the background frame, 2nd and 5th columns are selected frames, 3rd and 6th columns are BgSub output, 4th and 7th columns are pixel-to-pixel subtraction between selected frame and background frame.

the proposed model generalization ability, we capture frames from live public IP cameras (unseen during training). We show the results of traditional pixel-to-pixel background subtraction to show that on the pixel-level there are differences between the background frame and every other frame as shown in Fig. 8. However, our proposed model learns to perform background subtraction on the pixel level. Hence, the model ignores the pixel difference and focuses on the object difference.

5   Conclusion

In this paper, we propose a moving object detection model based on background subtraction and compiled a large-scale dataset. The proposed work addresses the problem of the limited performance of other models due to the lack of a large number of labeled frames existence. Our model demonstrated superior performance on the CDNet dataset, outperforming the current state-of-the-art model by 1.6%. This performance improvement is mainly due to the compiled dataset. These results show the necessity of a large-scale labeled dataset in the field of moving object detection. As the visual results show that the model is able to detect objects from out-of-domain frame sequences, in future work, we will collect a real-world large-scale dataset from public IP cameras and use the proposed model to pseudo-label the dataset and make it publicly available.

References

1. Abdelpakey, M.H., Shehata, M.S.: DomainSiam: domain-aware siamese network for visual object tracking. In: Bebis, G., et al. (eds.) ISVC 2019. LNCS, vol. 11844, pp. 45–58. Springer, Cham (2019). https://doi.org/10.1007/978-3-030-33720-9_4
2. Akilan, T., Wu, Q.J.: sEnDec: An improved image to image CNN for foreground localization. IEEE Trans. Intell. Transp. Syst. **21**(10), 4435–4443 (2019)

3. Ali, S., Shah, M.: Cocoa: tracking in aerial imagery. In: Airborne Intelligence, Surveillance, Reconnaissance (ISR) Systems and Applications III, vol. 6209, p. 62090D. International Society for Optics and Photonics (2006)
4. Aslani, S., Mahdavi-Nasab, H.: Optical flow based moving object detection and tracking for traffic surveillance. Int. J. Electr. Comput. Eng. **8**(12), 840–845 (2013)
5. Babaee, M., Dinh, D., Rigoll, G.: A deep convolutional neural network for video sequence background subtraction. Pattern Recogn. **76**, 635–649 (2018)
6. Bahnsen, C.H., Moeslund, T.B.: Rain removal in traffic surveillance: does it matter? IEEE Trans. Intell. Transp. Syst. **20**(8), 2802–2819 (2018)
7. Bianco, S., Ciocca, G., Schettini, R.: Combination of video change detection algorithms by genetic programming. IEEE Trans. Evol. Comput. **21**(6), 914–928 (2017)
8. Bouwmans, T.: Recent advanced statistical background modeling for foreground detection-a systematic survey. Recent Patents Comput. Sci. **4**(3), 147–176 (2011)
9. Bouwmans, T., Javed, S., Sultana, M., Jung, S.K.: Deep neural network concepts for background subtraction: a systematic review and comparative evaluation. Neural Netw. **117**, 8–66 (2019)
10. Braham, M., Droogenbroeck, M.V.: Deep background subtraction with scene-specific convolutional neural networks. In: 2016 International Conference on Systems, Signals and Image Processing (IWSSIP), pp. 1–4. IEEE (2016)
11. Braham, M., Piérard, S., Van Droogenbroeck, M.: Semantic background subtraction. In: 2017 IEEE International Conference on Image Processing (ICIP), pp. 4552–4556. IEEE (2017)
12. Camplani, M., Maddalena, L., Moyá Alcover, G., Petrosino, A., Salgado, L.: A benchmarking framework for background subtraction in RGBD videos. In: Battiato, S., Farinella, G.M., Leo, M., Gallo, G. (eds.) ICIAP 2017. LNCS, vol. 10590, pp. 219–229. Springer, Cham (2017). https://doi.org/10.1007/978-3-319-70742-6_21
13. Cuevas, C., Yáñez, E.M., García, N.: Labeled dataset for integral evaluation of moving object detection algorithms: LASIESTA. Comput. Vis. Image Underst. **152**, 103–117 (2016)
14. Deng, J., Dong, W., Socher, R., Li, L.J., Li, K., Fei-Fei, L.: ImageNet: a large-scale hierarchical image database. In: 2009 IEEE Conference on Computer Vision and Pattern Recognition, pp. 248–255. IEEE (2009)
15. ElTantawy, A., Shehata, M.S.: KRMARO: aerial detection of small-size ground moving objects using kinematic regularization and matrix rank optimization. IEEE Trans. Circuits Syst. Video Technol. **29**(6), 1672–1686 (2018)
16. Eltantawy, I., Valera, M.: Accelerated KRMARO: subspace background subtraction under dynamic environments. In: 2019 IEEE Winter Conference on Applications of Computer Vision (WACV), pp. 100–109. IEEE (2019)
17. Giraldo, J.H., Javed, S., Bouwmans, T.: Graph moving object segmentation. IEEE Trans. Pattern Anal. Mach. Intell. **01**, 1 (2020)
18. Giraldo, J.H., Javed, S., Sultana, M., Jung, S.K., Bouwmans, T.: The emerging field of graph signal processing for moving object segmentation. In: Jeong, H., Sumi, K. (eds.) IW-FCV 2021. CCIS, vol. 1405, pp. 31–45. Springer, Cham (2021). https://doi.org/10.1007/978-3-030-81638-4_3
19. Hu, H.N., et al.: Joint monocular 3D vehicle detection and tracking. In: Proceedings of the IEEE International Conference on Computer Vision, pp. 5390–5399 (2019)
20. Jodoin, J.P., Bilodeau, G.A., Saunier, N.: Urban tracker: multiple object tracking in urban mixed traffic. In: IEEE Winter Conference on Applications of Computer Vision, pp. 885–892. IEEE (2014)

21. Johnander, J., Chatterjee, A., Felsberg, M.: Generative probabilistic modeling for background subtraction: Moving beyond gaussian mixture models. In: Proceedings of the IEEE/CVF Conference on Computer Vision and Pattern Recognition, pp. 1559–1568 (2019)

22. Lee, S.h., Lee, G.c., Yoo, J., Kwon, S.: WisenetMD: motion detection using dynamic background region analysis. Symmetry **11**(5), 621 (2019)

23. Li, Y., Chang, H., Jiang, W., Zhang, X., Bao, L., Zhang, W.: Motion-guided attention for video salient object detection. In: Proceedings of the IEEE/CVF Conference on Computer Vision and Pattern Recognition, pp. 10050–10059 (2019)

24. Lim, I.C.Y., Phung, S.L.: FgSegNet V2: fully automated foreground segmentation using multi-scale feature fusion. IEEE Trans. Circuits Syst. Video Technol. **30**(12), 4682–4695 (2020)

25. Lim, L., Keles, H.: Foreground segmentation using a triplet convolutional neural network for multiscale feature encoding. arxiv 2018. arXiv preprint arXiv:1801.02225

26. Lin, T.Y., Goyal, P., Girshick, R., He, K., Dollár, P.: Focal loss for dense object detection. In: Proceedings of the IEEE International Conference on Computer Vision, pp. 2980–2988 (2017)

27. Lin, T.-Y., et al.: Microsoft COCO: common objects in context. In: Fleet, D., Pajdla, T., Schiele, B., Tuytelaars, T. (eds.) ECCV 2014. LNCS, vol. 8693, pp. 740–755. Springer, Cham (2014). https://doi.org/10.1007/978-3-319-10602-1_48

28. Mandal, M., Vipparthi, S.K.: An empirical review of deep learning frameworks for change detection: model design, experimental frameworks, challenges and research needs. IEEE Trans. Intell. Transp. Syst. **23**, 6101–6122 (2021)

29. Nakaya, Y., Harashima, H.: Motion segmentation based on similarity of spatial-temporal region. In: Proceedings of 3rd International Conference on Image Processing, vol. 2, pp. 378–382. IEEE (1994)

30. Oquab, M., et al.: DINOv2: learning robust visual features without supervision. Trans. Mach. Learn. Res. J. **2024**, 1–31 (2024). https://dblp.org/db/journals/tmlr/tmlr2024.html

31. Osman, I., Abdelpakey, M., Shehata, M.S.: TransBlast: self-supervised learning using augmented subspace with transformer for background/foreground separation. In: Proceedings of the IEEE/CVF International Conference on Computer Vision, pp. 215–224 (2021)

32. Osman, I., Eltantawy, A., Shehata, M.S.: Task-based parameter isolation for foreground segmentation without catastrophic forgetting using multi-scale region and edges fusion network. Image Vis. Comput. **113**, 104248 (2021)

33. Osman, I., Shehata, M.S.: Few-shot learning network for moving object detection using exemplar-based attention map. In: 2022 IEEE International Conference on Image Processing (ICIP), pp. 1056–1060. IEEE (2022)

34. Osman, I.I., Shehata, M.S.: MODSiam: moving object detection using siamese networks. In: 2020 IEEE Canadian Conference on Electrical and Computer Engineering (CCECE), pp. 1–6. IEEE (2020)

35. Osman, I.I., Shehata, M.S.: Mody-Net: moving object detection using multiscale output ensemble y-network. IEEE Can. J. Electr. Comput. Eng. **44**(4), 491–496 (2021)

36. Perazzi, F., Pont-Tuset, J., McWilliams, B., Van Gool, L., Gross, M., Sorkine-Hornung, A.: A benchmark dataset and evaluation methodology for video object segmentation. In: Proceedings of the IEEE Conference on Computer Vision and Pattern Recognition, pp. 724–732 (2016)

37. Pouzet, L., Garnier, L., Strugarek, J., Bascle, B.: Robust video object segmentation using region-wise classification. Sig. Process. Image Commun. **29**(10), 1185–1195 (2014)
38. Rosten, E., Drummond, T.: Machine learning for high-speed corner detection. In: Leonardis, A., Bischof, H., Pinz, A. (eds.) ECCV 2006. LNCS, vol. 3951, pp. 430–443. Springer, Heidelberg (2006). https://doi.org/10.1007/11744023_34
39. Scheiner, N., et al.: Seeing around street corners: Non-line-of-sight detection and tracking in-the-wild using doppler radar. In: IEEE/CVF Conference on Computer Vision and Pattern Recognition (CVPR) (2020)
40. Shen, X., Yu, J., Zeng, Q.: Moving object detection in the presence of complex background. Pattern Recogn. Lett. **34**(9), 1035–1044 (2013)
41. St-Charles, P.L., Bilodeau, G.A., Bergevin, R.: SuBSENSE: a universal change detection method with local adaptive sensitivity. IEEE Trans. Image Process. **24**(1), 359–373 (2014)
42. St-Charles, P.L., Bilodeau, G.A., Bergevin, R.: A self-adjusting approach to change detection based on background word consensus. In: 2015 IEEE Winter Conference on Applications of Computer Vision, pp. 990–997. IEEE (2015)
43. Tezcan, O., Ishwar, P., Konrad, J.: BSUV-Net: a fully-convolutional neural network for background subtraction of unseen videos. In: Proceedings of the IEEE/CVF Winter Conference on Applications of Computer Vision, pp. 2774–2783 (2020)
44. Vaswani, N., Bouwmans, T., Javed, S., Narayanamurthy, P.: Robust subspace learning: robust PCA, robust subspace tracking, and robust subspace recovery. IEEE Signal Process. Mag. **35**(4), 32–55 (2018)
45. Wang, N., Shelley, M., Heitz, G., Perona, P.: Interactive dynamic video segmentation using convolutional neural networks. In: 2017 IEEE Winter Conference on Applications of Computer Vision (WACV), pp. 1–9. IEEE (2017)
46. Wang, Y., Jodoin, P.M., Porikli, F., Konrad, J., Benezeth, Y., Ishwar, P.: CDnet 2014: An expanded change detection benchmark dataset. In: Proceedings of the IEEE Conference on Computer Vision and Pattern Recognition Workshops, pp. 387–394 (2014)
47. Xu, N., et al.: Youtube-VOS: a large-scale video object segmentation benchmark. arXiv preprint arXiv:1809.03327 (2018)
48. Zhang, X., Ma, J., Zhang, F.: Fast moving object detection with a network based on encoder-decoder structure. In: Proceedings of the IEEE/CVF Conference on Computer Vision and Pattern Recognition Workshops, pp. 344–352 (2019)
49. Zhou, S., Li, C., Chan, K.C., Loy, C.C.: Propainter: Improving propagation and transformer for video inpainting. In: Proceedings of the IEEE/CVF International Conference on Computer Vision, pp. 10477–10486 (2023)

Physically Interpretable Probabilistic Domain Characterization

Anaïs Halin[1]([✉])[iD], Sébastien Piérard[1][iD], Renaud Vandeghen[1][iD],
Benoît Gérin[2][iD], Maxime Zanella[2,4][iD], Martin Colot[3][iD], Jan Held[1][iD],
Anthony Cioppa[1][iD], Emmanuel Jean[5][iD], Gianluca Bontempi[3][iD],
Saïd Mahmoudi[4], Benoît Macq[2][iD], and Marc Van Droogenbroeck[1][iD]

[1] Montefiore Institute, University of Liège (ULiège), Liège, Belgium
{anais.halin,s.pierard}@uliege.be
[2] Catholic University of Louvain (UCLouvain), Louvain-la-Neuve, Belgium
[3] Université Libre de Bruxelles (ULB), Brussels, Belgium
[4] University of Mons (UMons), Mons, Belgium
[5] Multitel Research and Innovation Centre, Mons, Belgium

Abstract. Characterizing domains is essential for models analyzing dynamic environments, as it allows them to adapt to evolving conditions or to hand the task over to backup systems when facing conditions outside their operational domain. Existing solutions typically characterize a domain by solving a regression or classification problem, which limits their applicability as they only provide a limited summarized description of the domain. In this paper, we present a novel approach to domain characterization by characterizing domains as probability distributions. Particularly, we develop a method to predict the likelihood of different weather conditions from images captured by vehicle-mounted cameras by estimating distributions of physical parameters using normalizing flows. To validate our proposed approach, we conduct experiments within the context of autonomous vehicles, focusing on predicting the distribution of weather parameters to characterize the operational domain. This domain is characterized by physical parameters (absolute characterization) and arbitrarily predefined domains (relative characterization). Finally, we evaluate whether a system can safely operate in a target domain by comparing it to multiple source domains where safety has already been established. This approach holds significant potential, as accurate weather prediction and effective domain adaptation are crucial for autonomous systems to adjust to dynamic environmental conditions.

Keywords: Domain Characterization · Distribution Prediction · Normalizing Flows · Simulation-Based Inference · Domain Adaptation · Weather · Autonomous Vehicles · Operational Design Domain

---

A. Halin and S. Piérard—Equal contributions.

1   Introduction

Advances in computer vision allow widespread camera monitoring, but diverse weather conditions lead to visually different data, sometimes impacting the performance of high-level tasks like object detection and surveillance. Current solutions often lack generalizability across multi-weather scenarios, highlighting the need for adaptive methods that can process visual data under diverse conditions.

More specifically, weather conditions significantly affect the perception capabilities of autonomous driving systems [34], particularly under harsh conditions, such as heavy rain or fog, which can compromise their ability to operate safely. Therefore, it is essential to develop reliable approaches to analyze the environment, irrespective of the weather conditions. Detecting critical circumstances such as extreme weather events allow the system to respond appropriately within and outside its *Operational Design Domain (ODD)*, defined by SAE International [42] as the "operating conditions under which a given driving automation system, or feature thereof, is specifically designed to function, including, but not limited to, environmental, geographical, and time-of-day restrictions, and/or the requisite presence or absence of certain traffic or roadway characteristics".

However, predicting exact weather conditions is challenging due to the many ambiguous cases. A single observation of the environment (*e.g.*, recorded by a vehicle-mounted camera) could be characterized by multiple sets of weather parameters. Figure 1 presents two synthetic images, generated using the *CARLA* software [6], that appear nearly identical but were captured under different weather conditions. This shows the ambiguities inherent to characterizing the domain based on a single observation (*i.e.*, image).

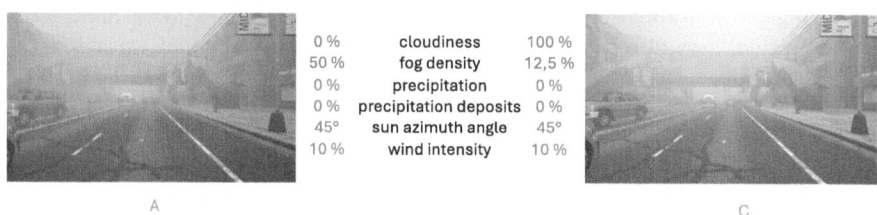

0 %	cloudiness	100 %
50 %	fog density	12,5 %
0 %	precipitation	0 %
0 %	precipitation deposits	0 %
45°	sun azimuth angle	45°
10 %	wind intensity	10 %

A   C

Fig. 1. The two synthetic images, A and C, generated using the *CARLA* software, appear almost identical, despite being acquired under very different weather conditions. This highlights the challenge of information loss from sensors like cameras when dealing with weather-related physical parameters. As a result, predictions that diverge from the ground truth in such ambiguous cases should not be penalized during evaluation.

Current methods predict weather conditions through regression or classification [17,18,26]. However, those approaches only produce a single crisp answer, lacking the insight about the ambiguities of characterizing the weather conditions. Figure 2 illustrates how the two scenarios from Fig. 1, when processed with an intermediate set of values, result in very different visual representations.

Fig. 2. Considering the images A and C from Fig. 1, generating an image B for the arithmetically averaged parameters, leads to an image very different from A and C. In other words, there are images i such that the probability $P(I = i, W = \hat{w}(i))$ of the pair (image, estimated weather) is zero when the estimated weather is the expected value of the weather knowing the image, $\hat{w}(i) = E[W|I = i]$, as done in regression. Working with distributions of weather parameters (as shown using a contour plot on the right-hand side) as proposed in this work, rather than predicting specific values for each parameter, *i.e.*, regression (as shown on the left-hand side), avoids this problem.

In this work, we propose a novel solution based on a probabilistic characterization of the weather. Historically, statistical inference relying on likelihood estimation was intractable when dealing with high dimensional data [4]. Recently, one class of density estimation techniques based on neural networks called normalizing flows [20, 32, 40] has become quite popular to solve this kind of problem. These models learn invertible transformations to go from complex distributions to more handy ones, and can therefore be used for modeling weather parameters from highly complex image and weather distributions.

These novel techniques allow us to express the domain of an image by a probabilistic distribution, *e.g.*, weather conditions, compared to current deterministic approaches. More specifically, we show on three consecutive tasks how various weather conditions can be predicted based (1) on a single color image acquired in front of a vehicle, (2) on a bag of color images (*absolute characterization*), and (3) how the current domain is related to arbitrarily chosen source domains (*relative characterization*).

Notions similar to *ODD* are very common in other fields where equivalent terms are used, such as *Operational Envelope* for maritime and *Operational Context* for railroad [46]. Additionally, other fields are facing domain shifts, such as in medical imagery from different acquisition devices [9]. We argue that our approach could serve a large range of practical applications. For example, determining the most suitable model for a given scenario within a fleet of lightweight AI models able to analyze the environment, as proposed in previous works [8,30,35], detecting significant domain transitions to collect new data for domain adaptation methods that rely on buffers or adaptable internal statistics [10,15,49–52] or activating adaptation mechanisms based on clustering [2,45,53,54].

We summarize our contributions as follows: (1) We propose a novel probabilistic methodology to characterize domains in the case of autonomous vehicles driving in various weather conditions. (2) We demonstrate that simulation-based inference (normalizing flows) is adequate to obtain distribution for weather parameters that are used to characterize the domain and compare different backbones for features extraction. (3) Based on this weather domain characterization, we show how to characterize a new target domain as a mixture of source models.

2   Related Work

2.1   Predicting Weather Conditions from Images

Prediction of weather conditions from images was first formulated as a single-label classification task (*e.g.*, sunny, cloudy, or foggy). In 2014, Lu *et al.* [28] proposed a binary classification task between sunny and cloudy weathers, using features extracted from visual cues such as the sky, shadows, reflections, contrast, and haze. Later, Guerra *et al.* [48] proposed a multi-class dataset extending the scope of the classification task to rain, snow, and fog. Recent works focus on optimizing *Convolutional Neural Network (CNN)* to obtain strong features for common and uncommon weather conditions [27,55]. However, weather conditions can hardly be represented by crisp classification due to its continuous nature.

To reach a more realistic description of intricate relations between weather conditions, recent methods simultaneously regress several physical parameters [17,18]. To increase interpretability, Li *et al.* [26] also assign cues of weather characteristics to each pixel. However, these methods still lack the ability to represent ambiguous scenarios (see Fig. 2). In this work, we predict the joint distribution of weather parameters by proposing a novel method based on normalizing flows.

2.2   Handling Weather Conditions for Autonomous Driving

Autonomous car driving systems need to be efficient under all weather conditions. Some methods propose to integrate a generalization step to the model to remove the environmental influences on the acquired images, by introducing a style layer inspired by images style-transfer neural networks [38] or through adversarial training [25]. Generalized features obtained from alignment of source and target domains tend to be suboptimal, as they do not consider the task. To improve the estimation, a solution is to add a domain adaptation step [24]. Jeon *et al.* [19] further improved their estimation by adding an unsupervised domain adaptation step after domain alignment.

Many studies have also highlighted the importance of a strong *ODD* definition to properly assess the ability of automatic driving systems to work in given conditions regarding weather, location, other vehicles on the road, state of the car sensors and many other environmental parameters [3,12,33]. Many strategies have been proposed to evaluate different situations according to specific evaluations of potential damage cost [23,44] and define the boundaries of the *ODD*. In this work, we characterize the domain by a probability distribution focusing on weather parameters in autonomous driving environment.

2.3   Probabilistic Modeling of Parameters Distributions

We propose to leverage recent observations in the *Simulation-Based Inference* (*SBI*) literature [4,47] for domain characterization. This literature has seen a rapid expansion thanks to new density estimation techniques in problems where likelihood estimation was often intractable, especially for high dimensional data. One class of these density estimation techniques based on neural networks is normalizing flows [32]. The principle consists in transforming an arbitrarily chosen distribution (*e.g.*, a Gaussian) into the desired distribution. Different types exist, such as the *Neural Spline Flow* (*NSF*) type [7] that can be used in two ways: (1) to determine the value of the *Probability Density Function* (*PDF*) at a given point and (2) to draw samples at random. Different techniques can be used to learn them, *e.g.* the *Neural Posterior Estimation* (*NPE*) technique [11,29].

   There are a few techniques to analyze the performance of models predicting distributions. *Coverage Plots* [14] show, objectively and quantitatively, whether the distribution prediction models are underconfident (*i.e.*, conservative), calibrated, or overconfident. Another technique, specific for parameter distributions (*e.g.*, weather parameters) predicted from an observation (*e.g.*, an image) and widespread in the field of *SBI*, is known as *Posterior Predictive Check* (*PPC*) [39]. It consists in drawing parameters at random from a predicted distribution, injecting these parameters into a simulator or physical system and comparing the resulting observations with the one from which the distribution was predicted. In this work, we leverage those analysis techniques for assessing the quality of our weather characterization models.

3   The Three Fundamental Tasks Behind the Physically Interpretable Probabilistic Domain Characterization

Our experiments are organized around three different tasks, involving a prediction of the distribution of weather conditions given some images acquired by color cameras placed in front of vehicles. Before elaborating on these tasks in Sect. 3.2, 3.3, and 3.4, we briefly introduce our framework in Sect. 3.1.

3.1   Framework

Mathematical Modeling. We denote the set of all possible values for the physical parameters (*e.g.* weather conditions) by $\mathbb{W}$ and the set of all observations of interest (*e.g.* images) by $\mathbb{I}$. We adopt the probability theory of Kolmogorov [21,22] and consider a measurable space (Ω, Σ) as well as the (generalized) random variables $W : \Omega \to \mathbb{W}$ for the physical parameters and $I : \Omega \to \mathbb{I}$ for the observation. Following the mathematical modeling introduced in [35], we consider the set $\mathbb{D}_{(\Omega,\Sigma)}$ of domains d in which there is a probability measure P_d on (Ω, Σ). We see the *ODD* of a given autonomous system as the set of domains in which it can be used safely, no matter if this has been established by design or by testing. Thus, $ODD \subseteq \mathbb{D}_{(\Omega,\Sigma)}$.

Table 1. In *CARLA*, the weather is controlled through 13 physical parameters. This table shows the range of values that we consider in our experiments and indicates, for each of them, if they are considered in our predictions.

parameter	range	predicted	parameter	range	predicted
cloudiness	0 to 100%	yes	fog distance	fixed to 0.75	no
fog density	0 to 100%	yes	fog falloff	fixed to 0.1	no
precipitation	0 to 100%	yes	mie scattering scale	fixed to 0.03	no
sun azimuth angle	$0°$ to $360°$	no	rayleigh scattering scale	fixed to 0.033	no
sun altitude angle	$-90°$ to $90°$	yes	scattering intensity	fixed to 1.0	no
wind intensity	0 to 100%	yes	wetness	fixed to 0.0	no
precipitation deposits	0 to 100%	yes			

Data. All our experiments are performed on data generated using the *CARLA* software, an open-source simulator for autonomous driving research [6]. All images are acquired by a simulated camera placed in front of a vehicle called *ego vehicle*. The weather is controlled through 13 real-valued parameters (see Table 1). There are 6 parameters for which we make predictions. Thus, $\mathbb{W} \subseteq \mathbb{R}^6$.

3.2   Task I: Predicting Distributions of Physical Parameters

The first task (see Fig. 3) consists in predicting how likely various physical parameters are, jointly. As these parameters cannot, in general, be measured directly, it is necessary to estimate the likelihood, *i.e.* the distribution of plausible values, based on an indirect observation.

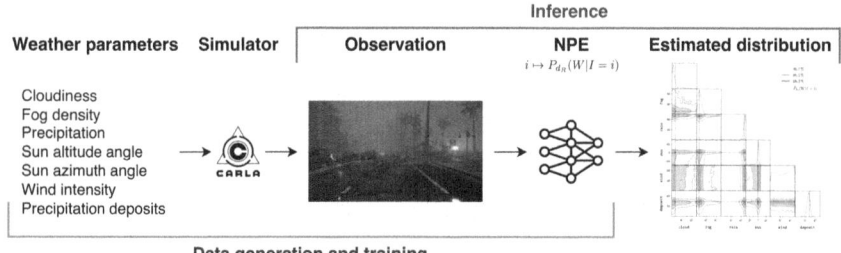

Fig. 3. Task I. The aim of the first task is, based on an image, to predict the joint distribution of the weather parameters. For this purpose, (1) we generate data using the *CARLA* software for uniformly distributed weather parameters (offline), (2) we train a *NPE* model using the learning set (*LS*) of our generated data (offline), and (3) we infer, given an image from the test set (*TS*), the estimated weather distribution (normalizing flow) and show the result on a corner plot.

Case Study. We study here the particular case in which the physical parameters are relative to the weather conditions and the observations are color images $i \in \mathbb{I}$ acquired by cameras placed in front of vehicles. We aim at learning, offline, a deep learning model $i \mapsto \hat{P}_{d_R}(W|I = i)$, where d_R denotes a domain of reference in which the probability measure on (Ω, Σ) is P_{d_R}. This domain is arbitrarily chosen in such a way that P_{d_R} has a large support.

Data. We use *CARLA* to generate a dataset with 635k images and the corresponding ground-truth values for the weather parameters. The dataset (Fig. 4) is split into a learning set (*LS*) with 600k images (500k for the training set and 100k for the validation set) and a test set (*TS*) with 35k images. Letting the model of the ego vehicle, the map, the number of pedestrians, and the number of vehicles vary brings a touch of diversity in our data.

Fig. 4. Excerpt of the images in our dataset generated with the *CARLA* simulator. The ground-truth weather parameters are drawn at random for each image, following a uniform distribution with the bounds given in Table 1.

Method. We consider three different (frozen) backbones to extract features from the input images: *ResNet-50* [13], *DINOv2* [31], and *CLIP* [37]. We use the libraries *LAMPE* [41] and *ZUKO* [40] to learn a model (of type *NPE*) and to manipulate the normalizing flows (of type *NSF*), respectively. All predicted weather distributions are posteriors relative to the weather priors in the *LS*. Note that *LAMPE* and *ZUKO* were not developed for domain characterization, but rather for *SBI*. Also, to the best of our knowledge, these libraries have only been used once with very high dimensional input data [47].

Evaluation and Results. Five different analyses are carried out.

1st analysis: histograms (Fig. 5b). Due to their physical meaning, the weather parameters are easily interpretable. This paves the way to a first, subjective, evaluation. We draw samples at random out of the predicted weather distributions $\hat{P}_{d_R}(W|I = i)$, for some images i arbitrarily chosen in *TS* and conduct a visual inspection of the histograms for the 6 marginals. The most credible weather parameters (*a.k.a.* highest density credibility sets, highest density regions, plausible sets, etc.) are highlighted for a credible level $l = 68.27\%$. For any input

weather parameter	ground-truth value
cloudiness (cloud)	19
fog density (fog)	8
precipitation (rain)	51
sun azimuth angle	90
sun altitude angle (sun)	54
wind intensity (wind)	43
precipitation deposits (deposit)	76

(a) An input image i and the corresponding ground-truth values.

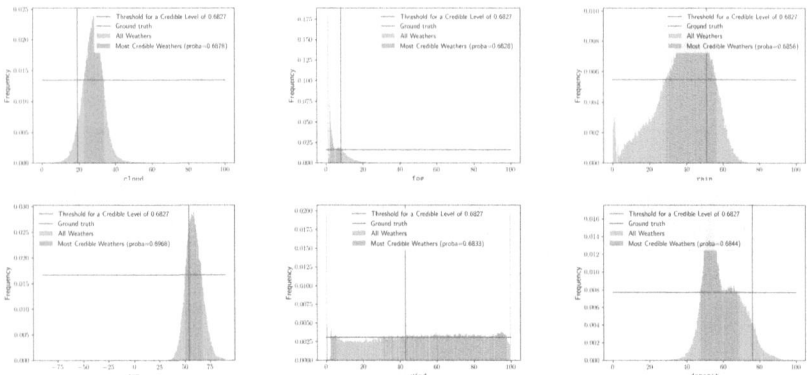

(b) Histograms for the 6 marginals of the predicted weather distribution for i.

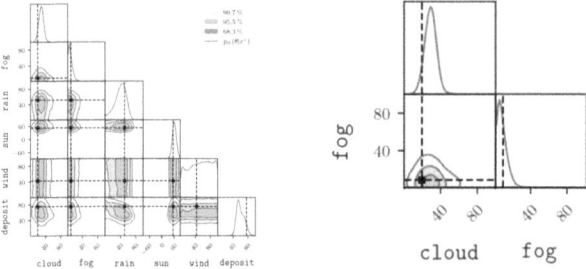

(c) Corner plot, with the ground-truth weather pinned (left: complete, right: zoomed).

(d) Posterior predictive check.

Fig. 5. Task I: results obtained, with the model learned with the *ResNet-50* backbone on 500k learning samples, for an arbitrarily chosen input image in the *TS*.

image i, these weathers are those such that the predicted *PDF* is above some threshold $t(i)$ and the predicted probability of the set is l [16].

2nd analysis: corner plots (Fig. 5c). A corner plot is a triangular array of plots. Those on the diagonal show the n marginals of a n-D distribution. Those below the diagonal depict the highest density credibility regions delimited at some arbitrarily chosen credibility levels (in this paper: 68.27%, 95.45%, and 99.73%), for each pair of parameters. We observe on the corner plots wide distributions, meaning that there is a large uncertainty for the weather parameters given an image. However, when analyzing the results, this uncertainty is explainable and the following analysis shows that this uncertainty is not excessive, *i.e.*, our models are not underconfident.

3rd analysis: posterior predictive checks (Fig. 5d). Once a weather distribution is predicted for an observation (input image), it is possible to (1) draw weather parameters vectors at random from it and then (2) to inject these vectors in *CARLA*, keeping all other parameters unchanged and immobilizing the vehicles and pedestrians, to obtain new images in order to finally (3) compare those with the initial input image. One cannot expect to obtain identical images as, in *CARLA*, the traffic lights continue to run, the rain continues to fall, the plants move with the wind, and pedestrians' poses are not perfectly frozen. Putting this aside, we observe that most retrieved images are very similar to the input image. We conclude that our models are not underconfident.

4th analysis: coverage plots (Fig. 6). Coverage plots show how the expected coverage varies with the credible level [14]. By definition, the expected coverage, at the credible level $l \in [0, 1]$, is the probability that the ground-truth weather belongs to the most credible weathers at level l. A method estimating the distribution of weathers based on an image is said underconfident (*i.e.*, conservative), calibrated, or overconfident at level l when the expected coverage at level l is, respectively, $> l$, $= l$, or $< l$. We observe that our models are all overconfident. The best calibrated model is the one that we obtained with 50k learning samples and *ResNet-50* as backbone.

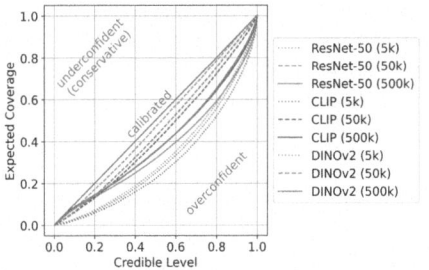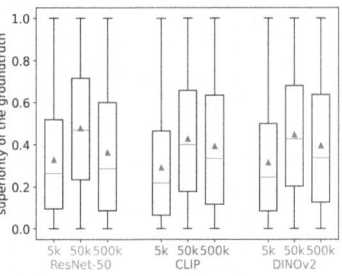

Fig. 6. Task I: comparison of our 9 models on the overall *TS*. Left: coverage plot. Right: box-and-whisker plots for the π statistic. Both analyses agree that the model learned with the *ResNet-50* backbone on 50k learning samples is preferable.

5th analysis: superiority of the ground truth (Fig. 6). We also determine to what extent the ground truth is more credible than the other weathers. This is the proportion π of weathers that can be drawn at random from the normalizing flow and that are predicted as having a *PDF* value (*i.e.*, likelihood) lower or equal than the one from the ground truth. The higher π is, the better it is, but approaching 1.0 is notably very challenging. We estimated π for each image of the *TS* and made box-and-whisker plots for the 9 models. This analysis is complementary to the coverage plot in the sense that a model can be perfectly calibrated while still presenting room for improvement. That being said, we are in the particular case in which this analysis leads to the same conclusion as the coverage plot: the model based on the backbone *ResNet-50* and learned from 50k samples is preferable to the others.

3.3   Task II: Absolute Domain Characterization

The second task (see Fig. 7) consists in characterizing a domain in an easy-to-interpret way. For this, we opt for a distribution of physical parameters estimated based on a bag (*a.k.a.* multiset) of observations.

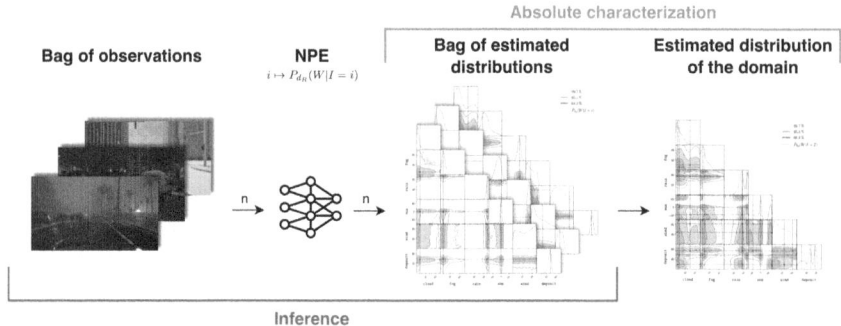

Fig. 7. Task II. The aim of the second task is to obtain an absolute characterization of a domain of interest. We process with the *NPE* each observation of a bag of observations of the domain of interest to obtain the bag of estimated distributions corresponding to the observations, then we obtain the distribution of weather parameters for the domain of interest by averaging the individual distributions of the bag.

Case Study. We characterize a domain $d \in \mathbb{D}_{(\Omega,\Sigma)}$ by the estimated distribution $\hat{P}_d(W)$ of weather conditions based on a real-valued bag b (multiset) of arbitrarily weighted images acquired by cameras placed in front of vehicles. In the following, we denote the weight (multiplicity) of the image $i \in b$ by $\omega(i)$. The images can either originate from a unique vehicle or from several, in case of vehicle-to-vehicle (*V2V*) communications. We want to establish $b \mapsto \hat{P}_d(W|B = b)$.

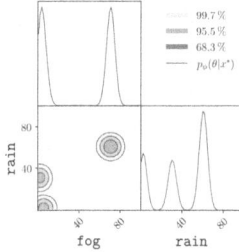  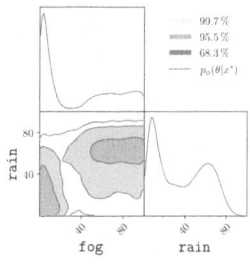

Fig. 8. Task II: example of result. Left: corner plot showing an arbitrarily chosen distribution of weather conditions, $P_d(W)$ (ground truth). Center: a bag b of images generated using $CARLA$ with weather conditions drawn from $P_d(W)$. Right: corner plot showing the estimated likelihood of the weather conditions $\hat{P}_d(W)$ based on b.

Data. We consider a bag b of 1,000 equally weighted images generated in maps already used in Task I. The weather parameters have been drawn at random from an arbitrary distribution $P_d(W)$ for which we fixed all parameters but two, `fog density` and `precipitation`, that follow a mixture of Gaussians.

Method. We use the NPE model developed for task I and build our solution for task II on top of it. We implement the following estimator for $P_d(W)$: $\hat{P}_d(W) = \sum_{i \in b} \omega(i) \hat{P}_{d_R}(W|I = i)$. The motivation for this estimator is threefold. (1) It is straightforward to evaluate the probability density function of $\hat{P}_d(W)$ and to draw weather conditions $w \in \mathbb{W}$ at random, following $\hat{P}_d(W)$, as we can do it with the normalizing flow $P_{d_R}(W|I = i)$. This will be valuable in our third task. (2) This estimator is useful for linear temporal filtering, *e.g.*, when one wants to weight more the recently acquired images than the old ones. (3) Finally, this estimator is fully justifiable under the assumptions that $P_d(W|I) = P_{d_R}(W|I)$ and $\omega(i) = P_d(I = i)$ as, in this case, $P_d(W) = \int_i \omega(i) P_{d_R}(W|I = i) di$.

Evaluation and Results. Our result is shown in Fig. 8. We observe that the three modes of the ground-truth weather distribution $P_d(W)$ are within the highest density regions of the prediction $\hat{P}_d(W)$. We made the same observation for many other ground-truth distributions (results omitted due to the limited space). We stress the fact that $\hat{P}_d(W)$ differs significantly from $P_d(W)$ can be explained by the inherent loss of information resulting from the use of color cameras, as already discussed in the introduction (*cf* Fig. 1).

3.4 Task III: Relative Domain Characterization

The third task (see Fig. 9) consists in characterizing any target domain $d_T \in \mathbb{D}_{(\Omega, \Sigma)}$, relatively, *w.r.t.* to some arbitrarily chosen source domains d_{S_1}, d_{S_2}, $\ldots d_{S_{n_S}} \in \mathbb{D}_{(\Omega, \Sigma)}$. Our motivation for this task originates from the importance of knowing if a system implementing a given many-to-one domain adaptation method can operate safely in the target domain d_T when it is known to operate safely in the source domains $d_{S_1}, d_{S_2}, \ldots d_{S_{n_S}}$.

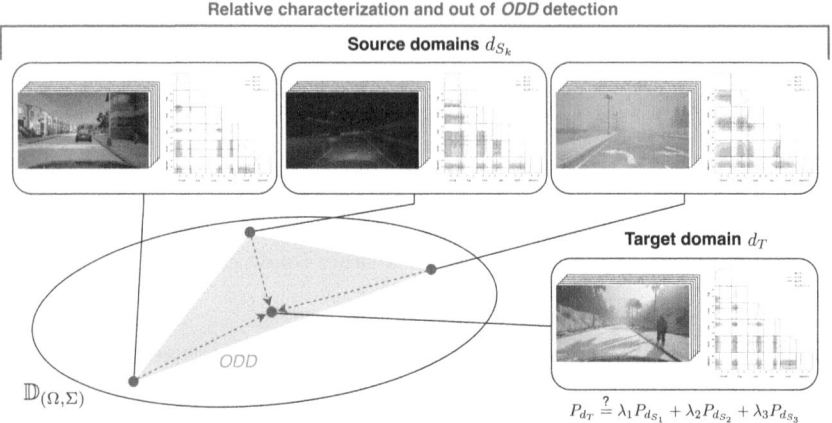

Fig. 9. Task III. The aim of the third task is twofold: (1) having a relative characterization of a target domain d_T based on source domains d_{S_k} and (2) detecting when this target domain d_T is out of the *ODD*.

Case Study. We discuss the case of *Mixture Domain Adaptation (MDA)* in which the system adapts to any target domain for which $P_{d_T} = \sum_{k=1}^{n_S} \lambda_k P_{d_{S_k}}$, with $\sum_{k=1}^{n_S} \lambda_k = 1$ and $\lambda_k \geq 0 \, \forall k$ (this is the *mixture assumption*). In this framework, the *ODD* is the convex hull of $\{P_{d_{S_1}}, P_{d_{S_2}}, \ldots P_{d_{S_{n_S}}}\}$. An autonomous car implementing *MDA* is expected to drive safely when $P_{d_T} \in ODD$. So, our goal in this third task is to determine if the mixture assumption holds and if so, what the values of the mixture weights λ_k are.

MDA supports many potential applications in the high-level tasks of the *Sense* pillar of the *Sense-Plan-Act* model [43]. While Mansour studied it generically [30], the application to the two-class classification task has been studied in [36], and the application to the semantic segmentation task of images acquired by vehicle-mounted cameras has been studied in [35]. These latter two works put an important emphasis on on-the-fly applicability and provide mathematically proven exact solutions. However, a critical limitation of these works is the need for the mixture weights to be known at adaptation time. Here, we remove this limitation by introducing a method that determines these weights automatically.

Data. We consider 4 subsets of the *TS* that we created for Task I: b_0, b_1, b_2, and b_3, containing 176, 27, 21, and 26 images, respectively. In the bag b_k, the distribution of the ground-truth weather parameters follows a uniform distribution on $\mathbb{W}_k \subsetneq \mathbb{W}$. These sets are such that $\mathbb{W}_0 \cap (\mathbb{W}_1 \cup \mathbb{W}_2 \cup \mathbb{W}_3) = \emptyset$.

Method. We characterize the source and target domains with the method presented for Task II, using the same weather distribution predictive model for all domains. We define the mean squared gap between the target domain and the mixture of the source domains as:

$$\delta(\hat{\lambda}_1, \ldots \hat{\lambda}_{n_S}) = \int_{\mathbb{W}} \left[\hat{P}_{d_T}(W = w) - \sum_{k=1}^{n_S} \hat{\lambda}_k \hat{P}_{d_{S_k}}(W = w) \right]^2 \hat{P}_{d_T}(W = w) dw$$

$$\simeq \frac{1}{n_W} \sum_i^{n_W} \left[\hat{P}_{d_T}(W = w_i) - \sum_{k=1}^{n_S} \hat{\lambda}_k \hat{P}_{d_{S_k}}(W = w_i) \right]^2 \quad (1)$$

with $\{w_i\}_{i=1}^{n_W} \sim \hat{P}_{d_T}(W)$. We aim at finding the values of $\hat{\lambda}_1, \ldots \hat{\lambda}_{n_S}$ that minimize $\delta(\hat{\lambda}_1, \ldots \hat{\lambda}_{n_S})$. This is a constrained least squares problem in which the constraints are $\sum_{k=1}^{n_S} \hat{\lambda}_k = 1$ and $\hat{\lambda}_k \geq 0 \,\forall k$. In order to use the *CVXPY* library [1,5] to solve our problem, we converted our original problem into a convex quadratic programming problem, with the same constraints.

Evaluation and Results. The goal of this experiment is twofold.

1. To show that it is possible to recover the mixture weights needed for the *MDA* technique presented in [35]. If the probability measure in the target domain is a mixture of the probability measures in the source domains, then we expect our characterization of the target domain to be a mixture of our characterizations for the source domains, with the same weights, as $P_{d_T} = \sum_{k=1}^{n_S} \lambda_k P_{d_{S_k}} \Rightarrow P_{d_T}(W) = \sum_{k=1}^{n_S} \lambda_k P_{d_{S_k}}(W)$.
2. To show that it is possible to detect when the target domain is not a mixture of the source domains, which means that the target domain is out of the *ODD* for the *MDA* technique presented in [35]. If our characterization of the target domain is not a mixture of our characterizations for the source domains, then we expect that the probability measure in the target domain is not a mixture of the probability measures in the source domains, as we have $P_{d_T}(W) \neq \sum_{k=1}^{n_S} \lambda_k P_{d_{S_k}}(W) \Rightarrow P_{d_T} \neq \sum_{k=1}^{n_S} \lambda_k P_{d_{S_k}}$.

To achieve this goal, we perform experiments in which we characterize target domains relatively to $n_S = 3$ source domains. However, to study the robustness of our method, we let the probability measures in the target domains d_T be mixtures not only of the probability measures in the source domains, but also in some domains that are out of the *ODD*. We use the bags b_1, b_2, and b_3 for the source domains d_{S_1}, d_{S_2}, and d_{S_3}, respectively. We also use the bag b_0 for domains $d_{\notin ODD}$ that are, by construction, guaranteed to be out of the *ODD*. Putting this into equations, we have $P_{d_T} = (1 - \eta)P_{d_{\in ODD}} + \eta P_{d_{\notin ODD}}$ with $P_{d_{\in ODD}} = \lambda_1 P_{d_{S_1}} + \lambda_2 P_{d_{S_2}} + \lambda_3 P_{d_{S_3}}$. The quantity $\eta \in [0, 1]$ is interpreted as a proportion of noise, which is the keystone to assess the robustness of our method.

In each experiment, the images in b_1, b_2, and b_3 are equally weighted, whereas those in b_0 are randomly weighted following a uniform distribution. The method introduced in Task II for the absolute characterization of domains is applied for d_{S_1}, d_{S_2}, d_{S_3}, and d_T. We arbitrarily chose $(\lambda_1, \lambda_2, \lambda_3) = (0.2, 0.3, 0.5)$. The mixture weights are optimized on $n_W = 16$ points.

The overall experiment consists in (1) choosing a target domain at random, as described here-above, (2) computing its absolute characterization with the method of Task II, (3) executing the weight estimation algorithm, and (4) reporting both the mean square gap $\delta(\hat{\lambda}_1, \ldots \hat{\lambda}_{n_S})$ and the Euclidean distance d_E between $(\hat{\lambda}_1, \hat{\lambda}_2, \hat{\lambda}_3)$ and $(\lambda_1, \lambda_2, \lambda_3)$. This experiment has been performed 330 times (30 times for 11 values of η). The results are shown in Fig. 10. The mean square gap achievable by chance is shown (baseline). Note that the target domain belongs to the ODD only when $\eta = 0$. We observe that d_E is negligible when $d_T \in ODD$ and that the mean square gap is effective to detect when $d_T \notin ODD$.

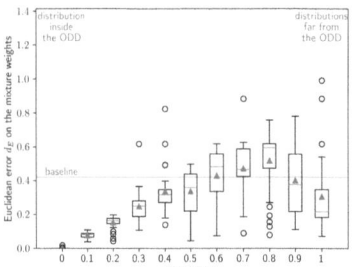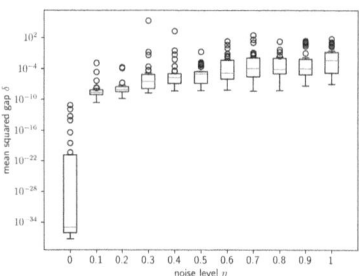

Fig. 10. Task III: results obtained when the noise level η sweeps the $[0, 1]$ interval. Left: box-and-whisker plots showing the distributions of d_E. Right: box-and-whisker plots showing the distributions of $\delta(\hat{\lambda}_1, \ldots \hat{\lambda}_{n_S})$. These plots show (1) that we can recover the relationship between the target domain and the source domains, under the mixture assumption, when the target domain belongs to the ODD and (2) that the $\delta(\hat{\lambda}_1, \ldots \hat{\lambda}_{n_S})$ can be used to determine when the target domain belongs to the ODD.

4   Conclusion

In this work, we present a novel approach to characterize domains by estimating the distribution of physical parameters. Our method enhances interpretability, facilitates domain adaptation, and provides safeguards for systems operating outside their ODD. Our experiments, organized around three tasks, are performed in the particular, but very important, case of autonomous vehicles and demonstrate how to obtain an absolute characterization of a domain by predicting a distribution of weather parameters (1) using a single image acquired by a vehicle-mounted camera and (2) using a bag of images. Our experiments also demonstrate (3) how to obtain a relative characterization of a target domain based on arbitrarily chosen source domains. Our approach includes two types of domain characterization: absolute and relative. The relative characterization is particularly valuable for domain adaptation, allowing the expression of a target domain in terms of source domains and verifying whether the current domain is part of the *ODD*. This is important for autonomous driving systems as well as for other fields requiring interpretable and trustworthy domain adaptation.

Acknowledgments. This work has been made possible thanks to the *TRAIL* initiative (https://trail.ac). Part of it was supported by the Walloon region (Service Public de Wallonie Recherche, Belgium) under grant n°2010235 (ARIAC by DIGITALWALLONIA4.AI). J. Held and A. Cioppa are funded by the F.R.S.-FNRS (https://www.frs-fnrs.be/en/).

Disclosure of Interests. The authors have no competing interests to declare that are relevant to the content of this article.

References

1. Agrawal, A., Verschueren, R., Diamond, S., Boyd, S.: A rewriting system for convex optimization problems. J. Control Decis. **5**(1), 42–60 (2018). https://doi.org/10.1080/23307706.2017.1397554
2. Boudiaf, M., Mueller, R., Ayed, I.B., Bertinetto, L.: Parameter-free online test-time adaptation. In: IEEE/CVF Conference on Computer Vision and Pattern Recognition (CVPR), pp. 8334–8343. Institute of Electrical and Electronics Engineers (IEEE), New Orleans, Louisiana, USA (2022). https://doi.org/10.1109/CVPR52688.2022.00816
3. Colwell, I., Phan, B., Saleem, S., Salay, R., Czarnecki, K.: An automated vehicle safety concept based on runtime restriction of the operational design domain. In: 2018 IEEE Intelligent Vehicles Symposium (IV) (2018). https://doi.org/10.1109/IVS.2018.8500530
4. Cranmer, K., Brehmer, J., Louppe, G.: The frontier of simulation-based inference. Proc. Nat. Acad. Sci. (PNAS) **117**(48), 30055–30062 (2020). https://doi.org/10.1073/pnas.1912789117
5. Diamond, S., Boyd, S.: CVXPY: a Python-embedded modeling language for convex optimization. J. Mach. Learn. Res. **17**(83), 1–5 (2016)
6. Dosovitskiy, A., Ros, G., Codevilla, F., Lopez, A., Koltun, V.: CARLA: an open urban driving simulator. In: Annual Conference on Robot Learning. Proceedings of Machine Learning Research, vol. 78, pp. 1–16. Mountain View, California, USA (2017). https://proceedings.mlr.press/v78/dosovitskiy17a.html
7. Durkan, C., Bekasov, A., Murray, I., Papamakarios, G.: Neural spline flows. In: Advances in Neural Information Processing Systems (NeurIPS), vol. 32 (2019)
8. Gérin, B., et al.: Multi-stream cellular test-time adaptation of real-time models evolving in dynamic environments. In: IEEE/CVF Conference on Computer Vision and Pattern Recognition Workshops (CVPRW), vol. 33, pp. 4472–4482. Institute of Electrical and Electronics Engineers (IEEE), Seattle, Washington, USA (2024). https://doi.org/10.1109/CVPRW63382.2024.00450
9. Gérin, B., Zanella, M., Wynen, M., Mahmoudi, S., Macq, B., De Vleeschouwer, C.: Exploring viability of test-time training: Application to 3D segmentation in multiple sclerosis. In: IEEE Conference on Artificial Intelligence (CAI), vol. 34, pp. 557–562. Institute of Electrical and Electronics Engineers (IEEE), Singapore, Singapore (2024). https://doi.org/10.1109/CAI59869.2024.00110

10. Gong, T., Jeong, J., Kim, T., Kim, Y., Shin, J., Lee, S.J.: NOTE: robust continual test-time adaptation against temporal correlation. In: Advances in Neural Information Processing Systems (NeurIPS), vol. 35, pp. 27253–27266. Curran Associates, Inc. (2022). https://openreview.net/forum?id=E9HNxrCFZPV

11. Greenberg, D., Nonnenmacher, M., Macke, J.H.: Automatic posterior transformation for likelihood-free inference. In: International Conference on Machine Learning (ICML). Proceedings of Machine Learning Research, vol. 97, pp. 2404–2414 (2019). https://proceedings.mlr.press/v97/greenberg19a.html

12. Gyllenhammar, M., et al.: Towards an operational design domain that supports the safety argumentation of an automated driving system. In: European congress on embedded real time systems (ERTS), pp. 1–10. Toulouse, France (2020)

13. He, K., Zhang, X., Ren, S., Sun, J.: Deep residual learning for image recognition. In: IEEE International Conference on Computer Vision and Pattern Recognition (CVPR), pp. 770–778. Las Vegas, Nevada, USA (2016). https://doi.org/10.1109/CVPR.2016.90

14. Hermans, J., Delaunoy, A., Rozet, F., Wehenkel, A., Begy, V., Louppe, G.: A trust crisis in simulation-based inference? your posterior approximations can be unfaithful. arXiv abs/2110.06581 (2021). https://doi.org/10.48550/arXiv.2110.06581

15. Houyon, J., et al.: Online distillation with continual learning for cyclic domain shifts. In: IEEE/CVF Conference on Computer Vision and Pattern Recognition Workshops (CVPRW), vol. abs 2211 16234, pp. 2437–2446. Institute of Electrical and Electronics Engineers (IEEE), Vancouver, Canada (2023). https://doi.org/10.1109/CVPRW59228.2023.00242

16. Hyndman, R.J.: Computing and graphing highest density regions. Am. Stat. **50**(2), 120–126 (1996). https://doi.org/10.1080/00031305.1996.10474359

17. Ibrahim, M., Haworth, J., Cheng, T.: WeatherNet: recognising weather and visual conditions from street-level images using deep residual learning. ISPRS Int. J. Geo-Inf. **8**(12), 1–18 (2019). https://doi.org/10.3390/ijgi8120549

18. Introvigne, M., Ramazzina, A., Walz, S., Scheuble, D., Bijelic, M.: Real-time environment condition classification for autonomous vehicles. In: IEEE Intelligent Vehicles Symposium (IV). Institute of Electrical and Electronics Engineers (IEEE), Jeju Island, Republic of Korea (2024). https://doi.org/10.1109/iv55156.2024.10588692

19. Jeon, M., Seo, J., Min, J.: DA-RAW: Domain adaptive object detection for real-world adverse weather conditions. In: IEEE International Conference on Robotics and Automation (ICRA), vol. 17, pp. 2013–2020. Institute of Electrical and Electronics Engineers (IEEE), Yokohama, Japan (2024). https://doi.org/10.1109/ICRA57147.2024.10611219

20. Kobyzev, I., Prince, S.J., Brubaker, M.A.: Normalizing flows: An introduction and review of current methods. IEEE Trans. Pattern Anal. Mach. Intell. **43**(11), 3964–3979 (2021). https://doi.org/10.1109/TPAMI.2020.2992934

21. Kolmogorov, A.N.: Grundbegriffe der wahrscheinlichkeitsrechnung. In: Grundbegriffe der Wahrscheinlichkeitsrechnung, Ergebnisse der Mathematik und Ihrer Grenzgebiete, vol. 2, p. 62. Springer, Heidelberg (1933). https://doi.org/10.1007/978-3-642-49888-6

22. Kolmogorov, A.N.: Foundations of the Theory of Probability. Chelsea Publishing Company (1950). https://archive.org/details/foundationsofthe00kolm

23. Lee, C.W., Nayeer, N., Garcia, D.E., Agrawal, A., Liu, B.: Identifying the operational design domain for an automated driving system through assessed risk. In: IEEE Intelligent Vehicles Symposium (IV). Institute of Electrical and Electronics Engineers (IEEE) (2020). https://doi.org/10.1109/IV47402.2020.9304552

24. Lee, S., Kim, D., Kim, N., Jeong, S.G.: Drop to adapt: learning discriminative features for unsupervised domain adaptation. In: IEEE/CVF International Conference on Computer Vision (ICCV), pp. 91–100. Institute of Electrical and Electronics Engineers (IEEE) (2019). https://doi.org/10.1109/ICCV.2019.00018

25. Li, J., Xu, R., Ma, J., Zou, Q., Ma, J., Yu, H.: Domain adaptive object detection for autonomous driving under foggy weather. In: IEEE/CVF Winter Conference on Applications of Computer Vision (WACV), pp. 612–622. Institute of Electrical and Electronics Engineers (IEEE), Waikoloa, Hawaii, USA (2023). https://doi.org/10.1109/WACV56688.2023.00068

26. Li, X., Wang, Z., Lu, X.: A multi-task framework for weather recognition. In: ACM International Conference on Multimedia, pp. 1318–1326. Mountain View, California, USA (2017). https://doi.org/10.1145/3123266.3123382

27. Li, Z., Li, Y., Zhong, J., Chen, Y.: Multi-class weather classification based on multi-feature weighted fusion method. In: Earth and Environmental Science. Journal of Physics: Conference Series, vol. 558, pp. 1–13. IOP Publishing (2020). https://doi.org/10.1088/1755-1315/558/4/042038

28. Lu, C., Lin, D., Jia, J., Tang, C.K.: Two-class weather classification. IEEE Trans. Pattern Anal. Mach. Intell. **39**(12), 2510–2524 (2017). https://doi.org/10.1109/TPAMI.2016.2640295

29. Lueckmann, J.M., Goncalves, P.J., Bassetto, G., Öcal, K., Nonnenmacher, M., Macke, J.H.: Flexible statistical inference for mechanistic models of neural dynamics. In: Advances in Neural Information Processing Systems (NeurIPS), vol. 30, pp. 1–11. Curran Associates, Inc., Long Beach, California, USA (November 2017)

30. Mansour, Y., Mohri, M., Rostamizadeh, R.: Domain adaptation with multiple sources. In: Advances in Neural Information Processing Systems (NeurIPS), vol. 21, pp. 1041–1048. Vancouver, Canada (December 2008), https://papers.nips.cc/paper_files/paper/2008/hash/0e65972dce68dad4d52d063967f0a705-Abstract.html

31. Oquab, M., et al.: DINOv2: learning robust visual features without supervision. arXiv abs/2304.07193 (2023). https://doi.org/10.48550/arXiv.2304.07193

32. Papamakarios, G., Nalisnick, E., Rezende, D.J., Mohamed, S., Lakshminarayanan, B.: Normalizing flows for probabilistic modeling and inference. J. Mach. Learn. Res. **22**(57), 1–64 (2021). http://jmlr.org/papers/v22/19-1028.html

33. Pappalardo, G., Caponetto, R., Varrica, R., Cafiso, S.: Assessing the operational design domain of lane support system for automated vehicles in different weather and road conditions. J. Traffic Transp. Eng. **9**(4), 631–644 (2022). https://doi.org/10.1016/j.jtte.2021.12.002

34. Perrels, A., Votsis, A., Nurmi, V., Pilli-Sihvola, K.: Weather conditions, weather information and car crashes. ISPRS Int. J. Geo-Inf. **4**(4), 1–23 (2015). https://doi.org/10.3390/ijgi4042681

35. Piérard, S., et al.: Mixture domain adaptation to improve semantic segmentation in real-world surveillance. In: IEEE/CVF Winter Conference on Applications of Computer Vision Workshops (WACVW), pp. 22–31. Institute of Electrical and Electronics Engineers (IEEE), Waikoloa, Hawaii, USA (2023). https://doi.org/10.1109/WACVW58289.2023.00007

36. Piérard, S., Marcos Alvarez, A., Lejeune, A., Van Droogenbroeck, M.: On-the-fly domain adaptation of binary classifiers. In: Belgian-Dutch Conference on Machine Learning (BENELEARN), pp. 20–28. Brussels, Belgium (2014)

37. Radford, A., et al.: Learning transferable visual models from natural language supervision. In: International Conference on Machine Learning (ICML). Proceedings of Machine Learning Research, vol. 139, pp. 8748–8763. Proceedings of Machine Learning Research (2021). https://proceedings.mlr.press/v139/radford21a.html

38. Rebut, J., Bursuc, A., Perez, P.: StyleLess layer: improving robustness for real-world driving. In: IEEE/RSJ International Conference on Intelligent Robots and Systems (IROS), pp. 8992–8999. Institute of Electrical and Electronics Engineers (IEEE), Prague, Czech Republic (2021). https://doi.org/10.1109/IROS51168.2021.9636204

39. Robertson, C., Long, J.A., Nathoo, F.S., Nelson, T.A., Plouffe, C.C.F.: Assessing quality of spatial models using the structural similarity index and posterior predictive checks. Geogr. Anal. **46**(1), 53–74 (2014). https://doi.org/10.1111/gean.12028

40. Rozet, F., Divo, F., Schnake, S.: Zuko: normalizing flows in PyTorch. Software (2022). https://doi.org/10.5281/ZENODO.7625672, https://pypi.org/project/zuko

41. Rozet, F., Miller, B.K., Delaunoy, A.: LAMPE: likelihood-free amortized posterior estimation. Software (2021). https://doi.org/10.5281/ZENODO.8405782, https://pypi.org/project/lampe

42. SAE International: Taxonomy and definitions for terms related to driving automation systems for on-road motor vehicles. Tech. Rep. SAE Standard J3016 202104, Society of Automobile Engineers, Warrendale, PA, USA (2021). https://doi.org/10.4271/J3016_202104

43. Shoker, A., Yasmin, R., Esteves-Verissimo, P.: Savvy: Trustworthy autonomous vehicles architecture. In: Symposium on Vehicles Security and Privacy (VehicleSec), pp. 1–7. San Diego, California, USA (2024). https://www.ndss-symposium.org/ndss-paper/auto-draft-465/

44. Sun, C., Deng, Z., Chu, W., Li, S., Cao, D.: Acclimatizing the operational design domain for autonomous driving systems. IEEE Intell. Transp. Syst. Mag. **14**(2), 10–24 (2022). https://doi.org/10.1109/MITS.2021.3070651

45. Tang, H., Chen, K., Jia, K.: Unsupervised domain adaptation via structurally regularized deep clustering. In: IEEE/CVF Conference on Computer Vision and Pattern Recognition (CVPR), pp. 8722–8732. Institute of Electrical and Electronics Engineers (IEEE), Seattle, WA, USA (2020). https://doi.org/10.1109/CVPR42600.2020.00875

46. Tonk, A., Boussif, A.: Operational design domain or operational envelope; seeking a suitable concept for autonomous railway systems. In: European Safety and Reliability Conference, pp. 2104–2111. Research Publishing Services (2022). https://doi.org/10.3850/978-981-18-5183-4_S06-08-245-cd

47. Vasist, M., Rozet, F., Absil, O., Mollière, P., Nasedkin, E., Louppe, G.: Neural posterior estimation for exoplanetary atmospheric retrieval. Astron. Astrophys. **672**, A147 (2023). https://doi.org/10.1051/0004-6361/202245263

48. Villarreal Guerra, J.C., Khanam, Z., Ehsan, S., Stolkin, R., McDonald-Maier, K.: Weather classification: a new multi-class dataset, data augmentation approach and comprehensive evaluations of convolutional neural networks. In: NASA/ESA Conference on Adaptive Hardware and Systems (AHS), pp. 305–310. Institute of Electrical and Electronics Engineers (IEEE), Edinburgh, UK (2018). https://doi.org/10.1109/AHS.2018.8541482

49. Wang, D., Shelhamer, E., Liu, S., Olshausen, B., Darrell, T.: Tent: Fully test-time adaptation by entropy minimization. arXiv abs/2006.10726 (2020). https://doi.org/10.48550/arXiv.2006.10726

50. Wang, Q., Fink, O., Van Gool, L., Dai, D.: Continual test-time domain adaptation. In: IEEE/CVF Conference on Computer Vision and Pattern Recognition (CVPR), pp. 7191–7201. Institute of Electrical and Electronics Engineers (IEEE), New Orleans, Louisiana, USA (2022). https://doi.org/10.1109/CVPR52688.2022.00706

51. Wang, W., et al.: Dynamically instance-guided adaptation: A backward-free approach for test-time domain adaptive semantic segmentation. In: IEEE/CVF Conference on Computer Vision and Pattern Recognition (CVPR), pp. 24090–24099. Institute of Electrical and Electronics Engineers (IEEE), Vancouver, Canada (2023). https://doi.org/10.1109/CVPR52729.2023.02307

52. Yuan, L., Xie, B., Li, S.: Robust test-time adaptation in dynamic scenarios. In: IEEE/CVF Conference on Computer Vision and Pattern Recognition (CVPR), pp. 15922–15932. Institute of Electrical and Electronics Engineers (IEEE), Vancouver, Canada (2023). https://doi.org/10.1109/CVPR52729.2023.01528

53. Zanella, M., Ayed, I.B.: On the test-time zero-shot generalization of vision-language models: Do we really need prompt learning? In: IEEE/CVF Conference on Computer Vision and Pattern Recognition (CVPR), vol. 33, pp. 23783–23793. Institute of Electrical and Electronics Engineers (IEEE), Seattle, Washington, USA (2024). https://doi.org/10.1109/CVPR52733.2024.02245

54. Zanella, M., Gérin, B., Ayed, I.B.: Boosting vision-language models with transduction. arXiv abs/2406.01837 (2024). https://doi.org/10.48550/arXiv.2406.01837

55. Zhang, Y., Sun, J., Chen, M., Wang, Q., Yuan, Y., Ma, R.: Multi-weather classification using evolutionary algorithm on EfficientNet. In: IEEE International Conference on Pervasive Computing and Communications Workshops and other Affiliated Events (PerCom Workshops), pp. 546–551. Institute of Electrical and Electronics Engineers (IEEE), Kassel, Germany (2021). https://doi.org/10.1109/PerComWorkshops51409.2021.9430939

Analysis of Adapter in Attention of Change Detection Vision Transformer

Ryunosuke Hamada(✉) ⓘ, Tsubasa Minematsu ⓘ, Cheng Tang ⓘ,
and Atsushi Shimada ⓘ

Kyushu University, 744 Motooka, Nishi-ku, Fukuoka 819-0395, Japan
hamada.ryunosuke.769@s.kyushu-u.ac.jp,
minematsu.tsubasa.659@m.kyushu-u.ac.jp, tang@limu.ait.kyushu-u.ac.jp,
atsushi@ait.kyushu-u.ac.jp

Abstract. Vision Transformer (ViT) contributes to accurate change detection with robustness to background changes. However, retraining ViT requires a large amount of computation to adapt to unlearned scenes. This study investigates the addition of learnable parameters into change detection ViT to reduce the computational complexity of retraining. We introduce MLP as an adapter as an addition to the attention output and the residual connection of the change detection ViT and apply LoRA method to the change detection ViT. We evaluate the retraining of additional parameter models for various background changes and analyze proper setting of additional parameters to adapt the target scenes. Introducing MLP and LoRA to change detection ViT improves the accuracy for the target scenes without competition between two additional parameter methods.

Keywords: Change detection · Vision Transformer · Adapter

1 Introduction

Accurate change detection methods are needed for surveillance systems using security cameras. Background subtraction methods can be used for the change detection task, which extracts regions where there is a scene that does not exist regularly. The background is defined as the scene that is regularly observed in the image. Background subtraction detects the foreground by removing the background from the input image. One of the challenges of background subtraction is background changes such as dynamic background, shadow, and illumination change.

Recently, Vision Transformer (ViT) [7] has demonstrated highly accurate performance in several vision tasks. Change detection methods using ViT have also been studied. Wang et al. [24] proposed TransCD using ViT for change detection. TransCD inputs two images into a ViT and computes the feature representation of each image. It detects changes by taking the difference between two feature representations. In addition to being robust to background changes

M. Cho et al. (Eds.): ACCV 2024 Workshops, LNCS 15482, pp. 36–51, 2025.
https://doi.org/10.1007/978-981-96-2641-0_3

that occur with fixed cameras, such as dynamic backgrounds and illumination changes, TransCD does not detect changes where different types of backgrounds are compared, such as those caused by PTZ camera scenes.

Change detection using ViT achieved high accuracy on trained datasets. However, the pre-trained change detection ViT can demonstrate lower performance on untrained scenes. Fine-tuning is often used to improve the accuracy on untrained scenes by retraining the model with a small number of untrained scene images. However, fine-tuning all parameters of ViT is computationally expensive because of the large dimension size. As a method to reduce the amount of calculation during fine-tuning ViT, adapter [19] was proposed. Adapter is a small number of parameters added to the model. We can handle target datasets with a low amount of calculation by tuning only the adapter for the target datasets because of its small size. ViT with adapters was effective in image classification tasks.

When adding an adapter for change detection ViT, it is necessary to consider the appropriate processing block in ViT for adding additional parameters. The knowledge that the adapter stores by retraining depends on the location of adapter, and affects the performance of the adapter model. For instance, Adaptformer [3], which introduced an adapter into the Multi-Layer Perceptron (MLP) block, enabled better image recognition than that of Visual Prompt Tuning [12], which introduced additional parameters into input tokens. When training a change detection model with an adapter, it is undesirable that the adapter only acquires knowledge about a specific object in the training dataset.

To keep the ability for change detection in ViT with adapters, we analyze the way to apply an adapter to acquire knowledge about changes in the training dataset. We introduce additional learnable parameters to attention, the residual connection, and attention weights in the change detection ViT and evaluate the contribution of each introduction method. We introduce MLP-based adapter for attention and the residual connection, and also introduce the existing additional parameter method to change detection ViT; LoRA [11]. We analyze the appropriate additional parameter method for change detection ViT by comparing the results of retraining with untrained scenes and changing the dimension size of additional parameters.

2   Related Works

Various background modeling have been proposed for handling background changes such as illumination changes and dynamic backgrounds. Conventional background subtraction methods represent the background that is stationary in the input image with a background model created using the median [17] and statistics [15,20] of pixels in the background image, and detect regions that deviate from the background as changes. Thereafter, background feature extraction using convolutional neural networks (CNNs) was proposed in [6]. Because the background subtraction method using CNNs enabled highly accurate change detection, various methods have been studied [1]. However, CNNs have problems such that their locality, which depends on the size of the receptive field, limits the range of images that can be referenced to obtain a feature representation. As the model for computing global feature representation from the entire image, ViT [7] was proposed and enabled highly accurate image classification. Following the emergence of ViT, their applications to change detection have been studied. Wang et al. [24] proposed TransCD as a model to introduce ViT to change detection through background subtraction. It inputs two images into ViT and computes the feature representation of each image. Chen et al. [2] proposed the change detection model of ViT that inputs the concatenated token sequences of each input image to the ViT encoder to model the temporal and spatial features between images. Inspired by the success of the Swin Transformer [16] in image classification, Zhang et al. [25] proposed SwinUNet for change detection. TransBlast [18] incorporates the inductive bias obtained from the CNN into the Transformer-based model training. Self-supervised learning using the extended loss function by subspace learning contributed to robust background/foreground separation even with training using limited labeled data. GIBS-Net [5] introduced global information of input images calculated by ViT into BSUV-Net [21] and improved the performance for unseen scenes. GIBS-Net also shows that it is possible to perform change detection with good accuracy even when we reduce the number of layers in GIBS-Net and computational complexity.

ViT provides highly accurate results for each task. However, it requires enormous computational complexity for fine-tuning because it has a huge number of parameters. Additionally, changing the parameters of a trained model may result in forgetting learned knowledge. In response to these issues, Rebuffi et al. proposed an adapter [19], which is a small number of parameters added to the model. The model fine-tunes the adapter module only without changing the parameters of the trained model. It is possible to improve the accuracy with low computation complexity by fine-tuning the adapter for untrained scenes and retaining pre-trained knowledge. Methods such as Adapter that add parameters for fine-tuning inside a model have been applied to various deep-learning tasks. Studies have been conducted on which modules of the model are appropriate to add parameters to. For image classification, AdaptFormer [3] introduced an adapter to the MLP ViT block, making it possible to store knowledge for a long token sequence of input. In the field of natural language processing, by adding an adapter module in two places after the attention and FeedForward blocks in Bert [14], it has become possible to handle multiple tasks with fewer calculations [10]. The knowledge and functionality of the adapter model depend on the location and shape of the adapter. Depending on the role we intend to add to a model, we need to verify where it is appropriate to introduce an adapter.

2.1   The Baseline Change Detection ViT

We selected TransCD [24] as a baseline change detection ViT. The reason is that it is a highly accurate change detection model with ViT and its internal processing is not changed from the first proposed ViT [7]. TransCD computes the feature representations of two images; the background image and the input image,

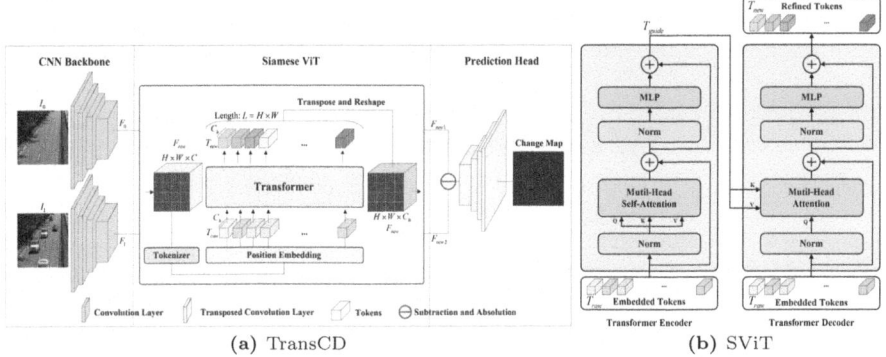

(a) TransCD   (b) SViT

Fig. 1. Overview of TransCD [24] and the SViT in TransCD

using a CNN and Siamese Vision Transformer (SViT). Thereafter it computes the difference between the obtained feature representations. Figure 1a shows the overview of the TransCD model.

First, each image is input to a CNN, and the features of the images are extracted from the feature map, which is converted into input tokens by dividing the map into patches. Positional encoding is added to input tokens, which is the positional information of the patch within the image. Next, each input token sequence, T_{raw}^1 and T_{raw}^2 is input to the SViT, two ViTs that share weights. The SViT models the features of each image from the feature map while acquiring the global relationship between each token and computes the two feature maps.

Figure 1b shows the overview of the SViT. In the encoder, an attention block computes the Multi-Head Self-Attention (MSA) of encoder input Z_{l-1} passed through Layer-Normalization (LN) and adds the residual connection (Eq. (2)). The MLP block computes the MLP of the attention block output Z_{l-1} passed through LN and adds the residual connection (Eq. (3)). Equations (2) and (3) are repeated k times and we obtain the output T_{guide}. In the decoder, the attention block computes the Multi-Head Attention (MA) of encoder input Y_{l-1} passed through LN and encoder output T_{guide}, and adds the residual connection (Eq. (6)). The MLP block computes Y_l similar to the encoder (Eq. (7)). Equations (6) and (7) are repeated k times and we obtain output T_{new}. Finally, TransCD computes the feature difference of two feature maps, T_{new}^1 and T_{new}^2, and converts it into a change map in the prediction head.

Encoder

$$Z_0 = T_{\mathrm{raw}} \tag{1}$$

$$Z_l' = \mathrm{MSA}(\mathrm{LN}(Z_{l-1})) + Z_{l-1}, \tag{2}$$

$$Z_l = \mathrm{MLP}(\mathrm{LN}(Z_l')) + Z_l', \tag{3}$$

$$T_{\mathrm{guide}} = \mathrm{LN}(Z_{k-1}) \tag{4}$$

`Decoder`

$$Y_0 = (T_{\text{guide}}, T_{\text{raw}}) \tag{5}$$

$$Y_l' = \text{MA}(T_{\text{guide}}, \text{LN}(Y_{l-1})) + Y_{l-1} \tag{6}$$

$$Y_l = \text{MLP}(\text{LN}(Y_l')) + Y_l' \tag{7}$$

$$T_{\text{new}} = \text{LN}(Y_{k-1}), \tag{8}$$

where $l = 1, 2, ..., k - 1$ is the number of layers.

3   Introducing the Additional Parameter to a Change Detection ViT

3.1   Analysis of ViT in Change Detection

We analyze the role of the internal processing of TransCD to gain insights into the appropriate places to introduce an adapter. A previous study on TransCD [24] revealed that TransCD can detect changes with high accuracy owing to the representation of image features. Therefore, in this analysis, we focused on the attention block in the SViT decoder, where TransCD computes the feature representation. The attention block comprises two main types of processes, the MA layer and residual connection. We analyze the role that these two processes play in change detection. We amplified the outputs of the attention layer or residual connection in TransCD by applying an appropriate scalar. Next, we input the untrained dataset to the amplified-TransCD. We compared the differences in change detection results between amplified TransCD and baseline TransCD. We estimated the amplified function has a role in dealing with the differences.

We used a TransCD pre-trained with Change Detection.Net 2014 dataset (CDNet-2014) [8] as the baseline model. We selected People-and-Foliage in SBM-Net [13] as untrained scenes. Background and Input in Fig. 2 show the background and input images in People-and-Foliage. People in the input image are the detection targets, and plants and a car in the background image are background objects. Baseline in Fig. 2 shows the result of change detection by the baseline TransCD. The result images show foreground pixels as white regions. The baseline model failed to detect some of the foreground human regions.

After that, we amplified the output of attention and residual connection at various rates and observed the change detection results. In Eq. (6), we multiplied MA by a scalar value to amplify the output of attention. The value implies the amplification rate of attention. We also multiplied Y_{l-1} by a scalar value to amplify the residual connection. We used various scalar ranges from 1.0 to 5.0. Res-amp, Att-amp, and Res-att-amp in Fig. 2 show the results of change detection by the model with various amplifications. Res-amp is the result of the model amplifying residual connection by 1.5. In Res-amp, the accuracy of foreground detection improved compared to that of the baseline model. This indicates that amplification of the residual connection is effective in suppressing defects in

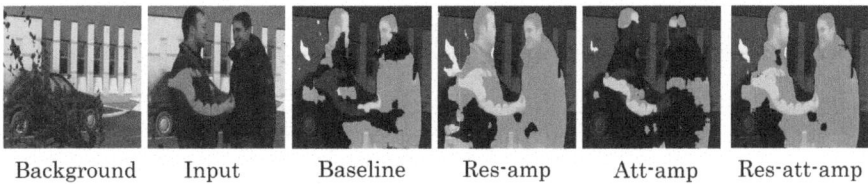

| Background | Input | Baseline | Res-amp | Att-amp | Res-att-amp |

Fig. 2. Results of change detection with amplification

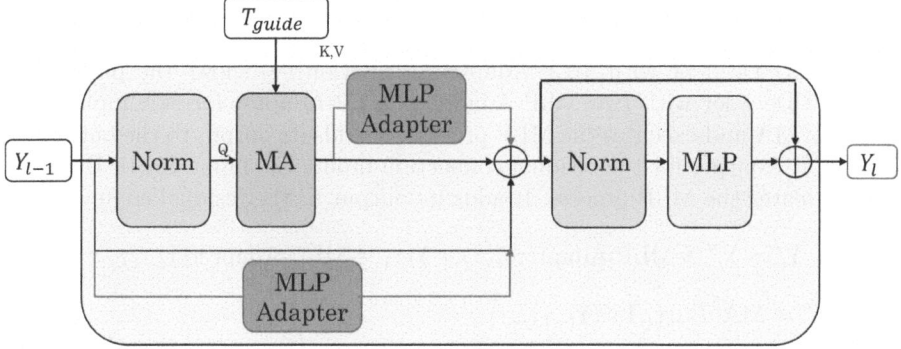

Fig. 3. The model of TransCD Decoder with MLP-Adapter

foreground detection. However, a new false detection of the plants occurred at the left of the image. Att-amp is the result of the model amplifying the attention by 1.5. Att-amp detected fewer foregrounds than the baseline model, but it suppressed the false detection of the plants. This result shows that amplifying attention effectively suppresses false detection of the background. Res-att-amp in Fig. 2 shows the results of change detection by the model amplifying attention and the residual connection by 1.5. Compared to Res-amp, the result had no false detection of the background caused by amplification of the residual connection. From these results, attention suppresses the false detection of background changes and the residual connection improves the foreground detection. Therefore, we expected that introducing learnable parameters to attention and the residual connection would contribute to adapting to the target scenes.

3.2 The Additional Parameters for Change Detection ViT

We introduce the learnable parameters for retraining to various functions of ViT and evaluate the effectiveness of each introduction method. We propose 2 types of introduction; MLP blocks in attention output and the residual connection, and LoRA layer in attention weights.

First, we propose to introduce MLP block as adapter to TransCD. We call the MLP block MLP-Adapter. We introduce MLP-Adapter for two functions; Multihead Attention (MA) and the residual connection. Figure 3 shows the model

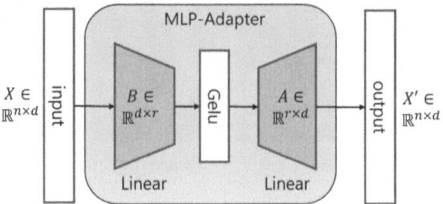

Fig. 4. The model of MLP-Adapter

of TransCD Decoder with MLP-Adapters. Equation (9) shows the process in TransCD Decoder with two MLP-Adapters. MLP-Adapter for MA inputs the output of MA and executes the MLP process. It adds its output to the output of MA. MLP-Adatper for the residual connection inputs the input of l th Decoder and calculates the MLP process. It adds its output to the residual connection.

$$Y_l' = Y_l'' + \text{MLP-Adapter}(Y_l'') + Y_{l-1} + \text{MLP-Adapter}(Y_{l-1}), \qquad (9)$$

where $Y_l'' = \text{MA}(T_{\text{guide}}, \text{LN}(Y_{l-1}))$.

MLP-Adapter consists of three step processes. Figure 4 shows the detail of MLP-Adapter and Eq. (10) shows the calculation in MLP-Adapter. MLP-Adapter has the internal dimension size r as a hyperparameter. First, it processes the input $X \in \mathbb{R}^{n \times d}$, where n is the number of patches and d is the hidden dimension size, by the first linear layer and resizes the hidden dimension from d to r by multiplying $B \in \mathbb{R}^{d \times r}$. Second, the activation function Gelu [9] processes the output of the first linear layer. Third, the second linear layer inputs the output of the activation function and outputs d dimension matrix $X' \in \mathbb{R}^{n \times d}$ the same size as the input of MLP-Adapter by multiplying $A \in \mathbb{R}^{r \times d}$.

$$\text{MLP-Adapter}(X) := \text{Gelu}(XB)A \qquad (10)$$

Second, we apply LoRA to change detection ViT. LoRA introduces LoRA layer to the weight of the model. LoRA layer consists in two matrixes; one is the matrix initialized to a normal distribution and the other is the zero matrix. LoRA can learn additional knowledge about the target domain. We introduce LoRA layer to attention weights of change detection ViT and evaluate the effectiveness of LoRA in the change detection tasks.

4   Experiments

First, we evaluate the effectiveness of various additional parameter models for change detection ViT. Second, we show the experiments to analyze the suitable hyperparameter of MLP-Adapter. Third, we introduce scalar parameter adapter to change detection ViT and analyze the propriety of introducing adapter to attention and the residual connection.

As the baseline, we used the TransCD pre-trained with CDNet-2014 [23]. It includes SViT with four encoder layers and four decoder layers. The input of the baseline model is a token sequence generated by splitting an image into 16×16 patches.

Every model was retrained and evaluated for each dataset independently. The training and evaluation datasets comprised the target datasets, and they are made to be different images. Training images are from the first half of the dataset, whereas evaluation images are from the second half of the dataset.

4.1 Experiments of MLP-Adapter

We evaluated tuning various additional parameter models by untrained scenes. We constructed three additional parameter models of TransCD; MLP-Adapter, LoRA, and MLP+LoRA. We analyzed the effectiveness of these models by comparing each other and the fine-tuned baseline model. In MLP-Adapter, the internal dimensions for both MA and the residual connection are four. We selected the best combination of internal dimensions because we make fair comparisons with LoRA. We also show the analysis of suitable combination of internal dimensions at Sect. 4.2. In LoRA, we introduced four-dimensional LoRA layers to attention key, query, and value weights. In MLP+LoRA, we introduced both two MLP-Adapters and LoRA to TransCD. We set the internal dimensions of MLP-Adapter to four and the rank of LoRA to four; They are the same setting as MLP-Adapter and LoRA. We executed retraining for 10 training datasets; Lasiesta I_IL_02 [4], LIMU Light switch on/off in Indoor Scene[1], and BMC Real [22] Video001 to Video009 except for Video003. We excluded Video003 due to its small number of images. We set the number of epochs to 500, and the learning rate to 0.0002. We set dropout in tuning parameters in each experiment setting. We calculated Precision, Recall, and F1 score of each dataset in each method.

Table 1 summarizes the accuracy of retraining by each method for each dataset. Figure 5 shows the change detection results of each method. First, we compared the results of MLP-Adapter with ones of LoRA. MLP-Adapter resulted in higher accuracy in three datasets, especially in Video001. The average F1 score of MLP-Adapter was Almost same as that of LoRA; 0.0296 lower than LoRA. We consider MLP-Adapter is an effective additional parameter model as LoRA from the above results. However, MLP-Adapter failed training in Video004. We should consider that MLP-Adapter has some scenes where it is not effective.

Second, we analyzed MLP+LoRA by comparing other methods. The average F1 of MLP+LoRA was the highest in additional parameter models. MLP+LoRA resulted in higher F1 than MLP-Adapter in seven datasets and higher F1 than LoRA in four datasets. MLP+LoRA improved accuracy in Video004 where MLP-Adapter failed training and got the highest F1 in LightSwitch. MLP+LoRA did not reduce accuracy significantly in any datasets. These results show that MLP-Adpater and LoRA do not compete with each other. Therefore, introducing

---

[1] https://limu.ait.kyushu-u.ac.jp/dataset/en/.

Table 1. Results of retraining for various datasets by each method

Dataset	MLP-Apdater			LoRA		
	F1	Precision	Recall	F1	Precision	Recall
Lasiesta	**0.9288**	0.9573	0.9019	0.9228	0.9393	0.9069
LightSwitch(LIMU)	0.2273	0.1464	0.5079	0.2305	0.1358	0.7611
Video001(BMC Real)	**0.6884**	0.7851	0.6130	0.5060	0.4790	0.5363
Video002(BMC Real)	0.5435	0.8270	0.4047	0.5648	0.8610	0.4203
Video004(BMC Real)	0.1664	0.9246	0.0914	0.4734	0.8980	0.3215
Video005(BMC Real)	0.1637	0.0911	0.8044	0.1697	0.0948	0.8066
Video006(BMC Real)	0.6996	0.8891	0.5767	0.7446	0.9129	0.6287
Video007(BMC Real)	0.4911	0.4382	0.5585	0.4890	0.4060	0.6148
Video008(BMC Real)	0.7605	0.7885	0.7344	0.7934	0.7959	0.7910
Video009(BMC Real)	0.6360	0.8878	0.4955	0.7072	0.8469	0.6070
Average	0.5305	0.6735	0.5688	0.5601	0.6370	0.6394
Dataset	MLP+LoRA			Fine-Tuning		
	F1	Precision	Recall	F1	Precision	Recall
Lasiesta	0.9227	0.9667	0.8826	0.9243	0.9655	0.8865
LightSwitch(LIMU)	**0.6221**	0.8205	0.5009	0.4133	0.3205	0.5820
Video001(BMC Real)	0.6713	0.7087	0.6376	0.6301	0.6491	0.6121
Video002(BMC Real)	0.5630	0.8400	0.4234	**0.6828**	0.8222	0.5838
Video004(BMC Real)	0.3038	0.9441	0.1810	**0.3061**	0.8930	0.1847
Video005(BMC Real)	0.1016	0.0543	0.7931	**0.3500**	0.2409	0.6399
Video006(BMC Real)	0.7242	0.9086	0.6020	**0.8061**	0.8960	0.7326
Video007(BMC Real)	0.5313	0.5818	0.4889	**0.6051**	0.7130	0.5255
Video008(BMC Real)	**0.8073**	0.8057	0.8090	0.7917	0.7940	0.7894
Video009(BMC Real)	0.6848	0.8473	0.5747	**0.7402**	0.8420	0.6603
Average	0.5932	0.7478	0.5893	**0.6250**	0.7136	0.6197

both MLP-Adapter and LoRA contributes to the improvement of accuracy of retraining.

Table 2 shows the number of tuned parameters in each method. The amount of parameters in MLP-Adapter was about 0.17% of those of Fine-tuning. Even in the larger additional parameter model MLP+LoRA, the number of tuned parameters was less than 0.4% of that of Fine-tuning. Therefore, introducing MLP and LoRA to change detection ViT contributes to the reduction of computational complexity while improving the accuracy of change detection ViT.

Table 2. The number of tuned parameters of each method

	MLP-Adapter	LoRA	MLP+LoRA	Fine-tuning
Number of parameter	18464	24576	43040	11158211

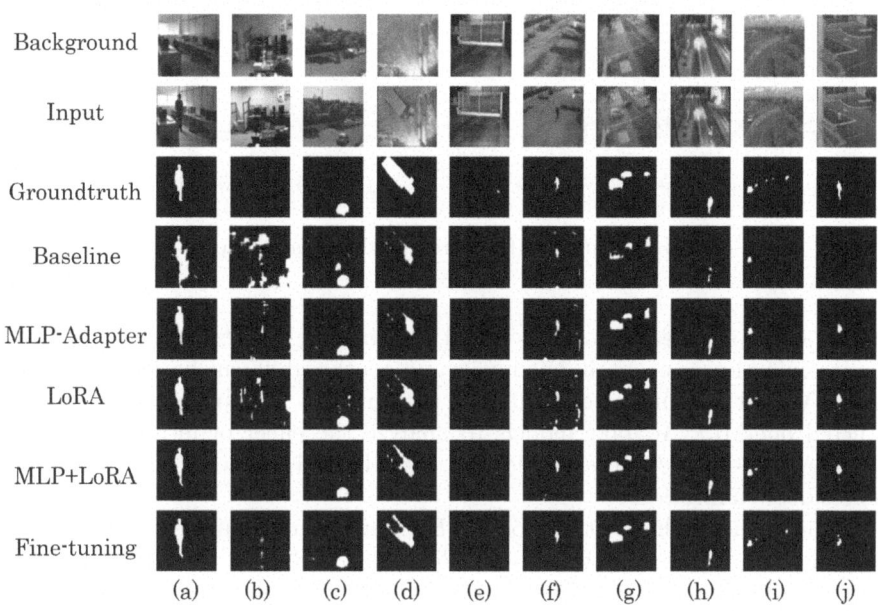

Fig. 5. The results of retraining by various methods (a)Lasiesta, (b)LIMU LightSwitch, (c)Video001, (d)Video002, (e)Video004, (f)Video005, (g)Video006, (h)Video007, (i)Video008, (j)Video009 Baseline means the result by baseline TransCD pre-trained by CDNet-2014

4.2   Analysis of Internal Dimension of MLP-Adapter

MLP-Adapter has an internal dimension size as a hyperparameter. In this section, we analyze the suitable dimension size of two MLP-Adapters; for MA and the residual connection. We selected 1, 2, 4, 16, and 64 as the candidate of internal dimension size. These internal dimensions are smaller than the input dimension of MLP-Adapter. We intended to store effective knowledge about the target scene in smaller parameters such as LoRA [11]. We executed retraining various internal dimension size MLP-Adapter for some datasets and evaluated the accuracy for target scenes. We retrained MLP-Adapter models with two different settings for the number of training images. We selected BMC Real Video008 as the training dataset. It includes simple dynamic background; swaying plants. We selected it because the baseline model cannot detect changes accurately despite the simplicity of background change. First, we retrained MLP-Adapter models with many training images as the experiments in Sect. 4.1; the first half of the dataset. Second, we retrained MLP-Adapter models with a few

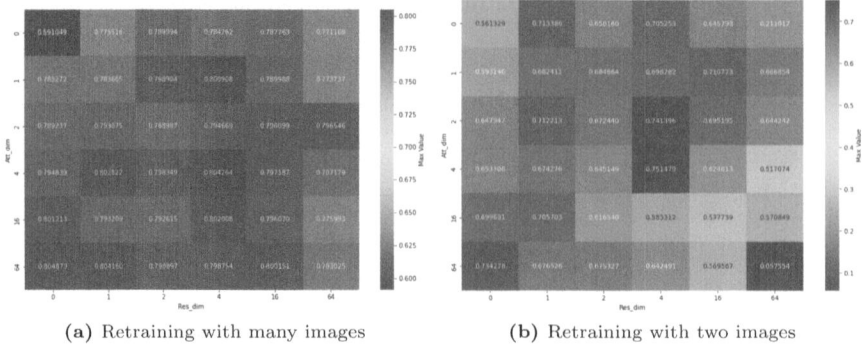

(a) Retraining with many images   (b) Retraining with two images

Fig. 6. Heatmap of F1 score for BMC real Video008

images as few-shot learning. We selected two training images from each dataset.
Two training images contain one image and another different background scene
image. We set learning late 0.0002 and retrained in 100 epochs. We also constructed one-MLP-Adapter models to analyze the effect of introduced MLP-
Adapter. One-MLP-Adapter model has one MLP-Adapter either for MA or for
the residual connection.

Figure 6 shows the heatmap of the F1 score by retraining various internal
dimension MLP-Adapters with BMC Real Video008. The vertical axis means
the internal dimension of MLP-Adapter for MA, and the horizontal axis means
that of MLP-Adapter for the residual connection. 0 dimension in each axis means
introducing no MLP-Adapter to TransCD; for example, the (0,4) model has only
one MLP-Adapter for the residual connection, not one for MA. (0,0) means the
result of baseline TransCD. The heatmap scale of Fig. 6a is different from that
of Fig. 6b.

Figure 6a shows the result of retraining with many images. All MLP-Adapter
models with every combination of dimensions improved the F1 score from the
baseline model. There was no significant difference in F1 between the different
dimension sizes; The gap between the best F1 (64,0) and the worst F1 (0,64) was
less than 0.035. Figure 6b shows the result of retraining with two images. MLP-
Adapter with combination of small-dimension sizes such as (4,4) and (2,4) had
higher F1 scores than the combination of large-dimension sizes such as (16,16)
and (64,64); (64,64) model failed to improve accuracy from the baseline model.

From these results, when retraining with many images, MLP-Adapter with
every dimension size demonstrated better performance than the baseline. In contrast, when retraining with two images, MLP-Adapter with the small-dimension
size improved the performance stably, while some of the MLP-Adapter with
large-dimension size failed to improve accuracy. In addition, retraining small-
dimension size of MLP-Adapter needs smaller computation than large-dimension
size of MLP-Adapter. Therefore, the small dimension MLP-Adapter is suitable
for retraining. However, the smallest dimension size is not always the best combination; the best combination in retraining with two images was (4,4), not

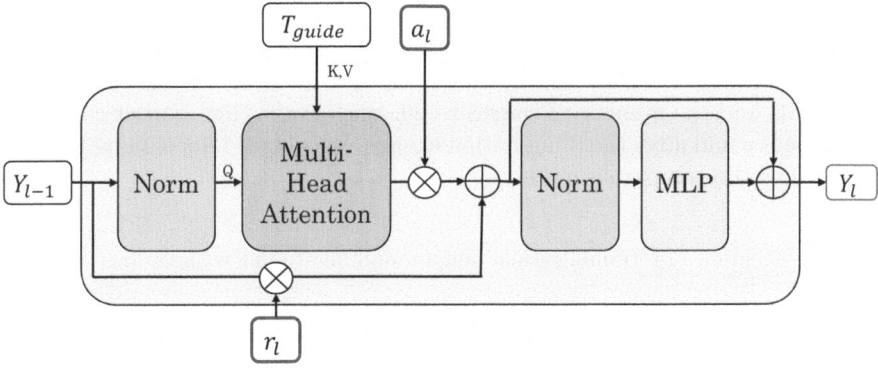

Fig. 7. Model of TransCD decoder with scalar adapters

such as (1,1). Therefore, we need to select an appropriate combination of two MLP-Adapters depending on the target scenes and retraining method.

4.3 Introducing Scalar Parameter Adapter to Change Detection ViT

In this section, we analyze the contribution of adapter for MA and the residual connection. We introduced scalar parameters as the smallest adapter to MA and the residual connection. We confirmed whether the introduction scalar adapter to MA and the residual connection can get the knowledge of the target scene. Figure 7 shows the decoder with two scalar parameter adapters. We introduced a scalar parameter adapter a_l after MA and another scalar parameter adapter r_l to the residual connection. We multiplied the output of MA by a_l and multiplied the residual connection r_l. These processes mean introducing one dimension MLP-Adapter in Fig. 4 like Eq. (11).

$$\text{Adapter}(X) := cX \tag{11}$$

In this experiment, we retrained the scalar parameter adapter for untrained illumination change scenes. We also verified the difference in detection accuracy depending on the number and content of untrained datasets. We selected two datasets in terms of the difference in strength of illumination change. We used Lasiesta I_IL_02 and LIMU Camera Parameter Changes in Indoor Scene[2] as untrained datasets. Lasiesta includes weak illumination changes depending on the presence and absence of sunshades. LIMU CameraParameter includes strong illumination changes depending on the light switch. We used the following four types of datasets as concerns the number of datasets and their content, 100-imgs, 2-imgs, Dark, and Bright. 100-imgs has one hundred images containing both dark and bright scenes. 2-imgs has one dark scene and one bright scene. Dark is one

---

[2] https://limu.ait.kyushu-u.ac.jp/dataset/en/.

dark scene. Bright is one bright scene. Tuning with 100-imgs implies adjusting with enough training images as conventional fine-tuning. We used 2-imgs as a few-shot learning, and Dark and Bright as one-shot learning. We set the learning rate to 0.1, and the number of epochs to 20. Each evaluation dataset contained images before and after the illumination change. We also did fine-tuning baseline TransCD as the comparing method.

Table 3. Accuracy of retraining scalar adapter and fine-tuning with various setting of dataset

| Method | Datasets | Laciesta | | | sLIMU CameraParameter | | |
		F1	Precision	Recall	F1	Precision	Recall
baseline	-	0.5703	0.4632	0.7418	0.112	0.062	0.607
Fine-tuning	100-imgs	**0.9412**	0.9556	0.9271	**0.889**	0.897	0.881
Fine-tuning	2-imgs	0.9100	0.9301	0.8907	0.297	0.483	0.214
Fine-tuning	Dark	0.7359	0.6284	0.8877	0.180	0.103	0.713
Fine-tuning	Brignt	0.8866	0.9298	0.8472	0.387	0.482	0.323
Scalar-tuning	100-imgs	0.7643	0.8493	0.6947	0.227	0.217	0.238
Scalar-tuning	2-imgs	**0.7768**	0.9623	0.6513	0.327	0.491	0.246
Scalar-tuning	Dark	0.7119	0.7891	0.6485	0.134	0.105	0.185
Scalar-tuning	Bright	0.7726	0.9561	0.6482	**0.402**	0.700	0.282

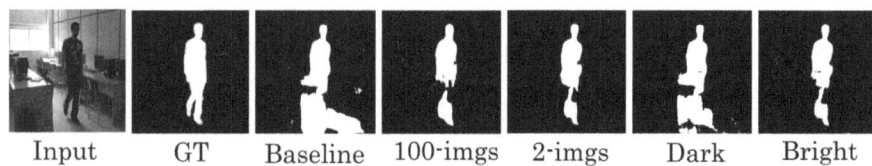

Input   GT   Baseline   100-imgs   2-imgs   Dark   Bright

Fig. 8. Results of change detection to Lasiesta by retraining scalar adapters

Table 3 summarizes the accuracy of change detection to Lasiesta and LIMU CameraParamter by fine-tuning and by tuning scalar adapter with the four datasets. In the results of Lasiesta, tuning scalar adapter with both training datasets recorded a higher F1 score than the baseline model. Scalar-tuning with 2-imgs had the best accuracy in Scalar-tuning, followed by Scalar-tuning with Bright which achieved an almost similar accuracy. Scalar-tuning with Dark resulted in an approximately 0.06 lower F1 score than the other Scalar-tuning. Figure 8 shows the detection results of each Scalar-tuning. Scalar-tuning except for Dark scene did not detect illumination changes and shadows on the floor.

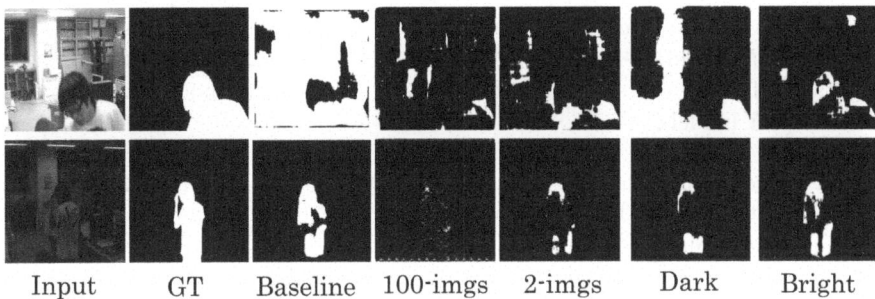

| Input | GT | Baseline | 100-imgs | 2-imgs | Dark | Bright |

Fig. 9. Results of change detection to LIMU CameraParameter by retraining scalar adapters

In the results of LIMU CameraParamter, Scalar-tuning except for with Dark improved the F1 score by more than 0.1. Scalar-tuning with Dark exhibited little improvement in the F1 score. Compared to fine-tuning, Scalar-tuning with 100-imgs was significantly less accurate than fine-tuning. However, Scalar-tuning with 2-imgs and Bright had a better F1 score than that of fine-tuning. For all datasets, the accuracy of the scalar adapter models was inferior to fine-tuning. Comparing the result of 100-imgs and that of Bright in LIMU CameraParameter, training with multiple datasets reduced the accuracy of the scalar parameter adapter model. Training with 1 scene could improve accuracy for the training scene. Figure 9 shows the result of change detection to LIMU CameraParameter. The baseline detected the brightness of bright input mistakenly. However, training with Bright suppressed to detect the brightness. From the results, a scalar adapter can get the knowledge of one illumination change and improve the change detection for the learned scene. Bright dataset has only one bright scene, so the scalar adapter may get the knowledge of the bright scene without competition by dark scenes.

5   Conclusion

This study analyzed the introduction of the various learnable parameters to change detection ViT. MLP-Adapter for MA and the residual connection and LoRA effectively improve accuracy for untrained scenes by tuning additional parameters, and the combination of MLP-Adapter and LoRA is available. The small internal dimension size is suitable for MLP-Adapter. From the experiments with scalar parameters, we showed that the introduction of adapter to MA and the residual connection is reasonable. Adapter needs more than two dimension sizes of parameters to adapt multiple scenes before and after background scenes. In the future, we will construct an efficient method for selecting and introducing appropriate additional parameters for the target scene. In this study, we used pre-trained TransCD and tuned additional parameters with the pre-trained TransCD. We should evaluate training TransCD from scratch to the target scene and the limits of the methods of additional parameters. In addition, we compared MLP-Adapter to LoRA and Fine-tuning only. To gain deeper insights, we further evaluate the difference between our method and other previous adapter models.

Acknowledgements. This work was supported by JST CREST Grant Number JPMJCR22D1, JSPS KAKENHI Grant Number JP22H00551, and JST, PRESTO Grant Number JPMJPR236A, Japan.

References

1. Bouwmans, T., Javed, S., Sultana, M., Jung, S.K.: Deep neural network concepts for background subtraction:a systematic review and comparative evaluation. Neural Netw. **117**, 8–66 (2019). https://doi.org/10.1016/j.neunet.2019.04.024, https://www.sciencedirect.com/science/article/pii/S0893608019301303
2. Chen, H., Qi, Z., Shi, Z.: Remote sensing image change detection with transformers. IEEE Trans. Geosci. Remote Sens. **60**, 1–14 (2022). https://doi.org/10.1109/TGRS.2021.3095166
3. Chen, S., et al.: AdaptFormer: adapting vision transformers for scalable visual recognition. Adv. Neural. Inf. Process. Syst. **35**, 16664–16678 (2022)
4. Cuevas, C., Yáñez, E.M., García, N.: Labeled dataset for integral evaluation of moving object detection algorithms: LASIESTA. Comput. Vis. Image Underst. **152**, 103–117 (2016)
5. Cui, H., Lv, Z., Yuan, T., Feng, C., Shan, X.: GIBS-Net: unseen video background subtraction with global information. In: 2023 IEEE 11th Joint International Information Technology and Artificial Intelligence Conference (ITAIC), vol. 11, pp. 125–130 (2023). https://doi.org/10.1109/ITAIC58329.2023.10408949
6. Culibrk, D., Marques, O., Socek, D., Kalva, H., Furht, B.: A neural network approach to Bayesian background modeling for video object segmentation. In: VISAPP (2006)
7. Dosovitskiy, A., et al.: An image is worth 16×16 words: transformers for image recognition at scale. arXiv preprint arXiv:2010.11929 (2020)
8. Goyette, N., Jodoin, P.M., Porikli, F., Konrad, J., Ishwar, P.: Changedetection.net: a new change detection benchmark dataset. In: 2012 IEEE Computer Society Conference on Computer Vision and Pattern Recognition Workshops, pp. 1–8 (2012). https://doi.org/10.1109/CVPRW.2012.6238919
9. Hendrycks, D., Gimpel, K.: Gaussian error linear units (GELUs). arXiv preprint arXiv:1606.08415 (2016)
10. Houlsby, N., et al.: Parameter-efficient transfer learning for NLP. In: International Conference on Machine Learning, pp. 2790–2799. PMLR (2019)
11. Hu, E.J., et al.: LORA: low-rank adaptation of large language models. arXiv preprint arXiv:2106.09685 (2021)
12. Jia, M., et al.: Visual prompt tuning. In: European Conference on Computer Vision, pp. 709–727 (2022)
13. Jodoin, P.M., Maddalena, L., Petrosino, A., Wang, Y.: Extensive benchmark and survey of modeling methods for scene background initialization. IEEE Trans. Image Process. **26**(11), 5244–5256 (2017). https://doi.org/10.1109/TIP.2017.2728181

14. Kenton, J.D.M.W.C., Toutanova, L.K.: BERT: pre-training of deep bidirectional transformers for language understanding. In: Proceedings of NAACL-HLT, pp. 4171–4186 (2019)
15. Lin, H.H., Liu, T.L., Chuang, J.H.: A probabilistic SVM approach for background scene initialization. In: Proceedings. International Conference on Image Processing, vol. 3, pp. 893–896 (2002). https://doi.org/10.1109/ICIP.2002.1039116
16. Liu, Z., et al.: Swin transformer: hierarchical vision transformer using shifted windows. In: Proceedings of the IEEE/CVF International Conference on Computer Vision, pp. 10012–10022 (2021)
17. McFarlane, N.J., Schofield, C.P.: Segmentation and tracking of piglets in images. Mach. Vis. Appl. **8**, 187–193 (1995)
18. Osman, I., Abdelpakey, M., Shehata, M.S.: TransBlast: Self-supervised learning using augmented subspace with transformer for background/foreground separation. In: 2021 IEEE/CVF International Conference on Computer Vision Workshops (ICCVW), pp. 215–224 (2021). https://doi.org/10.1109/ICCVW54120.2021.00029
19. Rebuffi, S.A., Bilen, H., Vedaldi, A.: Learning multiple visual domains with residual adapters. In: Advances in Neural Information Processing Systems, vol. 30 (2017)
20. Stauffer, C., Grimson, W.: Adaptive background mixture models for real-time tracking. In: Proceedings. 1999 IEEE Computer Society Conference on Computer Vision and Pattern Recognition (Cat. No PR00149), vol. 2, pp. 246–252 (1999). https://doi.org/10.1109/CVPR.1999.784637
21. Tezcan, O., Ishwar, P., Konrad, J.: BSUV-Net: a fully-convolutional neural network for background subtraction of unseen videos. In: Proceedings of the IEEE/CVF Winter Conference on Applications of Computer Vision, pp. 2774–2783 (2020)
22. Vacavant, A., Chateau, T., Wilhelm, A., Lequièvre, L.: A benchmark dataset for outdoor foreground/background extraction. In: Park, J.-I., Kim, J. (eds.) ACCV 2012. LNCS, vol. 7728, pp. 291–300. Springer, Heidelberg (2013). https://doi.org/10.1007/978-3-642-37410-4_25
23. Wang, Y., Jodoin, P.M., Porikli, F., Konrad, J., Benezeth, Y., Ishwar, P.: CDNet 2014: an expanded change detection benchmark dataset. In: 2014 IEEE Conference on Computer Vision and Pattern Recognition Workshops, pp. 393–400 (2014). https://doi.org/10.1109/CVPRW.2014.126
24. Wang, Z., Zhang, Y., Luo, L., Wang, N.: TransCD: scene change detection via transformer-based architecture. Opt. Express **29**(25), 41409–41427 (2021). https://doi.org/10.1364/OE.440720, https://opg.optica.org/oe/abstract.cfm?URI=oe-29-25-41409
25. Zhang, C., Wang, L., Cheng, S., Li, Y.: SwinSUNet: pure transformer network for remote sensing image change detection. IEEE Trans. Geosci. Remote Sens. **60**, 1–13 (2022). https://doi.org/10.1109/TGRS.2022.3160007

Supervised Domain Adaptation with Disjoint Label Spaces for Fine-Grained Classification

Enrico Krohmer[1,2,3]([✉]) [ID], Stefan Wolf[1,2,3] [ID], and Jürgen Beyerer[1,2,3] [ID]

[1] Fraunhofer IOSB, Institute of Optronics, System Technologies and Image
Exploitation, Fraunhoferstrasse 1, 76131 Karlsruhe, Germany
enricokrohmer@icloud.com
[2] Vision and Fusion Lab, Karlsruhe Institute of Technology KIT,
Vincenz-Prießnitz-Straße 3, 76131 Karlsruhe, Germany
[3] Fraunhofer Center for Machine Learning, Sankt Augustin, Germany

Abstract. Domain adaptation scenarios commonly assume that the label spaces of the source and target domains are either equal or share a common set of classes. However, in fine-grained classification settings, it is likely that the common label set is empty. Therefore, we approach a supervised domain adaptation scenario where the label spaces of the source and target domains are available but disjoint during training. The classifier is tasked with generalizing to the complete target domain where classes are not only from the target label space but also from the source label space. We introduce a novel CycleGAN variant, FCCGAN, which translates source images into target-stylized images that preserve their class-specific features. To further encourage the classifier to learn domain-invariant representations, we pre-train the classifier exclusively on the target domain and then employ supervised contrastive learning on source, target, and target-stylized images. We demonstrate that this framework outperforms existing domain adaptation methods in a fine-grained classification task under the disjoint label space assumption. Code and supplementary material is available at: https://github.com/enricokrohmer/sda_dls.

Keywords: Fine-Grained Classification · Supervised Domain Adaptation · Disjoint Label Spaces

1 Introduction

Fine-grained classification tasks like vehicle make and model recognition have been dominated by data-intensive deep learning methods recently. While good results have been achieved under the assumption of enough data available, the acquisition of images and image labels are an enormous task for fine-grained classification. Particularly, fine-grained classification requires large amount of training data due to the specific class-distinguishing features often being difficult

M. Cho et al. (Eds.): ACCV 2024 Workshops, LNCS 15482, pp. 52–68, 2025.
https://doi.org/10.1007/978-981-96-2641-0_4

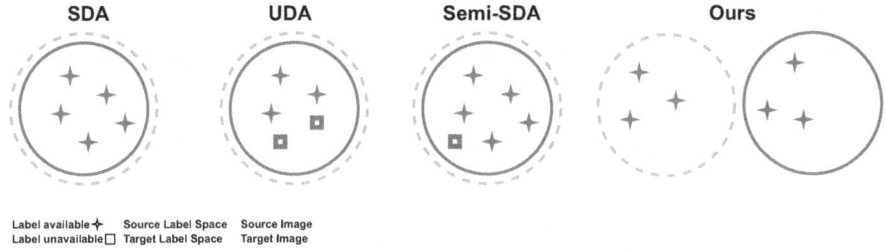

Fig. 1. Visualization of different domain adaptation scenarios. In our domain adaptation framework, we assume that the source and target domains have disjoint label spaces, contrasting with other approaches that assume identical label spaces. This means each domain contains images belonging only to its specific set of classes.

to extract. Additionally, the high specificity of the classes renders it challenging to find a large amount of appropriate samples for each class. For this reason, domain adaptation methods have been investigated for fine-grained classification which exploit domains with large amounts of well-annotated samples available as a source domain which provides information to learn class-distinguishing feature. Typical examples of source domains for fine-grained domain adaptation are web images of vehicles from car-selling sites or field guides for the classification of birds [9,37,38,42]. In domain adaptation scenarios, the final classification is performed in a target domain different from the source domain. While the source domain has abundant annotated samples available, the target domain shows some form of limited availability of data. However, the existing fine-grained domain adaptation studies only explore unsupervised or semi-supervised domain adaptation scenarios which consider a common label space for the source and target domain. This includes the presence of samples for all classes in the target domain, a hard to fulfill claim for fine-grained classification with a high number of classes, particularly, assuming the lack of labels in unsupervised domain adaptation.

Thus, we approach a different domain adaptation scenario with disjoint label spaces in the source and target domain, not requiring the same classes to be present in the source and target domain. Nonetheless, the final classifier is tasked to perform classification on target domain images for the join of the source and the target label space. So, it has to support all classes. To still be a viable scenario, we consider the presence of labels in the target domain necessary and thus, it is a supervised domain adaptation scenario. This enables a more flexible usage of domain adaptation. We can exploit a supervised dataset in the target domain, which is already commonly available for most relevant domains, and additional classes can be added to the classification solely based on samples from the source domain.

For example, for fine-grained vehicle classification in surveillance scenarios, data privacy limits the possibility to acquire data in public space drastically. However, the constant introduction of new vehicle models by the manufacturers

requires regular update of training datasets and thus, acquiring new data. Therefore, we aim to exploit the limited available data best possible. For the target
domain of surveillance images, annotated datasets are publicly available [32,39].
So, the main challenge is to extend it with new vehicle models which we solve
with a disjoint label spaces domain adaptation scenario.

Banatt *et al.* [1] explored a similar scenario and showed that the disjoint
label space leads to significant negative transfer. Thus, the classification performance on the source classes drops below random guessing while also existing
domain adaptation methods not proving to be effective in this setting. However, the authors also showed that the negative transfer can be reduced when
a small proportion of target domain samples of the source classes are added to
the training. We show that a similar effect can be achieved when target-stylized
source samples are used to approximate the target domain distribution of the
source classes. We approach the scenario with a novel domain adaptation framework which we essay in this study for a fine-grained vehicle classification task
with synthetic images as source domain and real surveillance images as target
domain. First on, we perform a pre-training on the target domain only including
the target label space which focuses on the extraction of domain-specific task-
important features. Afterwards, we extend the classification label space by the
source domain, in our case, a synthetic dataset of surveillance images of vehicles [31]. Samples of the target domains for the new classes are not required,
neither unsupervised nor supervised. Our novel method enhances the Cycle-
GAN [46] to tackle the challenges of applying it for fine-grained domain adaption.
While image-to-image translation is a common pattern for domain adaptation,
particularly synthetic-to-real adaptation, it has yet to be investigated for fine-
grained domain adaptation and its specific challenges. CycleGAN is well capable
of generating images of the target domain with the meta-class being well reconstructed. Nonetheless, it lacks the ability to reconstruct fine-grained details such
as the logo of the vehicle make. Thus, we propose a feature consistency loss
which imposes recovering fine-grained details important for classification. With
our feature-consistent CycleGAN, we can transform images from the synthetic
source domain to the real target domain while preserving fine details in the car.
Based on a model pre-trained only on the target images, we use the target-
stylized source images as training data combined with the real target images.
Thus, we have target images available for all classes, for some classes actual
real target images and for some classes pseudo target images which have been
target-stylized from source images. We confirm the impact of this enhancement
quantitatively and qualitatively. Additional to the classification loss, we apply
a supervised contrastive loss [15] to diminish remaining feature information loss
between the original source images and the target-stylized counterpart.

Our contributions can be summarized as follows:

1. We are the first to explore domain adaptation methods in depth for a domain
   adaptation scenario with disjoint label spaces, indicating the deficiencies of
   the existing methods for this scenario.

2. We propose a novel feature-consistency loss enhancing CycleGAN [46] by improving the capability to distinguish task-relevant details in the image which need to be transferred and domain-relevant details that should be adapted to the new domain.
3. We propose the integration of the supervised contrastive loss [15] for the classification training to diminish the remaining feature information loss between target-stylized source images and their original counterparts.

2   Related Work

Domain Adaptation Scenarios. Multiple domain adaptation scenarios have been proposed to address varying assumptions about label spaces and availability of labels as data availability can differ by application. Regarding assumptions about the label space, these scenarios include a closed-set, multiple open-set, a partial and a universal domain adaptation scenario. In the traditional closed-set scenario, the label spaces of the source and the target domains are assumed to be identical. However, due to applications of domain adaptation are commonly data-limited, especially concerning the presence of classes in datasets, more flexible scenarios were proposed and investigated. Busto and Gall [4] propose the open-set domain adaptation scenario assuming the existence of a set of common classes and additionally samples with an unspecified class for the source and the target domain. However, samples without a specified class are not required to be recognized by their class but recognizing them as an unknown class sample is enough. Saito *et al.* [28] propose a more stringent variant by removing the availability of unknown class samples in the source domain during training. Still, samples of unknown classes in the target domain have to be distinguished from known classes. Cao *et al.* [5] and Zhang *et al.* [44] introduce the partial domain adaptation scenario, where the target label space is a subset of the source label space. Universal domain adaptation [41] includes private classes in the source and the target domain beside a set of common classes. However, all of these scenarios consider private classes as not to be distinguished. Additionally, none of these scenarios consider completely separate label spaces. Nonetheless, both conditions are common in a fine-grained domain adaptation scenario. Thus, we consider disjoint label spaces with the join of the source and target classes to be distinguished in the target domain.

Domain adaptation has been approached in unsupervised, semi-supervised and supervised settings. In unsupervised domain adaptation, a large amount of unlabeled images of the target domain is leveraged for the adaptation process [2, 8,18,35]. Semi-supervised domain adaptation introduces labels for some of the images of the target domain [8,40]. In supervised domain adaptation, all samples from the target domain are labeled [13,16,25,26,34]. To preserve a challenging aspect in supervised domain adaptation scenarios, only a few samples per class in the target domain are used during training. In contrast, in the supervised domain adaptation scenario we consider, the challenge arises not from a scarcity of samples but from a disjoint label space with a complete lack of images for the source classes in the target domain.

Domain Adaptation Methods. Following the categorization of Wang and Deng [36], there are three general categories of domain adaptation methods. Discrepancy-based methods perform a fine-tuning on the target data to minimize the shift between the domains based on a criterion. The criterion can be a class label, either unsupervised with a pseudo label [45] or supervised [34], a statistic criterion like maximum mean discrepancy [10, 20] or Kullback-Leibler divergence [43] or an architecture criterion like an adaptive batch normalization [17]. Adversarial-based approaches aim to induce domain confusion by either using generative methods [3] which transform images from source domain to target domain or non-generative methods [8]. Reconstruction-based approaches expect a reconstruction process to generate features which are invariant to domain differences while maintaining class-discriminative properties. They can be based on an encoder-decoder combination [11] or an adversarial reconstruction [46]. Our approach can be categorized as a hybrid approach of a generative adversarial-based approach and a discrepancy-based approach based on a class criterion. While semantic consistency losses have been proposed to retain task-important details for generative domain adaptation approaches [3, 14], as we show in our experiments, they still lack the capability of retaining details on a level necessary for fine-grained classification. Thus, we introduce a feature consistency loss.

In the field of fine-grained domain adaptation, methods are based on the approaches for regular domain adaptation and adjusted for fine-grained classification tasks. A common pattern is to exploit coarse-grained labels to improve the domain alignment process in the expectation that the coarse-grained labels can be recognized more consistently across domains [9, 37]. Wang *et al.* [38] propose the integration of a spatial self-attention module to extract more relevant features for the fine-grained classification. Yu *et al.* [42] propose a new adversarial domain adaptation approach based on label switching which better retains fine-grained features. However, in the field of fine-grained domain adaptation, generative adversarial domain adaptation methods are yet to be investigated. To the best of our knowledge, we are the first to explore generative adversarial domain adaptation approaches for fine-grained domain adaptation.

3 Supervised Domain Adaptation with Disjoint Label Spaces

3.1 Problem Setting

We are given a labeled source domain S and a labeled target domain $T_{Full} = T \cup T^*$ with label space $\mathcal{Y} = \mathcal{Y}_1 \cup \mathcal{Y}_2$, where $\mathcal{Y}_1 \cap \mathcal{Y}_2 = \emptyset$ and $T \cap T^* = \emptyset$. However, S and T^* only contain data points whose labels are in $\mathcal{Y}_1$, and T only contains data points whose labels are in $\mathcal{Y}_2$. For simplicity, we will refer to $\mathcal{Y}_2$ as the label space of T, and $\mathcal{Y}_1$ as the label space of both S and T^*. This setting was first explored by [1] and a comparison of different domain adaptation scenarios can be seen in Fig. 1.

The goal of supervised domain adaptation with disjoint label spaces (SDA-DLS) is to train a classifier capable to generalize to T_{Full}, while only having access to training data from S and T.

3.2   Technical Challenges

Models that try to tackle fine-grained classification tasks need to learn good representations of each class to mitigate the low inter-class and high intra-class variance. As in traditional domain adaptation, the model additionally needs to bridge the domain gap between $\mathcal{S}$ and $\mathcal{T}^*$. Otherwise, the classifier may learn domain-specific features from $\mathcal{S}$ which hinders the model's ability to generalize to the target domain.

In SDA-DLS, a classifier is prone to learn domain-specific features not only from $\mathcal{S}$ but also from $\mathcal{T}$ to differentiate between classes from $\mathcal{Y}_1$ and $\mathcal{Y}_2$ during training. If the classifier must classify a sample from $\mathcal{T}^*$ during inference, the domain-specific features of the sample could mislead the classifier into thinking that the sample belongs to $\mathcal{T}$. As a result, the classifier is prone to confusing classes from $\mathcal{Y}_1$ with classes from $\mathcal{Y}_2$. This behavior can be seen as a special case of negative transfer as described by [1].

Addressing the issue of negative transfer within the current domain adaptation frameworks presents its own set of problems: Supervised Domain Adaptation (SDA) operates under the assumption that during training, the source and target domains share the same label space. Multiple SDA methods [25, 26] directly utilize label information on the target domain to derive a domain-invariant representation of samples that belong to the same class but originate from different domains. However, during the training phase of SDA-DLS no class is present in both source and target domain, due to the disjoint label spaces.

An alternative approach involves turning to Unsupervised Domain Adaptation (UDA) methods, which do not depend directly on label information. However, they still assume that the label spaces of the source and target domains are identical, as is typical in classical unsupervised domain adaptation, or that they share a common set of labels, as seen in universal domain adaptation [41].

4   Method

This section will introduce our novel approach for tackling the SDA-DLS scenario. We will incrementally introduce new components to a convolutional neural network that minimizes the standard cross-entropy loss $\mathcal{L}_{Task}$ on both the source and target domain. The framework consists of the following training stages:

1. Train a CycleGAN [46] to translate images from $\mathcal{S}$ to $\mathcal{T}$. Class-specific features are preserved after translation, as CycleGAN also optimizes the novel feature-consistency loss.
2. Pre-train a classifier C on only the target domain.
3. Train C on the source, target and translated source images in conjunction to a supervised contrastive loss [15].

4.1   Target-Only Pretraining

First, the classifier is pre-trained exclusively on the target domain. After pre-training is completed, we freeze the weights of the earlier layers and fine-tune

C on both domains. To accelerate the training process during the second phase, only a small subset of the target domain is utilized. We do not completely exclude target data to prevent C from forgetting the class-specific features of $\mathcal{Y}_2$.

During target-only pre-training (TO-PT), C is optimized to extract features that are specialized for fine-grained classification. Nonetheless, earlier layers are should still be able to extract more general features. By freezing these layers in the subsequent step, we maintain their generality and avoid that C learns domain-specific features from the source domain.

4.2   CycleGAN Recap

Our framework employs a CycleGAN [46] to translate images from the source domain $\mathcal{S}$ to the target domain $\mathcal{T}$ and vice versa. The CycleGAN framework consists of two GANs [12] with generators G and F, and discriminators $D_\mathcal{S}$ and $D_\mathcal{T}$. G is tasked with transforming images $x \in \mathcal{S}$ to images that adopt the style of images from $\mathcal{T}$. The discriminator $D_\mathcal{T}$ is tasked with differentiating between images from $\mathcal{T}$ and the transformed images $G(\mathcal{S})$. Conversely, the generator aims to deceive the discriminator into believing that the transformed images are indeed sampled from $\mathcal{T}$. A generator-discriminator pair optimizes the following adversarial loss function introduced by Goodfellow et $al.$ [12]:

$$\mathcal{L}_{gan}(G, D, \mathcal{S}, \mathcal{T}) = \mathbb{E}_{z \sim \mathcal{T}} \log D(z) + \mathbb{E}_{x \sim \mathcal{S}} \log(1 - D(G(x))) \qquad (1)$$

The loss function for the opposite direction is defined analogously. Additionally, CycleGAN requires that both generators should be inverse to each other such that $F(G(x)) \approx x$ and $G(F(z)) \approx z$, where $x \in \mathcal{S}, z \in \mathcal{T}$. This is achieved by G and F minimizing the cycle-consistency loss $\mathcal{L}_{cyc}$:

$$\mathcal{L}_{cyc}(G, F, \mathcal{S}, \mathcal{T}) = \mathbb{E}_{x \sim \mathcal{S}} \|F(G(x)) - x\|_1 \qquad (2)$$
$$+ \mathbb{E}_{z \sim \mathcal{T}} \|G(F(z)) - z\|_1$$

Combining Eq. (1) and Eq. (2) yields the final loss function of CycleGAN:

$$\mathcal{L}_{CycleGAN}(G, F, D_\mathcal{S}, D_\mathcal{T}, \mathcal{S}, \mathcal{T}) = \mathcal{L}_{gan}(G, D_\mathcal{T}, \mathcal{S}, \mathcal{T}) \qquad (3)$$
$$+ \mathcal{L}_{gan}(F, D_\mathcal{S}, \mathcal{T}, \mathcal{S})$$
$$+ \lambda_{cyc} \mathcal{L}_{cyc}(G, F, \mathcal{S}, \mathcal{T})$$

where λ_{cyc} is a hyperparameter. $\mathcal{L}_{CycleGAN}$ is optimized in the following minimax objective:

$$\min_{G,F} \max_{D_\mathcal{S}, D_\mathcal{T}} \mathcal{L}_{CycleGAN}(G, F, D_\mathcal{S}, D_\mathcal{T}, \mathcal{S}, \mathcal{T}) \qquad (4)$$

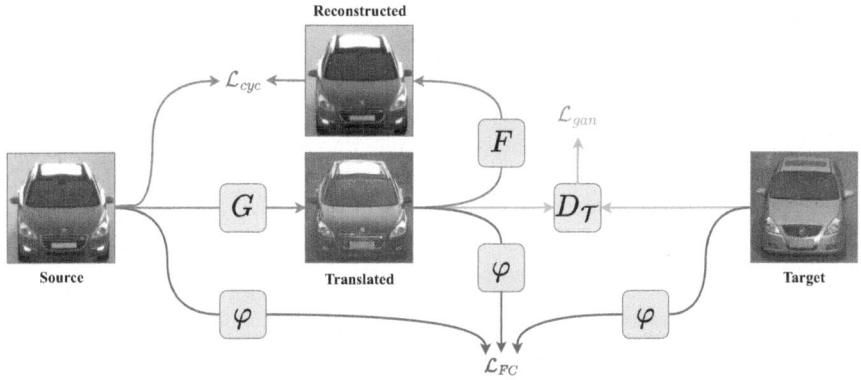

Fig. 2. Translation from $\mathcal{S}$ to $\mathcal{T}$ using FCCGAN. Green arrows show the data flow for the cycle-consistency loss $\mathcal{L}_{cyc}$. Orange arrows show the data flow for the gan loss. Purple arrows show the data flow for the feature-consistency loss. The translation in the opposite direction is performed analogously. (Color figure online)

4.3   Feature-Consistency Loss

Translating images via CycleGAN for SDA-DLS has a main caveat. First, due to the disjoint label spaces, the class-specific features in $\mathcal{S}$ are different from the class-specific features in $\mathcal{T}$. As a consequence, the discriminator could easily differentiate between translated and real images if the generator preserves all class-specific features from the input. To minimize $\mathcal{L}_{gan}$, the generators are forced to transform these class-specific features such that they align with the probability distribution of $\mathcal{T}$. In the worst case, label flipping occurs.

To preserve class-specific features after translation and prevent label flipping, we first train a classifier $h \circ \varphi$ by minimizing $\mathcal{L}_{task}$ on the source and target domain. h denotes the classification head of the classifier, and φ denotes its feature extractor. After training, we emit h and freeze the weights of φ. Second, CycleGAN is optimized using $\mathcal{L}_{CycleGAN}$ in conjunction to the novel feature-consistency loss $\mathcal{L}_{FC}$:

$$\mathcal{L}_{FC}(G, \varphi, \mathcal{S}, \mathcal{T}) = \mathbb{E}_{\substack{x \sim \mathcal{S}, \\ z \sim \mathcal{T}}} \Big[\|\varphi(G(x)) - \varphi(x)\|_2 - \|\varphi(G(x)) - \varphi(z)\|_2 + m \Big]_+ \quad (5)$$

At its core, $\mathcal{L}_{FC}$ employs the triplet-loss [29] with margin m where the translated image serves as the anchor, the input image as a positive sample, and an image from the opposite domain as a negative sample. As φ is frozen, the only way to minimize $\mathcal{L}_{FC}$ is by G optimizing $\mathcal{L}_{FC}$ by itself. Therefore, the class-specific features of translated images have to be similar to the features of their original counterpart and distant to class-specific features from $\mathcal{Y}_2$.

We employ the feature-consistency loss for both generators. This leads to the full loss function of our CycleGAN variant, named Feature-Consistent CycleGAN (**FCCGAN**):

$$\mathcal{L}_{FCCGAN}(G, F, D_{\mathcal{S}}, D_{\mathcal{T}}, \varphi, \mathcal{S}, \mathcal{T}) = \mathcal{L}_{CycleGAN}(G, F, D_{\mathcal{S}}, D_{\mathcal{T}}, \mathcal{S}, \mathcal{T}) \quad (6)$$
$$+ \lambda_{FC}\mathcal{L}_{FC}(G, \varphi, \mathcal{S}, \mathcal{T})$$
$$+ \lambda_{FC}\mathcal{L}_{FC}(F, \varphi, \mathcal{T}, \mathcal{S})$$

where λ_{FC} is hyperparameter. A visualization of FCCGAN training is shown in Fig. 2.

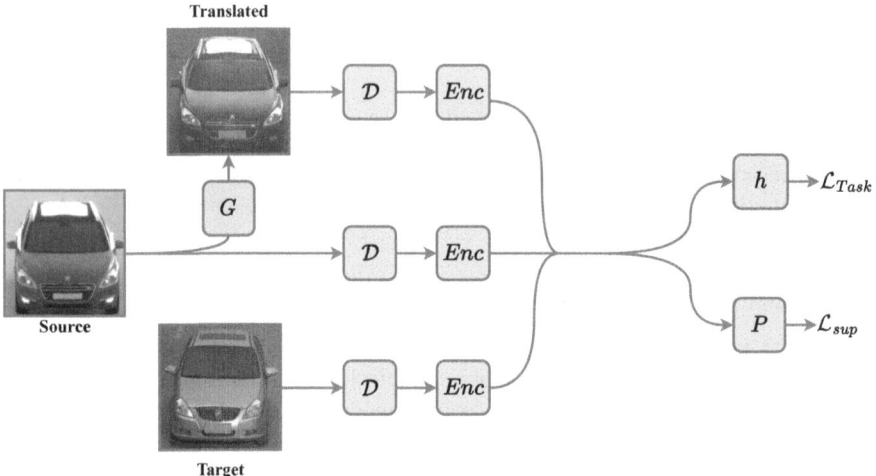

Fig. 3. Visualization of our modified supervised contrastive learning framework. *Enc* denotes the feature extractor of the classifier with classification head h. First, source images are translated via the FCCGAN generator G. Second, the data-augmentation module $\mathcal{D}$ is applied to each image, before being fed to *Enc*. The features are then either fed into h and P to calculate the cross-entropy loss $\mathcal{L}_{Task}$ and the supervised contrastive loss $\mathcal{L}_{sup}$, respectively.

4.4   Feature-Level Alignment

Translated images generated by FCCGAN have no guarantee that they preserve class-specific features perfectly. This implies a trade-off between training with source and translated images: Source images contain all class-specific features but there exists a domain gap between $\mathcal{S}$ and $\mathcal{T}^*$. Images from $G(\mathcal{S})$ only approximate class-specific features but mitigate the domain gap by adapting the style of the target domain. To leverage the advantages of both translated and original data, we present a modification of the supervised contrastive learning framework developed by Khosla *et al.* [15].

Enc denotes the feature extractor of the classifier C. Equivalent to Khosla *et al.* we introduce a data augmentation module $\mathcal{D}$ and a simple feed-forward network P. $\mathcal{D}$ applies data augmentations to input images. P projects feature

Table 1. Comparison between the proposed method and the baselines. All values are rounded to the third decimal.

Model	SSB → CCSV		CCSV → SSB	
	$F_1(\mathcal{Y}_1)$	$F_1(\mathcal{Y}_2)$	$F_1(\mathcal{Y}_1)$	$F_1(\mathcal{Y}_2)$
Direct Transfer	0.660	0.965	0.751	0.978
CycleGAN [46]	0.782	0.968	0.861	0.986
DANN [8]	0.827	0.908	0.629	0.965
DAN [20]	0.895	0.962	0.858	0.985
CyCADA [14]	0.808	0.966	0.800	0.983
Ours	**0.946**	**0.985**	**0.906**	**0.991**
Full Target	0.990	0.971	0.999	0.999

maps produced by Enc into a lower dimensional space. Additionally, the output of P gets normalized.

Given a batch of N source image-label pairs $\{x_i, y_i\}_{i=1}^{N}$ and a batch of target image-label pairs $\{z_i, y_i\}_{i=1}^{N}$. First, we generate $\{G(x_i), y_i\}_{i=1}^{N}$ using the FCC-GAN source-to-target generator. Finally, we calculate the supervised contrastive loss [15] using the concatenated batch of all image-label pairs $\{a_i, y_i\}_{i=1}^{3N}$:

$$\mathcal{L}_{sup} = -\sum_{i=1}^{3N} \frac{1}{|A(i)|} \sum_{j \in A(i)} \log \frac{\exp(p_i \cdot p_j/\tau)}{\sum_{k \neq i} \exp(p_i \cdot p_k/\tau)} \tag{7}$$

Here, $p_i = P(Enc(\mathcal{D}(a_i)))$, $A(i) = \{k \in \mathbb{N} : y_i = y_k, 1 \leq k \leq 3N, k \neq i\}$, τ is a positive temperature parameter, and $\cdot$ denotes the inner product. $\mathcal{L}_{sup}$ forces features within the same class to be close to each other and features from different classes to be distant. Therefore, $\mathcal{L}_{sup}$ encourages Enc to find a domain-invariant representation for source images and their translated counterparts, as they share their label and thus need to be closely aligned in the feature space. Additionally, $\mathcal{L}_{sup}$ aids in addressing the challenges of fine-grained classification tasks: First, it mitigates the low inter-class variance, as images from different classes need to be well separated. Second, it mitigates the high intra-class variance, as samples within the same class need to be close to each other in the feature space.

After target-only pre-training the classifier can minimize $\lambda_{sup}\mathcal{L}_{sup}$ in addition to $\mathcal{L}_{task}$, where λ_{sup} is a hyperparameter. Note that $\mathcal{L}_{task}$ is optimized on source, target and translated source images. A visualization of the proposed method can be seen in Fig. 3.

5 Experiments

5.1 Setup

Datasets. We use the fine-grained vehicle classification datasets **Synset Boulevard** (SSB) [31] and **CompCars Surveillance** (CCSV) [39], specifically employing the Bayer-Bad Configuration of SSB. Both datasets feature surveillance-type vehicle images captured during daytime. There exists a domain gap between SSB and CCSV, as SSB is entirely synthetically generated where as CCSV is a real-world dataset which additionally contains nighttime images. The source and target datasets contain 156 and 281 classes, respectively, with each class denoting a distinct vehicle model. We report results for both adaptation directions, *i.e.* from SSB to CCSV and vice versa. For both directions, we set $\mathcal{Y}_1$ to the 21 classes from SSB and CCSV that match by vehicle model and year. The remaining images in the source domain are omitted, as their classes are not present in the target domain, precluding the possibility of evaluating them. The remaining classes within the respective target domain are designated as $\mathcal{Y}_2$. This leads to 260 classes in $\mathcal{Y}_2$ for SSB to CCSV and 135 classes for CCSV to SSB. Therefore, $\mathcal{S}$ comprises images from the source dataset with classes from $\mathcal{Y}_1$, $\mathcal{T}^*$ comprises images from target dataset with classes from $\mathcal{Y}_1$, and $\mathcal{T}$ comprises the remaining images in the target dataset.

Evaluation Details. We report the F1-Score on $\mathcal{T}_{Full}$ for each method. More specifically, we first compute the class-wise metrics on $\mathcal{T}_{Full}$ and average them based on their associated label spaces. During training, we periodically evaluate each method on a small subset of $\mathcal{T}_{Full}$. This subset serves as our validation set, with the remaining samples comprising the test set. For each method, we report the results computed on the test set for the model instance that achieved the highest F1-Score in the source domain during validation.

Baselines. We report results for a classifier trained on $\mathcal{S}$ and $\mathcal{T}$, a classifier trained on $\mathcal{T}_{Full}$, and a classifier trained on $G(\mathcal{S})$ and $\mathcal{T}$. G denotes the source-to-target generator of a vanilla **CycleGAN** [46]. Additionally, we report results for the following state-of-the-art UDA methods: Domain-Adversarial Neural Networks (**DANN**) [8], Deep Adaptation Networks (**DAN**) [20], and Cycle-Consistent Adversarial Domain Adaptation (**CyCADA**) [14]. We modified each method so that they can be directly applied to the SDA-DLS setting. More specifically, the classifier of each method now minimizes the cross-entropy loss on both the source and target domains, instead of only on the source domain as in the UDA setting.

Implementation. All methods were implemented using the open-source deep learning framework PyTorch [27]. For the classifier and FCCGANs feature extractor we employ a **ConvNeXt-Tiny** [19] that was pre-trained on ImageNet [7]. Features of the classifier are extracted after the final pooling layer. For classifiers pre-trained on the target domain, we freeze the first three stages. For all other cases, only one stage is frozen. FCCGAN utilizes the generator and discriminator architecture from Zhu *et al.* [46]. We replaced Eq. (1) with a least-square objective [24], and the discriminator is updated based on a history of generated images rather than the most recent one [30]. Both methods were employed by Zhu *et al.* [46] to stabilize CycleGAN training. P is implemented as a feed-forward network with one hidden layer of size 2048 and an output size of 128. We set $\lambda_{cyc} = 10$, $\lambda_{FC} = 10$, the margin of $\mathcal{L}_{FC}$ to 0.5, $\lambda_{sup} = 0.5$, and $\tau = 0.1$.

Training. Each model is trained for 60 epochs using AdamW [22], with betas set to 0.9 and 0.999, and an initial learning rate of 0.0002. We employ cosine annealing [21] without restarts for learning rate scheduling. The weight decay is set to 0.05 for FCCGAN training and to 0.1 for classifier training. For FCCGAN, we set the batch size to 4 images per domain. Classifiers are trained with a batch size of 32. For FCCGAN, we resize each image to 276×276 pixels, then randomly crop them to a size of 256×256, and apply a random flip to each image with a probability of 0.5. For classifier training, we resize each image to 256×256, randomly crop them to 224×224, and apply RandAugment [6]. If a classifier is trained with translated images, the data augmentations are applied post-translation. Within our supervised contrastive learning framework, RandAugment functions as the data-augmentation module $\mathcal{D}$. Additionally, for classifier training, we employ label smoothing [33] with a smoothing parameter of 0.1. We use the same hyperparameters for both directions.

5.2   Results

Source	FCCGAN	CycleGAN	CyCADA	Target

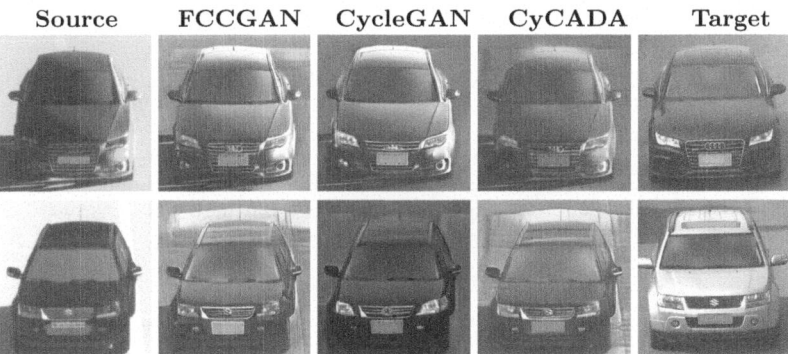

Fig. 4. Images from the source, target and translated source domain. Images in each row belong to the same class.

Classification Results. The classification results are shown in Table 1. The Full Target model serves as an upper bound as it represents the optimal scenario where training data for all of $\mathcal{T}_{Full}$ is available. Therefore, it achieves the highest F1-Score for classes in $\mathcal{Y}_1$ for both directions. Regardless of the direction, the direct transfer model significantly underperforms on $\mathcal{Y}_1$. A minor drop in performance is observed on $\mathcal{Y}_2$. This observation aligns with the findings of [1].

For classes in $\mathcal{Y}_2$, our method achieves the highest F1-Score in both directions. On SSB to CCSV, our classifier is even able to outperform the classifier trained on the full target domain. This could be attributed to reduced overfitting, as the initial stages are frozen after TO-PT. The UDA methods and the direct transfer approach maintain a relatively high F1-Score for classes from $\mathcal{Y}_2$, as there exists no domain gap between classes from $\mathcal{Y}_2$ during training and inference.

DANN, however, experiences a significant performance decline on $\mathcal{Y}_2$ in both directions. On CCSV to SSB, DANN additionally underperforms to the direct transfer model on $\mathcal{Y}_1$. This indicates that domain adversarial training is not well-suited for the SDA-DLS setting.

Our method significantly outperforms all other domain adaptation methods by combining feature-consistent image translation with domain-invariant feature learning.

DAN achieves the second highest F1-Score on $\mathcal{Y}_1$ for SSB to CCSV and third highest for CCSV to SSB. This suggests that discrepancy-based approaches, such as our method and DAN, are viable strategies for tackling SDA-DLS.

Even though CyCADA employs a semantic consistency loss to preserve features after translation, it fails to match the performance of CycleGAN on CCSV to SSB and only marginally improves over CycleGAN on the opposite direction. This could be due to the second phase of CyCADA training, which employs domain adversarial training. We refer to Fig. 4 for a comparison of image translations for CycleGAN, CyCADA, and FCCGAN.

Table 2. Ablation study on SSB to CCSV. All values rounded to the third decimal.

TO-PT	CycleGAN	FCCGAN	SupCon	$F_1(\mathcal{Y}_1)$	$F_1(\mathcal{Y}_2)$
-	-	-	-	0.660	0.965
✓	-	-	-	0.915	0.983
✓	✓	-	-	0.816	0.980
✓	-	✓	-	0.929	0.984
✓	-	✓	✓	**0.946**	**0.985**

Ablation Study. We systematically evaluate each component introduced by our framework. The results are shown in Table 2. The introduction of TO-PT significantly improves the classifier's ability to generalize to $\mathcal{T}_{Full}$, as it prevents the classifier from learning domain-specific features. However, if the classifier is additionally trained with source images translated by a vanilla CycleGAN, this improvement is mitigated, as class-specific features are lost after translation. Consequently, the classifier is not able to learn a representation that generalizes to $\mathcal{T}^*$. FCCGAN manages to close the domain gap while preserving class-specific features, thanks to the novel feature-consistency loss. Therefore, the combination of FCCGAN and TO-PT yields an improvement over using TO-PT alone. Nonetheless, the best results are achieved by the complete framework that employs supervised contrastive learning. The classifier learns a domain-invariant representation of source classes, as source and target-stylized images are forced to reside near each other in the feature space. Additionally, the low inter-class variance of fine-grained datasets is mitigated, as samples from distinct classes are forced to be distant from each other.

6 Conclusion

6.1 Summary

In this paper, we approach the supervised domain adaptation scenario with disjoint label spaces for fine-grained classification. We propose FCCGAN, a variant of CycleGAN, designed to translate images in a way that adapts the style of the target domain while preserving class-specific features. To further encourage the classifier to learn domain-invariant representations, we employ supervised contrastive learning and target-only pre-training. Target-only pre-training prevents the classifier from learning domain-specific information from the source domain. Supervised contrastive learning ensures that images from the source and target domains, as well as translated source images, are positioned close to one another in the feature space if they belong to the same class and distanced if not. Operating under the disjoint label space assumption, we evaluate our proposed framework on a fine-grained classification task and demonstrate that it significantly outperforms existing domain adaptation methods.

6.2 Limitations and Future Work

This paper explored four state-of-the-art unsupervised domain adaptation approaches for SDA-DLS. We further encourage the investigation of existing domain adaptation approaches, because UDA methods like DAN [20] are still able to perform relatively well under these new challenges.

Our method was only tested on a fine-grained vehicle classification task. We expect our method to perform well on other datasets *e.g.* birds [37] and aircrafts [23]. However, benchmarking requires fine-grained datasets with a large enough shared label space. Additionally, the domain gap should not involve significant perspective changes, as these can lead to disparities in class-specific features across domains. For example, evaluating on CompCars Web [39] was not feasible because SSB and CCSV include only frontal vehicle images, whereas CompCars Web features vehicles from various perspectives, such as the side and back. This mismatch would mean that for the same vehicle model class-specific features in the source domain might not cover all class-specific features in the target domain. Therefore, the curation of new fine-grained datasets is needed to further generate insights for the supervised domain adaptation with disjoint label spaces scenario.

References

1. Banatt, E., Rajendran, V., Packer, L.: Target domain data induces negative transfer in mixed domain training with disjoint classes. arXiv preprint arXiv:2303.01003 (2023)
2. Boqing, G., Yuan, S., Fei, S., Grauman, K.: Geodesic flow kernel for unsupervised domain adaptation. In: CVPR, pp. 2066–2073. IEEE, Providence, RI (2012). https://doi.org/10.1109/CVPR.2012.6247911
3. Bousmalis, K., Silberman, N., Dohan, D., Erhan, D., Krishnan, D.: Unsupervised Pixel-Level Domain Adaptation with Generative Adversarial Networks. In: CVPR, pp. 95–104. IEEE, Honolulu, HI (2017). https://doi.org/10.1109/CVPR.2017.18
4. Busto, P.P., Gall, J.: Open set domain adaptation. In: ICCV, pp. 754–763. IEEE, Venice (2017). https://doi.org/10.1109/ICCV.2017.88
5. Cao, Z., Long, M., Wang, J., Jordan, M.I.: Partial transfer learning with selective adversarial networks. In: CVPR, pp. 2724–2732. IEEE, Salt Lake City, UT (2018). https://doi.org/10.1109/CVPR.2018.00288
6. Cubuk, E.D., Zoph, B., Shlens, J., Le, Q.V.: RandAugment: practical automated data augmentation with a reduced search space. In: CVPRW, pp. 702–703 (2020). https://doi.org/10.1109/cvprw50498.2020.00359
7. Deng, J., Dong, W., Socher, R., Li, L.J., Li, K., Fei-Fei, L.: ImageNet: a large-scale hierarchical image database. In: CVPR, pp. 248–255 (2009). https://doi.org/10.1109/CVPR.2009.5206848
8. Ganin, Y., et al.: Domain-adversarial training of neural networks. JMLR **17**(59), 1–35 (2016)
9. Gebru, T., Hoffman, J., Fei-Fei, L.: Fine-grained recognition in the wild: a multitask domain adaptation approach. In: 2017 IEEE International Conference on Computer Vision (ICCV), pp. 1358–1367. IEEE, Venice (2017). https://doi.org/10.1109/ICCV.2017.151
10. Ghifary, M., Kleijn, W.B., Zhang, M.: Domain adaptive neural networks for object recognition. In: Pham, D.-N., Park, S.-B. (eds.) PRICAI 2014. LNCS (LNAI), vol. 8862, pp. 898–904. Springer, Cham (2014). https://doi.org/10.1007/978-3-319-13560-1_76
11. Glorot, X., Bordes, A., Bengio, Y.: Domain adaptation for large-scale sentiment classification: a deep learning approach. In: ICML, pp. 513–520 (2011)
12. Goodfellow, I., et al.: Generative adversarial networks. Commun. ACM **63**(11), 139–144 (2020). https://doi.org/10.1145/3422622
13. Hedegaard, L., Sheikh-Omar, O.A., Iosifidis, A.: Supervised domain adaptation: a graph embedding perspective and a rectified experimental protocol. IEEE TIP **30**, 8619–8631 (2021). https://doi.org/10.1109/TIP.2021.3118978
14. Hoffman, J., et al.: CyCADA: Cycle-consistent adversarial domain adaptation. In: ICML, pp. 1989–1998 (2018)
15. Khosla, P., et al.: Supervised contrastive learning. In: NeurIPS, pp. 18661–18673 (2020). https://doi.org/10.48550/arXiv.2004.11362
16. Koniusz, P., Tas, Y., Porikli, F.: Domain adaptation by mixture of alignments of second-or higher-order scatter tensors. In: CVPR, pp. 7139–7148. IEEE, Honolulu, HI (2017). https://doi.org/10.1109/CVPR.2017.755
17. Li, Y., Wang, N., Shi, J., Hou, X., Liu, J.: Adaptive batch normalization for practical domain adaptation. PR **80**, 109–117 (2018). https://doi.org/10.1016/j.patcog.2018.03.005

18. Liu, M.Y., Tuzel, O.: Coupled generative adversarial networks. In: Lee, D., Sugiyama, M., Luxburg, U., Guyon, I., Garnett, R. (eds.) NeurIPS. vol. 29. Curran Associates, Inc. (2016). https://proceedings.neurips.cc/paper_files/paper/2016/file/502e4a16930e414107ee22b6198c578f-Paper.pdf

19. Liu, Z., Mao, H., Wu, C.Y., Feichtenhofer, C., Darrell, T., Xie, S.: A ConvNet for the 2020s. In: CVPR, pp. 11976–11986 (2002). https://doi.org/10.1109/cvpr52688.2022.01167

20. Long, M., Cao, Y., Wang, J., Jordan, M.: Learning transferable features with deep adaptation networks. In: ICML, pp. 97–105 (2015)

21. Loshchilov, I., Hutter, F.: SGDR: stochastic gradient descent with warm restarts (2017). https://doi.org/10.48550/arXiv.1608.03983

22. Loshchilov, I., Hutter, F.: Decoupled weight decay regularization (2019). https://doi.org/10.48550/arXiv.1711.05101

23. Maji, S., Rahtu, E., Kannala, J., Blaschko, M., Vedaldi, A.: Fine-grained visual classification of aircraft (2013). https://doi.org/10.48550/arXiv.1306.5151

24. Mao, X., Li, Q., Xie, H., Lau, R.Y., Wang, Z., Paul Smolley, S.: Least squares generative adversarial networks. In: ICCV, pp. 2794–2802 (2017). https://doi.org/10.1109/iccv.2017.304

25. Motiian, S., Jones, Q., Iranmanesh, S., Doretto, G.: Few-shot adversarial domain adaptation. In: NeurIPS, vol. 30 (2017)

26. Motiian, S., Piccirilli, M., Adjeroh, D.A., Doretto, G.: Unified deep supervised domain adaptation and generalization. In: ICCV, pp. 5715–5725 (2017). https://doi.org/10.1109/iccv.2017.609

27. Paszke, A., et al.: Pytorch: an imperative style, high-performance deep learning library. In: NeurIPS, vol. 32 (2019)

28. Saito, K., Yamamoto, S., Ushiku, Y., Harada, T.: Open set domain adaptation by backpropagation. In: Ferrari, V., Hebert, M., Sminchisescu, C., Weiss, Y. (eds.) ECCV 2018. LNCS, vol. 11209, pp. 156–171. Springer, Cham (2018). https://doi.org/10.1007/978-3-030-01228-1_10

29. Schroff, F., Kalenichenko, D., Philbin, J.: FaceNet: a unified embedding for face recognition and clustering. In: CVPR, pp. 815–823 (2015). https://doi.org/10.1109/CVPR.2015.7298682

30. Shrivastava, A., Pfister, T., Tuzel, O., Susskind, J., Wang, W., Webb, R.: Learning from simulated and unsupervised images through adversarial training. In: CVPR, pp. 2107–2116 (2017). https://doi.org/10.1109/cvpr.2017.241

31. Sielemann, A., Wolf, S., Roschani, M., Ziehn, J., Beyerer, J.: Synset boulevard: a synthetic image dataset for VMMR. In: 2024 IEEE International Conference on Robotics and Automation (ICRA) (2024)

32. Sochor, J., Špaňhel, J., Herout, A.: Boxcars: Improving fine-grained recognition of vehicles using 3-d bounding boxes in traffic surveillance. IEEE Trans. Intell. Transp. Syst. **20**(, 97–108 (2018). https://doi.org/10.1109/TITS.2018.2799228

33. Szegedy, C., Vanhoucke, V., Ioffe, S., Shlens, J., Wojna, Z.: Rethinking the inception architecture for computer vision. In: CVPR, pp. 2818–2826 (2016). https://doi.org/10.1109/cvpr.2016.308

34. Tzeng, E., Hoffman, J., Darrell, T., Saenko, K.: Simultaneous deep transfer across domains and tasks. In: ICCV, pp. 4068–4076. IEEE, Santiago, Chile (2015). https://doi.org/10.1109/ICCV.2015.463

35. Tzeng, E., Hoffman, J., Saenko, K., Darrell, T.: Adversarial discriminative domain adaptation. In: CVPR, pp. 2962–2971. IEEE, Honolulu, HI (2017). https://doi.org/10.1109/CVPR.2017.316

36. Wang, M., Deng, W.: Deep visual domain adaptation: s survey. Neurocomputing **312**, 135–153 (2018). https://doi.org/10.1016/j.neucom.2018.05.083
37. Wang, S., Chen, X., Wang, Y., Long, M., Wang, J.: Progressive adversarial networks for fine-grained domain adaptation. In: CVPR, pp. 9213–9222 (2020). https://doi.org/10.1109/cvpr42600.2020.00923
38. Wang, Y., Song, R.J., Wei, X.S., Zhang, L.: An adversarial domain adaptation network for cross-domain fine-grained recognition. In: 2020 IEEE Winter Conference on Applications of Computer Vision (WACV), pp. 1217–1225. IEEE, Snowmass Village, CO, USA (2020). https://doi.org/10.1109/WACV45572.2020.9093306
39. Yang, L., Luo, P., Change Loy, C., Tang, X.: A large-scale car dataset for fine-grained categorization and verification. In: CVPR, pp. 3973–3981 (2015). https://doi.org/10.1109/cvpr.2015.7299023
40. Yao, T., Yingwei Pan, Ngo, C.W., Houqiang Li, Tao, M.: Semi-supervised domain adaptation with subspace learning for visual recognition. In: CVPR. pp. 2142–2150. IEEE, Boston, MA, USA (2015). https://doi.org/10.1109/CVPR.2015.7298826
41. You, K., Long, M., Cao, Z., Wang, J., Jordan, M.I.: Universal domain adaptation. In: CVPR, pp. 2715–2724. IEEE, Long Beach, CA, USA (2019). https://doi.org/10.1109/CVPR.2019.00283
42. Yu, H., Jiang, R., Li, A.: Striking a balance in unsupervised fine-grained domain adaptation using adversarial learning. In: Li, G., Shen, H.T., Yuan, Y., Wang, X., Liu, H., Zhao, X. (eds.) KSEM 2020. LNCS (LNAI), vol. 12275, pp. 401–413. Springer, Cham (2020). https://doi.org/10.1007/978-3-030-55393-7_36
43. Yu, Q., Hashimoto, A., Ushiku, Y.: Divergence optimization for noisy universal domain adaptation. In: CVPR, pp. 2515–2524 (2021)
44. Zhang, J., Ding, Z., Li, W., Ogunbona, P.: Importance Weighted Adversarial Nets for Partial Domain Adaptation. In: CVPR, pp. 8156–8164. IEEE, Salt Lake City, UT (2018). https://doi.org/10.1109/CVPR.2018.00851
45. Zhang, X., Yu, F.X., Chang, S.F., Wang, S.: Deep transfer network: unsupervised domain adaptation (2015). https://doi.org/10.48550/ARXIV.1503.00591
46. Zhu, J.Y., Park, T., Isola, P., Efros, A.A.: Unpaired image-to-image translation using cycle-consistent adversarial networks. In: ICCV, pp. 2223–2232 (2017). https://doi.org/10.1109/iccv.2017.244

Leveraging Thermal Imaging for Robust Human Pose Estimation in Low-Light Vision

Mickael Cormier[1,3,4]([envelope]), Caleb Ng Zhi Yi[1], Andreas Specker[1,4], Benjamin Blaß[2], Michael Heizmann[1,3,4], and Jürgen Beyerer[1,3,4]

[1] Fraunhofer IOSB, Karlsruhe, Germany
{mickael.cormier,caleb.yi,andreas.specker,michael.heizmann,
jurgen.beyerer}@iosb.fraunhofer.de
[2] Stahl-Holding-Saar, Saarland, Germany
benjamin.blass@stahl-holding-saar.de
[3] Karlsruhe Institute of Technology, Karlsruhe, Germany
{mickael.cormier,michael.heizmann,jurgen.beyerer}@kit.edu
[4] Fraunhofer Center for Machine Learning, Munich, Germany

Abstract. Human Pose Estimation (HPE) is becoming increasingly ubiquitous, finding applications in diverse fields such as surveillance and worker safety, healthcare, sport and entertainment. Despite substantial research in HPE within the visible domain, there is limited focus on thermal imaging for HPE, primarily due to the scarcity and annotation difficulty of thermal data. Thermal imaging offers significant advantages, including better performance in low-light conditions and enhanced privacy, which can lead to greater acceptance of monitoring systems. In this work, we introduce LLVIP-Pose, an extension of the existing LLVIP dataset, to include 2D single-image pose estimation for aligned nighttime RGB and thermal images, containing approximately 26k annotated skeletons. We detail our annotation process and propose a novel metric for identifying and correcting poorly annotated skeletons. Furthermore, we present a comprehensive benchmark of top-down, bottom-up, and single-stage pose estimation models evaluated on both RGB and thermal images. Our evaluations demonstrate how pre-training on grayscale COCO data with data augmentation can benefit thermal pose estimation. The LLVIP-Pose dataset addresses the lack of thermal HPE datasets, providing a valuable resource for future research in this area. The pose annotations and baseline code are available on github: https://github.com/MickaelCormier/llvip-pose.

1 Introduction

Human pose estimation (HPE) constitutes a critical task within the domain of computer vision, aiming to predict human poses from image or video sources.

Supplementary Information The online version contains supplementary material available at https://doi.org/10.1007/978-981-96-2641-0_5.

M. Cho et al. (Eds.): ACCV 2024 Workshops, LNCS 15482, pp. 69–86, 2025.
https://doi.org/10.1007/978-981-96-2641-0_5

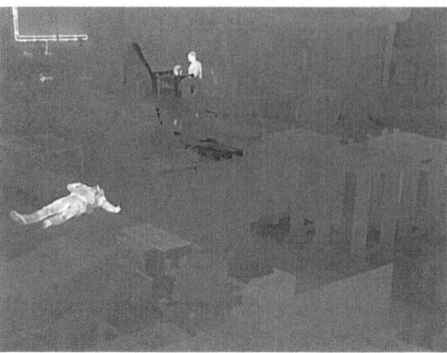

Fig. 1. Thermal image of a worker lying unconscious in a maintenance hall. Colleagues are not able to see him to call for help.

The ability to extract poses from visual data facilitates enhanced scene understanding, thereby garnering significant interest across various academic and industrial fields, including autonomous driving [28,36], fall detection in healthcare [1], human-computer interaction [16], robotics [45], and surveillance [9]. Specifically, in surveillance, HPE enables skeleton-based activity recognition, autonomously triggering alerts that necessitate immediate human intervention. This proves highly advantageous as it replaces vision-based activity recognition that may otherwise contain sensitive biometric data [9]. Furthermore, HPE may also be leveraged for worker safety through privacy-preserving monitoring systems, which ensure that the collected data cannot be misused against employees, thereby fostering a secure and ethical work environment. As shown in Fig. 1, detecting an injured worker or preventing collision with larger machines may prevent heavy injuries or fatalities. However, with enough context information, the identity of a person in thermal imaging may be inferred. Thus, HPE provides a second layer of anonymization. However, prevailing HPE techniques predominantly rely on images from the visible light spectrum, thereby rendering them susceptible to low-light conditions and obstructive weather phenomena [8,18].

To address these limitations, thermal infrared cameras present a promising alternative. These cameras exhibit substantial robustness against the aforementioned conditions. This robustness is evident in tasks such as person detection, where thermal-based predictions demonstrate superior accuracy compared to their vision-based counterparts. Nonetheless, the potential of thermal imagery in the context of HPE remains largely underexplored.

The primary impediment to progress in this area is the scarcity of thermal-based HPE datasets. The acquisition of such datasets comes at considerable costs due to the challenges associated with manually annotating poses, where annotating a single pose can take up to 60 s [10]. Furthermore, the absence of a benchmark to compare thermal-based with vision-based HPE complicates the assessment of the respective advantages and disadvantages of these modalities. Thus, the LLVIP dataset [18] emerges as a dataset of particular interest, offering aligned images from both modalities and designed to investigate the impact of low-light conditions on pedestrian detection. However, it lacks pose annotations and necessitates modifications to serve as a suitable benchmark for HPE.

In this work, we propose to extend the aforementioned dataset into a new benchmark dataset called LLVIP-Pose. To this aim, pose annotations are incorporated, where poses are shared between the two modalities. Since manual pose annotation demand high costs, we propose a semi-autonomous workflow to accelerate the pose annotation and validation processes, where previously each pose was annotated and validated by hand. For this purpose, pose predictions from HPE models are corrected by the annotators, eliminating the need to label a pose from scratch. To minimize the validation costs, an outlier detector based on the principle of anatomically correct skeleton structure leverages the relations between bone segments to highlight poorly annotated poses. This results in the contribution of 26k pose annotations for thermal imaging, which is by far the largest dataset for HPE in thermal imaging. Alongside the acquisition of this dataset, two studies are conducted to examine the viability and capabilities of thermal-based HPE models. First, our benchmark experiments compare the performance of thermal-based models against traditional RGB-based models featuring different multi-person pose estimation methodologies, e.g., top-down, bottom-up, and single-stage approaches. Next, generalization experiments are conducted on the thermal images of the LLVIP-Pose dataset with models pretrained on the COCO dataset to provide insight into the generalization capabilities of RGB-trained models for thermal applications.

The main contributions of this paper are summarized as follows:

- We propose LLVIP-Pose, the first large scale visible-infrared paired dataset for HPE in thermal imaging and low-light vision.
- We propose a semi-automated workflow to label and validate skeletons in still-images in LLVIP-Pose.
- We evaluate the experimental results of recent methods for HPE on LLVIP-Pose, and find that the dataset is a challenge for both HPE in RGB-low-light vision and in thermal imaging, although the latter shows impressive results.

2 Related Work

2.1 2D Human Pose Estimation

In HPE, human poses are predicted from images or videos by localizing keypoints (e.g., body parts or joints) and visualizing their connections based on a skeletal topology. Keypoint estimation is traditionally categorized into regression-based [22,27,33,40] or heatmap-based [7,35,41] methods. Regression-based methods predict keypoint coordinates directly, while heatmap-based methods infer them through multiple heatmaps, each representing the likelihood of a specific keypoint type. However, recents methods also use classification to perform discrete regression [23,25]. Top-down approaches [23,27,35,40,41,43] localize individuals first and then estimate their poses. In contrast, bottom-up approaches [5,7,15,31] localize keypoints first and then group them. Top-down methods offer better accuracy but are slower with more individuals, while bottom-up methods are faster but less accurate. One-stage pipeline approaches [25,26,42] generate pose candidates without intermediate steps, processing both global and local context simultaneously. They are computationally efficient but suffer from the same accuracy issues as bottom-up approaches. HPE is predominantly applied to RGB images, leveraging rich color and texture information. However, its application to infrared thermal images remains largely unexplored. A major reason for this gap is the absence of thermal image datasets for HPE.

2.2   RGB Datasets

With growing interest in applications for HPE [38], numerous large-scale datasets have gained prominence [2,3,11,21,24,34,37,44]. The COCO dataset [24], one of the most widely utilized, comprises over 200,000 images and 250,000 annotated poses. Similar to the MPII dataset [3], COCO features non-continuous images with common poses and frontal views. PoseTrack18 [2], based on the MPII dataset, includes continuous video frames capturing more complex real-life scenarios in controlled environments, such as sports events. COCO defines its own topology with 17 keypoints, including five on the head (nose, eyes, ears), which are challenging to detect in realistic scenarios with steeper camera angles. Consequently, the MPII and PoseTrack18 topologies simplify this by reducing head keypoints to two and three, respectively. In this work, we prefer this representation to facilitate manual annotation. These datasets primarily represent human poses in common and straightforward situations with favorable camera angles. OCHuman [44] and OCPose [44] address (self-)occlusion with frontal views in single, non-continuous images with two subjects. Crowd-Pose [21] features crowded scenarios in controlled environments like group photos or sports events. However, training on these datasets often transfers poorly to real-world surveillance scenarios, which involve steep camera angles, heavy (self-)occlusion, and dense crowds. Occlusion in crowded environments and complex poses remains a significant challenge in HPE [9] even for annotators labeling manually [10].

2.3   Thermal Infrared Datasets

The CAMEL dataset [14], modeled after the MOT challenge [11,20,29], includes 26 video sequences of RGB-thermal pairs, totaling around 23,437 annotated pairs, with 7,775 aligned. Captured at 336×256 pixels resolution and 30 fps in the LWIR band, it features indoor and outdoor urban environments under varying conditions and times. The KAIST dataset [17] offers 95,000 aligned RGB-thermal image pairs with 103,128 annotated bounding boxes for pedestrians, captured at 640×480 resolution and 20 fps in the LWIR band. It features outdoor scenarios with varying conditions and day times from a moving vehicle's perspective. The ThermalIM dataset [32] aims to infer past human motion via thermal residuals on objects. It consists of 783 video clips with thermal images at 384 × 288 pixels resolution and RGB at 1,920 × 1,080 pixels. It lacks bounding boxes, but provides 2D and 3D pose labels for RGB images, captured in three rooms with different actors, angles, and layouts. The OpenThermal-Pose dataset [19] includes 6,090 images with 14,315 annotated human instances. It provides bounding boxes and 17 keypoints, following COCO standards. The dataset spans various activities such as fitness exercises, multi-person interactions, and outdoor walking in different locations and weather conditions. However, the point of view is not suitable for surveillance contexts. The LLVIP dataset [18] addresses image fusion and low-light pedestrian detection, featuring 15,438 aligned image pairs in the LWIR band at a resolution of 1,280 × 1,024 pixels across 26 nighttime scenarios. It is possible to transfer labels from thermal images to RGB images due to alignment, delivering bounding boxes usable for both modalities.

3   LLVIP-Pose

We propose LLVIP-Pose, an HPE extension of the LLVIP visible-infrared paired dataset for low-light vision. The train-test split is shared with the original LLVIP dataset. After annotation and dataset cleaning, 26,135 bounding boxes with poses are provided. The final training set includes 6,853 images pairs (18,633 persons) and the test set contains 3,462 image pairs (7,502 persons).

In the remainder of this section, the annotation and validation of the human poses is described. First, a semi-automated annotation process is proposed. Second, a distance metric is proposed in order to identify outliers in the manual annotation which may contain annotation errors. Finally, the validation workflow is described.

3.1   Annotation Process

A semi-autonomous pipeline is proposed that uses an HPE model to reduce the labor-intensive process of labeling each keypoint individually. Instead, the focus is on correcting the predicted poses from the HPE model. Predicted poses from existing RGB-based pose estimators delivered suboptimal results due to the low visibility in the RGB images. Therefore, paid-annotators manually labeled poses until a sufficient number of poses was reached to train an HPE model. As a compromise between high quality predicted poses and fast training, HRNetw48-udp model [35] is used, which is a top-down HPE model pre-trained on the COCO dataset. Prior to prediction, the initial width and height of the bounding boxes were increased by a tenth of their width and height to ensure that all body parts are enclosed within the bounding box. These RGB-based pose predictions were visualized on thermal images and corrected by the annotators for their 2D image coordinates and visibility score. Subsequently, a thermal-based pose estimator is trained to provide predictions for the remaining more challenging sequences.

3.2   Validation Process

After the initial pose annotation phase, a validation phase is carried out for quality assurance. Validation often incurs high time costs, as each individual pose is reviewed manually.

Anthropometric Detection of Outliers. Based on anthropometric measurements, we propose to evaluate the viability of poses based on the proportions of various body segments. Drillis *et al.* [12] provided estimates of different body lengths as a percentage of the body height H, as depicted in Fig. 2. In this work, we re-contextualized the body segment lengths as ratios represented as the percentage of the shortest over the longest body segment. This new representation is necessary as the availability of body height H is not guaranteed in the poses, e.g., poses found at the edges of the image or occlusions blocking either the upper or lower body.

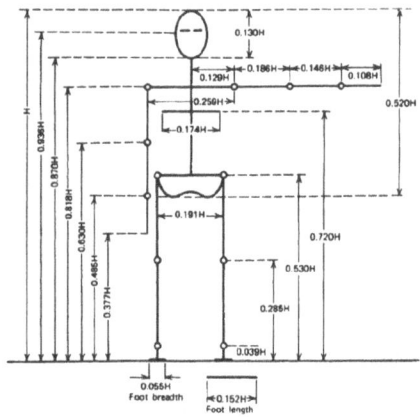

Fig. 2. Body segments with respect to body height H. (Source: [12,39])

As the anatomical structure of a human skeleton consists of multiple body parts interconnected together at the joints (keypoints), the length of a body part affects the length of its neighbors. Therefore, the use of the hip-shoulder, thigh-torso and torso-leg ratios is proposed as these rations share a common connection with the keypoints of the hip. Ratios including the arms were omitted, as their high degrees of freedom in 3D space are not fully captured in 2D images. In addition to the three ratios, the hip-shoulder-angle is included as an additional term to evaluate the viability of the poses, where the angle difference between both body segments is calculated. In the case of LLVIP, the angle difference is approximately zero, as the three main actions performed are standing, walking, and riding.

The evaluator features a distance metric d_i to score the individual poses, where a lower score indicates a low difference between the values derived from the poses x_i and the groundtruth values y_i found in Table 1.

Table 1. Groundtruth values used in the pose validator.

Ratio	Value
Hip-Shoulder	0.737
Hip-Shoulder-Angle	0.0
Thigh-Torso	0.851
Torso-Leg	0.587

The pose and distance metric are depicted in Eq. (1) and Eq. (2).

$$\text{Metric} = \frac{\sum_{i=1}^{n_{\text{values}}} d_i(x_i, y_i)}{n_{\text{values}}} \tag{1}$$

$$d_i(x_i, y_i) = \begin{cases} 1.0 - f(|x_i - y_i| \mid 0, \sigma_{|x_i-y_i|}) \cdot \sigma_{|x_i-y_i|}\sqrt{2\pi}, & \text{if } x_i \text{ is present} \\ 1.0 - f(|x_{\text{approx},i} - y_i| \mid 0, \sigma_{|x_i-y_i|}) \cdot \sigma_{|x_i-y_i|}\sqrt{2\pi}, & \text{if } x_i \text{ is not present} \end{cases} \tag{2}$$

The variables and parameters found in the metrics are:

- x_i: calculated values
- y_i: groundtruth values
- $x_{\text{approx},i}$: approximated values
- $\mu_{|x_i-y_i|}$: mean of the error terms between calculated and groundtruth values
- μ_{x_i}: mean of calculated values
- $\sigma_{|x_i-y_i|}$: standard deviation of the error terms between calculated and groundtruth values
- σ_{x_i}: standard deviation of calculated values
- n_{values}: number of values
- n_x : number of available values $(1 - n_{\text{values}})$

Through the distance metric d_i each calculated value x_i is evaluated against its respective groundtruth value y_i. Their error term z_i is calculated based on the Manhattan Distance and is set into an unnormalized Gaussian with the mean $\mu_{|x_i-y_i|}$ and the standard deviation $\sigma_{|x_i-y_i|}$ of the error term, as depicted by Eq. 3. The resulting pose metric score is the average of the distance metric for each value.

$$f(z \mid \mu_{|x_i-y_i|}, \sigma_{|x_i-y_i|}) = \frac{1}{\sigma_{|x_i-y_i|}\sqrt{2\pi}} \cdot e^{-\frac{1}{2}\left(\frac{z-\mu_{|x_i-y_i|}}{\sigma_{|x_i-y_i|}}\right)^2} \tag{3}$$

Poses are classified as an outlier, if the calculated pose metric score exceeds a threshold, which is calculated based on the percentage of the standard deviation σ_{x_i} over the mean μ_{x_i} for the different values. This representation reflects the variability of the values, where a high variability indicates an inconsistent quality of pose annotations resulting in a lower threshold. For the hip-shoulder-angle, the threshold is calculated as the percentage of the mean μ_{x_i} over the standard deviation σ_{x_i}, as $\mu_{x_i} \gg \sigma_{x_i}$. Formally, the threshold is computed as

$$\text{Threshold} = \frac{\sum_{i=1}^{n_{\text{ratios}}}\left(1 - \frac{\sigma_i}{\mu_i}\right)}{n_{\text{ratios}}} \tag{4}$$

The calculation of the values relies on the availability of keypoints. If a ratio is excluded from the overall pose metric, multiple different thresholds would need to be assigned for each pose within a sequence. Thus, the omission of any value is undesirable as it will introduce inconsistency in the pose metric. For this reason, we propose to approximate the missing ratios or angle from the distribution of known n_x values, where the error term of the missing ratio or angle is approximated based on the n_x error terms with respect to their respective standard deviation $\sigma_{|x_i-y_i|}$, as depicted in Eq. 5.

$$\frac{|x_{\text{approx},i} - y_i|}{\sigma_{|x_i-y_i|}} = \frac{\left(\sum_{n=1}^{n_x} \frac{|x_i-y_i|}{\sigma_{|x_i-y_i|}}\right)}{n_x} \tag{5}$$

Validation Workflow. After applying the outlier detector on the manually annotated sequences, a ranking based on the metric is used to prioritize the work of the human validators. Those are tasked with the correction of the detected outliers for each sequence in the dataset. The mean and the standard deviation of the error terms z_i are summarized in Fig. 3. We report the mean and standard deviation before and after the validation phase of the entire dataset to ascertain the effectiveness of outlier detection and improve the consistency of poses.

It is observed that the mean and standard deviation decrease for each value, showcasing the increase of quality as the individual error terms are converging towards zero.

In summary, instead of manually reviewing each pose and probably generating similar errors during the annotation process, the proposed distance metric is used to prioritize and objectively quantify the potential error from each pose estimation. Through multiple review iterations, the correction of a quarter to half of a sequence optimizes the overall metric by 0.135 points. In certain sequences, an optimization of 0.277 points is observed, strongly reducing the costs of validation while guaranteeing appropriate quality of annotation.

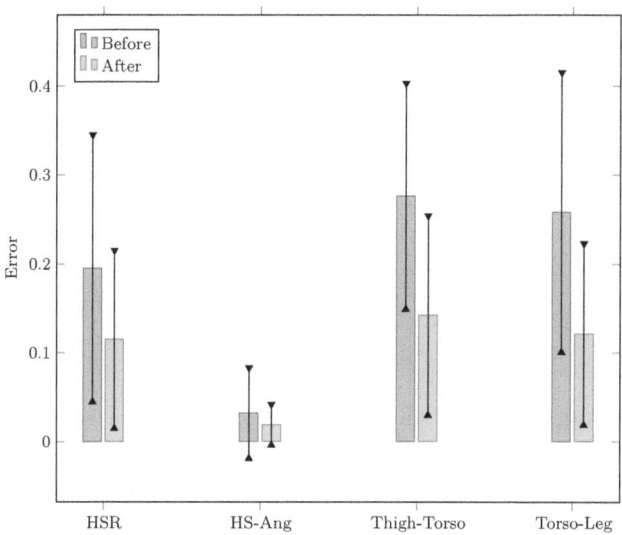

Fig. 3. Means and standard deviations of the error term before and after the validation phase. One can observe that both the mean and standard deviation decrease which proves the effectiveness of the proposed pose validator.

4   Experiments

This section details the experiments for pedestrian detection required for top-down models, the HPE experiments on LLVIP-Pose dataset for RGB and thermal images, and evaluates the results. Further experiments on grayscale pre-training and data-augmentation are also presented. The experiments are conducted on NVIDIA RTX Quatro 6000 24GB or Tesla V100-SXM2-32GB depending on their availability.

4.1   Pedestrian Detection

For our experiments, a YOLOX [13] object detection model is trained for person detection for each modality. These models are used to generate the detections for the evaluation of the top-down HPE models in the benchmark experiments. The YOLOX-l predictions with the same configuration setting as the LLVIP paper report worse predictions for both modalities as the baseline. The YOLOX-x versions achieved better detection results on RGB images but slightly lower detection performance on thermal images. The detailed results are presented in Table 2.

Table 2. Person detection results for LLVIP datasets. YOLOv5-l results are taken from the LLVIP paper. The YOLOX-l and YOLOX-x models are trained with the mmdetection toolbox [6]. Best model is highlighted in red. Second best is highlighted in blue.

Models	Thermal		RGB		Epoch	Lr
	AP	AP_{50}	AP	AP_{50}		
YOLOv5-l [18]	0.670	0.965	0.527	0.908	200	3.2×10^{-3}
YOLOX-l (ours)	0.652	0.959	0.529	0.906	200	3.2×10^{-3}
YOLOX-x (ours)	0.664	0.962	0.536	0.908	100	1.0×10^{-2}

4.2   Evaluation Setup

The Average Precision (AP) based on the object keypoint similarity (OKS) metric is used for evaluation on the LLVIP-Pose. In addition to AP and AP_{50}, the AP_M and AP_L are reported for medium and large bounding boxes as well as the AP_E^C, AP_M^C and AP_H^C for easy, medium and hard crowdedness levels based on [21].

4.3   Human Pose Estimation

Seven SOTA 2D HPE models with representatives for top-down (TD), bottom-up (BU) and one stage (OS) models are trained and evaluated on the newly acquired LLVIP-Pose dataset. To this aim, the mmpose toolbox [30] is used. See the supplementary materials for detailed training configurations. The models are trained with the same configuration with a batch size of 16 and weights from ImageNet, if provided. Otherwise, the weights are randomly initialized. To assess the effectiveness of thermal-based pose predictions against low-light RGB, each model is trained twice: once with thermal images and once with RGB images independently.

The evaluation results on the test set of the LLVIP-Pose dataset are reported in Table 3 and Table 4. As can be observed, the thermal-based models outperform the RGB-based models in all AP categories. The best model for both modalities is the ViTPose-h, for both groundtruth and predicted bounding boxes.

While considering the top-down approaches for both modalities, one notices a larger AP gap in the results between groundtruth and predicted bounding boxes for the RGB-based models. The switch from groundtruth to predicted bounding boxes results in a drop of approximately 0.015 AP and 0.05 AP for thermal- and RGB-based models, respectively. The lower drop of APs showcases the higher detection rate of individuals in thermal images and the difficulty of person detection in RGB images due to low-illumination. The thermal-based models exhibit higher AP_M, showcasing the effectiveness of thermal images for medium bounding boxes. However, the reported AP_M does not highlight the ability of thermal images for medium scale human instances due to the fact that

Table 3. Benchmarking results with AP for the different bounding box sizes (AP_M, AP_L). Best model is highlighted in red. Second best is higlighted in blue.

Models	Category	Groundtruth BBoxes				Predicted BBoxes (Thermal, AP: 0.664)			
Thermal									
		AP	AP_{50}	AP_M	AP_L	AP	AP_{50}	AP_M	AP_L
HRNetw48-udp [35]	TD	0.900	0.980	0.605	0.902	0.886	0.965	0.248	0.887
ViTPose-h [41]	TD	0.916	0.990	0.634	0.917	0.899	0.965	0.270	0.902
DeepPose-r50 [33]	TD	0.852	0.979	0.350	0.854	0.841	0.964	0.176	0.842
SimCC [23]	TD	0.877	0.979	0.457	0.879	0.867	0.964	0.217	0.869
DEKR [15]	BU					0.845	0.953	0.065	0.850
YOLOX-Pose-l [26]	OS					0.848	0.962	0.223	0.850
RTMO-l [25]	OS					0.855	0.961	0.243	0.858

Models	Category	Groundtruth BBoxes				Predicted BBoxes (RGB, AP: 0.536)			
RGB									
		AP	AP_{50}	AP_M	AP_L	AP	AP_{50}	AP_M	AP_L
HRNetw48-udp [35]	TD	0.643	0.914	0.176	0.647	0.591	0.866	0.087	0.594
ViTPose-h [41]	TD	0.681	0.937	0.253	0.684	0.632	0.879	0.142	0.634
DeepPose-r50 [33]	TD	0.596	0.906	0.153	0.599	0.547	0.854	0.064	0.549
SimCC [23]	TD	0.619	0.906	0.199	0.621	0.572	0.856	0.074	0.574
DEKR [15]	BU					0.562	0.854	0.010	0.566
YOLOX-Pose-l [26]	OS					0.569	0.867	0.128	0.571
RTMO-l [25]	OS					0.561	0.853	0.096	0.563

the scales of the person for medium and large bounding boxes are similar for both, as depicted in Fig. 4a.

Bottom-up and single-stage models exhibit similar trends, with higher APs reported for thermal images, comparable to results of the top-down models.

As shown in Table 4, thermal models report higher AP^C than RGB models for both predicted and groundtruth bounding boxes. Single-stage models with the exeption of ViTPose-h have higher AP_H^C than top-down approaches for both modalities. For the top-down approaches, two overlapping bounding boxes result in sub-optimal pose predictions where the background bounding box receives the pose for the foreground bounding box, leaving the background bounding box without any poses and the foreground bounding box with two poses. This occurrence is relevant to the thermal images on SimCC, where a higher AP_H^C is reported for predicted bounding boxes, as shown in Fig. 4b.

Qualitative results are shown in Fig. 5, showcasing pose predictions from the ViTPose models for RGB and thermal images, respectively. Qualitatively,

(a) Medium bounding boxes for image pair with pose predictions from ViTPose-h. Medium bounding boxes contain human instances with similar scale to instances of large bounding boxes but are cropped due to image boundary.

(b) Comparison of SimCC with groundtruth and predicted bounding boxes. Poses with groundtruth bounding boxes (left) report lower AP_H^C than predicted bounding boxes (right).

Fig. 4. Visualization of pose predictions for both medium bounding boxes (a) and hard crowding level (b).

Fig. 5. Qualitative results of ViTPose for RGB and thermal image pairs. For scenarios with low-illumination (e.g. top right) and high illumination changes (e.g. top-left, bottom-middle and bottom-right) thermal-based predictions exhibit higher accuracy as indicated by the higher pose heatmap values.

thermal- and RGB-based predictions prove similar results. However, for scenarios with low-illumination (e.g., top right) and high illumination changes (e.g., top-left, bottom-middle, and bottom-right) thermal-based predictions exhibit higher accuracy as indicated by the higher heatmap values. On the top right image pairs, thermal-based pose predictions capture the body orientations of the pedestrians situated in the dark, whereas RGB-based predictions failed to do so.

4.4 Further Experiments

Due to the scarce availability of annotated thermal data for HPE, leveraging RGB data is investigated. More precisely, experiments with COCO RGB pretraining and COCO reduced to grayscale for pretraining are conducted. The

Table 4. Benchmarking results with AP for the different crowding level(AP_E^C, AP_M^C, AP_H^C). Best model is highlighted in red. Second best is higlighted in blue.

		Thermal									
Models	**Category**	**Groundtruth BBoxes**					**Predicted BBoxes** (Thermal, AP: 0.664)				
		AP	AP_{50}	AP_E^C	AP_M^C	AP_H^C	AP	AP_{50}	AP_E^C	AP_M^C	AP_H^C
HRNetw48-udp [35]	TD	0.900	0.980	0.906	0.886	0.833	0.886	0.965	0.893	0.866	0.760
ViTPose-h [41]	TD	0.916	0.990	0.921	0.901	0.837	0.899	0.965	0.907	0.878	0.797
DeepPose-r50 [33]	TD	0.852	0.979	0.859	0.826	0.751	0.841	0.964	0.849	0.809	0.728
SimCC [23]	TD	0.877	0.979	0.884	0.853	0.721	0.867	0.964	0.876	0.839	0.724
DEKR [15]	BU						0.845	0.953	0.852	0.818	0.707
YOLOX-Pose-l [26]	OS						0.848	0.962	0.851	0.838	0.775
RTMO-l [25]	OS						0.855	0.961	0.859	0.845	0.797

		RGB									
Models	**Category**	**Groundtruth BBoxes**					**Predicted BBoxes** (RGB, AP: 0.536)				
		AP	AP_{50}	AP_E^C	AP_M^C	AP_H^C	AP	AP_{50}	AP_E^C	AP_M^C	AP_H^C
HRNetw48-udp [35]	TD	0.643	0.914	0.651	0.626	0.546	0.591	0.866	0.598	0.576	0.497
ViTPose-h [41]	TD	0.681	0.937	0.683	0.678	0.714	0.632	0.879	0.635	0.626	0.567
DeepPose-r50 [33]	TD	0.596	0.906	0.601	0.584	0.560	0.547	0.854	0.551	0.529	0.480
SimCC [23]	TD	0.619	0.906	0.626	0.601	0.575	0.572	0.856	0.578	0.556	0.508
DEKR [15]	BU						0.562	0.854	0.570	0.536	0.519
YOLOX-Pose-l [26]	OS						0.569	0.867	0.571	0.565	0.504
RTMO-l [25]	OS						0.561	0.853	0.561	0.565	0.527

pre-trained models are first evaluated directly on LLVIP-Pose on the 13 common keypoints between the COCO topology and the LLVIP (Posetrack18) topology. Each model is then fine-tuned on the LLVIP-Pose dataset and once more evaluated.

Further variants are trained using the Albumentation library [4] to apply pixel-wise augmentation techniques on the COCO training data. In this case, HSV augmentation and image inversion with a probability of $p = 1$ are used to further simulate thermal data. HSV augmentation is applied prior to grayscaling to provide more variations, followed by image inversion to imitate the appearance of thermal images. As a control, the same augmentations are applied to RGB images and an experiment combining both RGB and grayscale images is conducted.

For a representative overview, three models are trained, representing the three categories of HPE models. The results are reported in Table 5. Prior to fine-tuning, two main observations are reported. The first is the benefit of the HSV and image inversion augmentations to both RGB and grayscale images and the second is the lower *APs* of models trained with grayscale images. This observation showcases the domain gap between RGB and thermal images, in which grayscale images lack the characteristics of thermal images, but the loss of two distinct color channels hinders keypoint localization. The *APs* reported for "RGB + Grayscale" for DEKR and YOLOX-Pose further support this with the lowest *APs* recorded. HRNetw48-udp acts, however, as an outlier, where the mixture of RGB and grayscale images resulted in the higher *AP*.

After the fine-tuning using the LLVIP-Pose train set, the grayscale (monochromatic) models report higher *AP* compared to the RGB models. This observation shows how learning monochromatic features can assist the fine-tuning of thermal data, as the grayscale models even outperforms the "RGB + Augmentations" models. The sub-optimal performance of models trained with RGB and grayscale COCO images report the highest *AP* for YOLOX-Pose and the second highest *AP* for HRNetw48-udp and DEKR, trailing behind the *APs* for "Grayscale + Augmentations". This demonstrates the importance of RGB features alongside monochromatic features, as it outperforms the *APs* of the grayscale models. "Grayscale + Augmentations" exhibits the highest *APs* for

Table 5. Pre-training and fine-tuning results of three SOTA models with different augmentations. The tests are conducted on the LLVIP-Pose test dataset with the 13 mutual keypoints between the COCO and PoseTrack18 skeleton topology.

Models	Augmentations	Thermal							
		COCO Pre-Training				LLVIP Fine-Tuning			
		AP	AP_{50}	AP_M	AP_L	AP	AP_{50}	AP_M	AP_L
HRNetw48-udp [35]	RGB	0.626	0.813	0.435	0.626	0.930	0.990	0.720	0.930
	Grayscale	0.611	0.789	0.431	0.611	0.938	0.990	0.705	0.938
	RGB + Grayscale	0.631	0.814	0.466	0.633	0.939	0.990	0.744	0.939
	RGB + Aug	0.705	0.872	0.411	0.707	0.937	0.989	0.764	0.938
	Grayscale + Aug	0.739	0.896	0.384	0.742	0.941	0.990	0.786	0.942
	LLVIP-Pose (Scratch)	-	-	-	-	0.917	0.979	0.685	0.917
DEKR [15]	RGB	0.427	0.606	0.016	0.440	0.888	0.964	0.071	0.892
	Grayscale	0.400	0.573	0.012	0.417	0.894	0.964	0.067	0.898
	RGB + Grayscale	0.383	0.550	0.010	0.398	0.896	0.965	0.062	0.899
	RGB + Aug	0.540	0.719	0.007	0.553	0.892	0.963	0.061	0.897
	Grayscale + Aug	0.601	0.782	0.009	0.616	0.897	0.965	0.066	0.900
	LLVIP-Pose (Scratch)	-	-	-	-	0.862	0.952	0.070	0.866
YOLOX-Pose-l [26]	RGB	0.511	0.707	0.061	0.513	0.880	0.964	0.295	0.881
	Grayscale	0.496	0.684	0.045	0.499	0.881	0.972	0.287	0.882
	RGB + Grayscale	0.491	0.677	0.055	0.494	0.882	0.971	0.219	0.883
	RGB + Aug	0.598	0.799	0.049	0.600	0.881	0.972	0.245	0.883
	Grayscale + Aug	0.607	0.814	0.042	0.610	0.878	0.971	0.235	0.879
	LLVIP-Pose (Scratch)	-	-	-	-	0.863	0.962	0.222	0.864

HRNetw48-udp and DEKR, however, the lowest AP for YOLOX-Pose, despite showcasing the highest AP without fine-tuning.

5 Conclusions

In conclusion, this work addresses the limitations in HPE under low-light conditions by introducing the LLVIP-Pose dataset, a modification of the existing LLVIP dataset to include pose annotations for both visible and thermal imagery. Our dataset includes an extensive contribution of 26,135 pose annotations, significantly enhancing the resources available for thermal-based HPE research. By implementing a semi-autonomous workflow for pose annotation and validation, we have substantially reduced the manual effort and associated costs. Our benchmark experiments demonstrate that thermal-based HPE models can outperform traditional RGB-based models in challenging conditions, while our generalization experiments provide valuable insights into the transferability of pre-trained RGB models to thermal applications.

These findings underscore the potential of thermal imagery for robust and privacy-preserving human pose estimation, paving the way for future advancements in various applications such as surveillance, worker safety, and beyond. Future work may address the problem of generalization between different sensors, cameras, and perspectives. Ensuring that HPE models can robustly adapt to varying hardware and viewpoints will be crucial for the widespread adoption and effectiveness of these technologies in real-world scenarios.

References

1. Alam, E., Sufian, A., Dutta, P., Leo, M.: Real-time human fall detection using a lightweight pose estimation technique. In: Dasgupta, K., Mukhopadhyay, S., Mandal, J.K., Dutta, P. (eds.) International Conference on Computational Intelligence in Communications and Business Analytics, pp. 30–40. Springer (2023). https://doi.org/10.1007/978-3-031-48879-5_3
2. Andriluka, M., et al.: PoseTrack: a benchmark for human pose estimation and tracking. In: Proceedings of the IEEE Conference on Computer Vision and Pattern Recognition (CVPR) (2018)
3. Andriluka, M., Pishchulin, L., Gehler, P., Schiele, B.: 2D human pose estimation: New benchmark and state of the art analysis. In: Proceedings of the IEEE Conference on computer Vision and Pattern Recognition, pp. 3686–3693 (2014)
4. Buslaev, A., Iglovikov, V.I., Khvedchenya, E., Parinov, A., Druzhinin, M., Kalinin, A.A.: Albumentations: fast and flexible image augmentations. Information **11**(2) (2020). https://doi.org/10.3390/info11020125, https://www.mdpi.com/2078-2489/11/2/125
5. Cao, Z., Simon, T., Wei, S.E., Sheikh, Y.: Realtime multi-person 2D pose estimation using part affinity fields. In: Proceedings of the IEEE Conference on Computer Vision and Pattern Recognition, pp. 7291–7299 (2017)
6. Chen, K., et al.: MMDetection: Open MMLAB detection toolbox and benchmark. arXiv preprint arXiv:1906.07155 (2019)

7. Cheng, B., Xiao, B., Wang, J., Shi, H., Huang, T.S., Zhang, L.: HigherHRNet: scale-aware representation learning for bottom-up human pose estimation. In: Proceedings of the IEEE/CVF Conference on Computer Vision and Pattern Recognition, pp. 5386–5395 (2020)

8. Choi, Y., et al.: KAIST multi-spectral day/night data set for autonomous and assisted driving. IEEE Trans. Intell. Transp. Syst. **19**(3), 934–948 (2018)

9. Cormier, M., Clepe, A., Specker, A., Beyerer, J.: Where are we with human pose estimation in real-world surveillance? In: Proceedings of the IEEE/CVF Winter Conference on Applications of Computer Vision (WACV) Workshops, pp. 591–601 (2022)

10. Cormier, M., Röpke, F., Golda, T., Beyerer, J.: Interactive labeling for human pose estimation in surveillance videos. In: Proceedings of the IEEE/CVF International Conference on Computer Vision (ICCV) Workshops, pp. 1649–1658 (2021)

11. Dendorfer, P., et al.: Mot20: a benchmark for multi object tracking in crowded scenes. arXiv: 2003.09003 (2020)

12. Drillis, R., Contini, R.: Body segment parameters, New York University. Tech. rep, NY, Technical Report (1966)

13. Ge, Z., Liu, S., Wang, F., Li, Z., Sun, J.: YOLOX: exceeding yolo series in 2021. arXiv preprint arXiv:2107.08430 (2021)

14. Gebhardt, E., Wolf, M.: Camel dataset for visual and thermal infrared multiple object detection and tracking. In: 2018 15th IEEE International Conference on Advanced Video and Signal Based Surveillance (AVSS), pp. 1–6. IEEE (2018)

15. Geng, Z., Sun, K., Xiao, B., Zhang, Z., Wang, J.: Bottom-up human pose estimation via disentangled keypoint regression. In: Proceedings of the IEEE/CVF Conference on Computer Vision and Pattern Recognition, pp. 14676–14686 (2021)

16. Heindl, C., Ikeda, M., Stübl, G., Pichler, A., Scharinger, J.: Metric pose estimation for human-machine interaction using monocular vision. arXiv preprint arXiv:1910.03239 (2019)

17. Hwang, S., Park, J., Kim, N., Choi, Y., Kweon, I.S.: Multispectral pedestrian detection: Benchmark dataset and baselines. In: Proceedings of IEEE Conference on Computer Vision and Pattern Recognition (CVPR) (2015)

18. Jia, X., Zhu, C., Li, M., Tang, W., Zhou, W.: LLVIP: a visible-infrared paired dataset for low-light vision. In: Proceedings of the IEEE/CVF International Conference on Computer Vision, pp. 3496–3504 (2021)

19. Kuzdeuov, A., Taratynova, D., Tleuliyev, A., Varol, H.A.: OpenThermalPose: an open-source annotated thermal human pose dataset and initial yolov8-pose baselines. In: 2024 IEEE 18th International Conference on Automatic Face and Gesture Recognition (FG), pp. 1–8 (2024). https://doi.org/10.1109/FG59268.2024.10581992

20. Leal-Taixé, L., Milan, A., Reid, I., Roth, S., Schindler, K.: MOTChallenge 2015: towards a benchmark for multi-target tracking. arXiv:1504.01942 (2015)

21. Li, J., Wang, C., Zhu, H., Mao, Y., Fang, H.S., Lu, C.: CrowdPose: efficient crowded scenes pose estimation and a new benchmark. In: Proceedings of the IEEE/CVF Conference on Computer Vision and Pattern Recognition (CVPR) (2019)

22. Li, K., Wang, S., Zhang, X., Xu, Y., Xu, W., Tu, Z.: Pose recognition with cascade transformers. In: Proceedings of the IEEE/CVF Conference on Computer Vision and Pattern Recognition, pp. 1944–1953 (2021)

23. Li, Y., et al.: SIMCC: a simple coordinate classification perspective for human pose estimation. In: Avidan, S., Brostow, G., Cissé, M., Farinella, G.M., Hassner, T. (eds.) European Conference on Computer Vision, pp. 89–106. Springer (2022). https://doi.org/10.1007/978-3-031-20068-7_6

24. Lin, T.-Y., et al.: Microsoft COCO: common objects in context. In: Fleet, D., Pajdla, T., Schiele, B., Tuytelaars, T. (eds.) ECCV 2014. LNCS, vol. 8693, pp. 740–755. Springer, Cham (2014). https://doi.org/10.1007/978-3-319-10602-1_48
25. Lu, P., Jiang, T., Li, Y., Li, X., Chen, K., Yang, W.: RTMO: towards high-performance one-stage real-time multi-person pose estimation. arXiv preprint arXiv:2312.07526 (2023)
26. Maji, D., Nagori, S., Mathew, M., Poddar, D.: YOLO-pose: enhancing yolo for multi person pose estimation using object keypoint similarity loss. In: Proceedings of the IEEE/CVF Conference on Computer Vision and Pattern Recognition, pp. 2637–2646 (2022)
27. Mao, W., et al.: Poseur: direct human pose regression with transformers. In: In: Avidan, S., Brostow, G., Cissé, M., Farinella, G.M., Hassner, T. (eds.) European conference on computer vision, pp. 72–88. Springer (2022). https://doi.org/10.1007/978-3-031-20068-7_5
28. Martin, M., Popp, J., Anneken, M., Voit, M., Stiefelhagen, R.: Body pose and context information for driver secondary task detection. In: 2018 IEEE Intelligent Vehicles Symposium (IV), pp. 2015–2021 (2018).https://doi.org/10.1109/IVS.2018.8500523
29. Milan, A., Leal-Taixé, L., Reid, I., Roth, S., Schindler, K.: MOT16: a benchmark for multi-object tracking. arXiv:1603.00831 (2016)
30. MMPose-Contributors: Openmmlab pose estimation toolbox and benchmark. https://github.com/open-mmlab/mmpose (2020)
31. Newell, A., Huang, Z., Deng, J.: Associative embedding: end-to-end learning for joint detection and grouping. In: Advances in Neural Information Processing Systems, vol. 30 (2017)
32. Tang, Z., Ye, W., Ma, W.C., Zhao, H.: What happened 3 seconds ago? Inferring the past with thermal imaging. In: CVPR (2023)
33. Toshev, A., Szegedy, C.: DeepPose: human pose estimation via deep neural networks. In: Proceedings of the IEEE Conference on Computer Vision and Pattern Recognition, pp. 1653–1660 (2014)
34. Vendrow, E., Le, D.T., Cai, J., Rezatofighi, H.: JRDB-pose: a large-scale dataset for multi-person pose estimation and tracking. In: Proceedings of the IEEE/CVF Conference on Computer Vision and Pattern Recognition (CVPR), pp. 4811–4820 (2023)
35. Wang, J., et al.: Deep high-resolution representation learning for visual recognition. TPAMI 43, 3349–3364 (2019)
36. Wang, S., et al.: Skeleton-based traffic command recognition at road intersections for intelligent vehicles. Neurocomputing 501, 123–134 (2022). https://doi.org/10.1016/j.neucom.2022.05.107, https://www.sciencedirect.com/science/article/pii/S0925231222006944
37. Wang, X., et al.: PANDA: a gigapixel-level human-centric video dataset. In: 2020 IEEE/CVF Conference on Computer Vision and Pattern Recognition (CVPR), pp. 3265–3275 (2020). https://doi.org/10.1109/CVPR42600.2020.00333
38. Wei, S.E., Ramakrishna, V., Kanade, T., Sheikh, Y.: Convolutional pose machines. In: Proceedings of the IEEE Conference on Computer Vision and Pattern Recognition (CVPR) (2016)
39. Winter, D.A.: Biomechanics and Motor Control of Human Movement. Wiley (2009)
40. Xiao, B., Wu, H., Wei, Y.: Simple baselines for human pose estimation and tracking. In: Proceedings of the European Conference on Computer Vision (ECCV), pp. 466–481 (2018)

41. Xu, Y., Zhang, J., Zhang, Q., Tao, D.: ViTPose: simple vision transformer baselines for human pose estimation. Adv. Neural. Inf. Process. Syst. **35**, 38571–38584 (2022)
42. Yang, J., Zeng, A., Liu, S., Li, F., Zhang, R., Zhang, L.: Explicit box detection unifies end-to-end multi-person pose estimation. arXiv preprint arXiv:2302.01593 (2023)
43. Yang, S., Quan, Z., Nie, M., Yang, W.: Transpose: Keypoint localization via transformer. In: Proceedings of the IEEE/CVF International Conference on Computer Vision, pp. 11802–11812 (2021)
44. Zhang, S.H., et al.: Pose2Seg: detection free human instance segmentation. In: Proceedings of the IEEE/CVF Conference on Computer Vision and Pattern Recognition (CVPR) (2019)
45. Zimmermann, C., Welschehold, T., Dornhege, C., Burgard, W., Brox, T.: 3D human pose estimation in RGBD images for robotic task learning. In: 2018 IEEE International Conference on Robotics and Automation (ICRA), pp. 1986–1992. IEEE (2018)

U-ENHANCE: Underwater Image Enhancement Using Wavelet Triple Self-attention

Priyanka Mishra[1] ⓘ, Santosh Kumar Vipparthi[1(✉)] ⓘ,
and Subrahmanyam Murala[2] ⓘ

[1] CVPR Lab, Indian Institute of Technology Ropar, Rupnagar, India
{priyanka.20eez0010,skvipparthi}@iitrpr.ac.in
[2] CVPR Lab, SCSS, Trinity College Dublin, Dublin, Ireland

Abstract. Transformer-based methods have demonstrated remarkable performance in underwater image enhancement due to their ability to capture long-range dependencies, crucial for high-quality reconstruction of degraded images. However, existing Transformer-based techniques often treat all token similarities equally during self-attention, which can lead to the aggregation of irrelevant features, hampering clear image restoration. We propose **U-ENHANCE**, a novel Underwater image **Enhance**ment framework that integrates wavelet-based frequency decomposition with spatial domain attention to address these challenges. In particular, we introduce a Wavelet Triple Self-Attention (WTSA) mechanism that performs self-attention across three dimensions—horizontal, vertical, and channel-wise, effectively capturing multi-scale features critical for restoring fine details and structural integrity. Additionally, we design a Self-Calibrated Feedforward Network (SCFN) that refines feature representation by dynamically adjusting the receptive field, further enhancing spatial and frequency domain integration. Extensive experiments on underwater image enhancement benchmarks demonstrate that U-ENHANCE outperforms state-of-the-art methods by providing superior restoration of color, clarity, and structural details. The code is available at: https://github.com/Priyanka01mishra/UENHANCE.

Keywords: Underwater image enhancement · Wavelet transform · Transformer

1 Introduction

Underwater imagery plays a critical role in various applications, including marine biology research [39], underwater archaeology [1,9], environmental monitoring, offshore engineering, coastal border security. It also supports efficient fish farming [37] and the use of autonomous underwater vehicles (AUVs) [2,30,31] for ocean exploration and surveillance [52,54]. High-quality underwater images are

ⓒ The Author(s), under exclusive license to Springer Nature Singapore Pte Ltd. 2025
M. Cho et al. (Eds.): ACCV 2024 Workshops, LNCS 15482, pp. 87–104, 2025.
https://doi.org/10.1007/978-981-96-2641-0_6

essential for tasks such as species identification, habitat mapping, and structural inspections of underwater infrastructure like pipelines and shipwrecks [30]. Moreover, in robotic and autonomous systems, the effectiveness of underwater navigation and object detection heavily relies on the clarity of captured images. However, capturing clear and accurate underwater images is a significant challenge due to the unique properties of the aquatic environment. As light penetrates through water, it undergoes refraction, absorption, and scattering, resulting in several types of image degradation [51]. These effects lead to color distortion, where certain wavelengths of light (particularly red) are absorbed more quickly, causing a green or blue hue in images. Additionally, scattering caused by particles in the water results in reduced contrast, blurriness, and haziness, which obscure fine details and structural features. These factors make underwater images visually unappealing and can severely limit the performance of underwater robotic systems, which rely on accurate visual information for navigation and object detection. The restoration of underwater images is critical to overcoming these challenges, recovering lost details, increasing visibility, and improving the overall quality of underwater scenes [14,17,20]. Traditional image restoration techniques often rely on handcrafted priors and physical models to address specific degradations [16]. However, the diversity and complexity of underwater environments make it difficult to generalize these methods across different conditions. In contrast, deep learning-based approaches have gained popularity due to their ability to learn data-driven solutions that can adapt to various underwater conditions [6,7,13,26]. These methods have shown promise in improving the quality of underwater images by effectively addressing color distortion, low contrast, and blurriness. CNN-based networks have demonstrated outstanding results in underwater image enhancement because of their ability to capture local features through convolutional operations. However, these methods inherently lack the ability to model long-range dependencies and global context [47], crucial for managing the complex degradations found in underwater environments. CNNs primarily rely on fixed-sized kernels [5], which restrict their ability to capture multi-scale features effectively, and they often struggle to generalize across diverse underwater conditions. As a result, exploring Transformer-based methods, which excel in modelling long-range dependencies and global feature interactions [44], presents a more promising direction for underwater image enhancement [32,35]. While existing Transformer-based approaches have focused primarily on spatial domain attention, limited work has explored the integration of frequency domain information [19,45], which is crucial for fully addressing underwater degradations. Among the few methods that do incorporate frequency domain features, most rely on Fourier transforms [19,45]. However, Fourier-based methods, while useful for capturing global frequency components, lack the ability to localize features in both time (spatial) and frequency domains simultaneously. This limitation makes it challenging to accurately restore fine details and structural elements, which are essential in underwater scenes. In contrast, wavelet transforms offer a distinct advantage by providing multi-scale frequency representation with spatial localization [38,41]. Unlike Fourier transforms, which only capture global frequency information, wavelets allow for the decomposition of an image into both frequency and spatial components, making them highly effective in preserving localized details while also addressing global distortions. Given the clear benefits of wavelet transforms in providing both frequency domain and spatial domain localization, incorporating wavelet-based techniques into Transformer models for underwater image enhancement is a logical and advantageous step forward.

In this work, we propose an efficient transformer-based network, U-ENHANCE, for underwater image enhancement that effectively leverages wavelet-based frequency decomposition to capture both spatial and frequency domain features, significantly enhancing restoration quality. The key component of U-ENHANCE is the Wavelet Transformer Block (WTB), which integrates the Wavelet Triple Self-Attention (WTSA) mechanism and a Self-Calibrated Feedforward Network (SCFN). The WTSA captures essential multi-scale features by applying self-attention across three dimensions—horizontal, vertical, and channel-wise—thus effectively capturing long-range dependencies and fine details. Additionally, WTSA decomposes features into wavelet sub-bands to preserve both high-frequency details and low-frequency structures, ensuring comprehensive feature extraction for underwater image enhancement. Furthermore, SCGFN refines these features by dynamically adjusting the receptive field, resulting in improved feature representation in both spatial and frequency domains. By adaptively handling local and global features, SCFN enhances the model's ability to restore clear, sharp images from severely degraded underwater scenes. This novel combination of wavelet-based frequency analysis and transformer attention mechanisms enables U-ENHANCE to significantly outperform existing methods, achieving superior restoration quality across multiple underwater image benchmarks.

The main contributions of this paper are as follows:

- We propose a Wavelet Triple Self-Attention (WTSA) Module that decomposes input features into frequency sub-bands using Discrete Wavelet Transform (DWT) for better noise reduction and detail preservation. In this module, we further introduce a Triple Attention mechanism (horizontal, vertical, and channel self-attention) to reduce computational complexity and capture long-range dependencies essential for underwater image enhancement.
- We propose a Self-Calibrated Feedforward Network (SCFN) which dynamically adjusts the receptive field to capture richer spatial and inter-channel dependencies. This improves spatial and contextual information processing, leading to more discriminative feature representation and enhanced restoration performance.

Experimentation on synthetic and real-world datasets, along with ablation studies verify the effectiveness of the proposed method for underwater image enhancement.

2   Literature Survey

2.1   Underwater Image Enhancement

Underwater image enhancement (UIE) techniques are generally classified into three main categories: physical model-based methods, visual prior-based approaches, and deep learning-based strategies [10,21,32,33,42]. Physical model-based methods often employ prior knowledge to construct enhancement models, utilizing concepts such as attenuation curve priors [46], fuzzy priors [8], and water dark channel priors [34]. While these methods can be effective, their reliance on externally defined priors can limit scalability and robustness in complex and diverse underwater environments.

Recent advancements in deep learning have shown significant promise in UIE. To address the challenge of limited real-world underwater paired training data, many researchers have turned to Generative Adversarial Network (GAN)-based frameworks. Notable examples include UGAN [12], FUnIE-GAN [17] UIE-DAL [43], and Watergan [23], which generate synthetic training data to improve enhancement performance. Additionally, Semi-UIR [16] introduced a mean teacher-based semi-supervised network that effectively leverages unlabeled data to enhance model training.

Emerging research has started to explore the utilization of frequency domain properties in UIE, highlighting the significant potential of frequency-based methods. Spectroformer [19], for instance, exploits frequency characteristics through a hybrid Fourier-spatial upsampling technique to enhance the resolution of degraded image features. Similarly, WF-Diff [50] combines frequency domain analysis with diffusion models for image enhancement and adjustment. Recent developments in wavelet-pixel domain fusion, such as WPFNet [28], have demonstrated improved UIE by integrating wavelet and pixel domains. This fusion preserves fine details and enhances color fidelity while reducing noise more effectively than previous methods. However, these frequency-based approaches may introduce unintended interactions in irrelevant areas due to their computational complexity and the potential for processing overhead.

2.2   Transformers in Vision

Building on the transformative impact of Transformers in NLP and high-level vision applications, recent advancements have extended their use to image restoration tasks, where they have outperformed traditional CNN-based models by effectively capturing long-range dependencies [4]. Nevertheless, the quadratic complexity of standard self-attention presents challenges for processing high-resolution images. To mitigate this issue, [49] proposed a transformer architecture optimized for restoration tasks like image deraining, deblurring, and denoising by computing attention across the channel dimension, thereby reducing computational burden. Another approach is the use of window-based attention, exemplified by Uformer [47], which enhances local interactions within the Transformer architecture. SwinIR [25] also leverages window-based attention but incorporates a shift mechanism to facilitate better cross-window communication. Additional strategies have explored the application of Transformers with channel-wise and spatial-wise attention layers [32], or through the integration of both frequency and spatial domains for self-attention as seen in [19]. In contrast to these methods, we propose an adaptive sparse self-attention mechanism in the wavelet domain, aimed at reducing redundancy by focusing on the most relevant interactions.

3 Proposed Method

3.1 Overall Pipeline

The proposed U-ENHANCE framework, as illustrated in Fig. 1, begins by embedding an input RGB image, I into a feature space using the Input Projection module, which consists of a 3×3 standard convolution. This step produces low-level features, $X_0 \in H \times W \times C$ where, H, W, C refers to the height, width and channel dimensions, respectively that are passed into a hierarchical encoder-decoder backbone network, designed with four stages. Between stages, down-sampling and up-sampling are achieved through pixel-unshuffle and pixel-shuffle operations, enabling multi-scale representation of underwater-degradation effects. At the core of each block is the use of Wavelet Triple Self-Attention (WTSA), which replaces the conventional Transformer self-attention [44]. This WTSA mechanism introduces a three-dimensional co-computation of horizontal, vertical, and channel-wise self-attention, making feature aggregation more efficient. Additionally, each U-ENHANCE Block incorporates a Self-Calibrated Feedforward Network (SCFN) to ensure more effective feature refinement. After

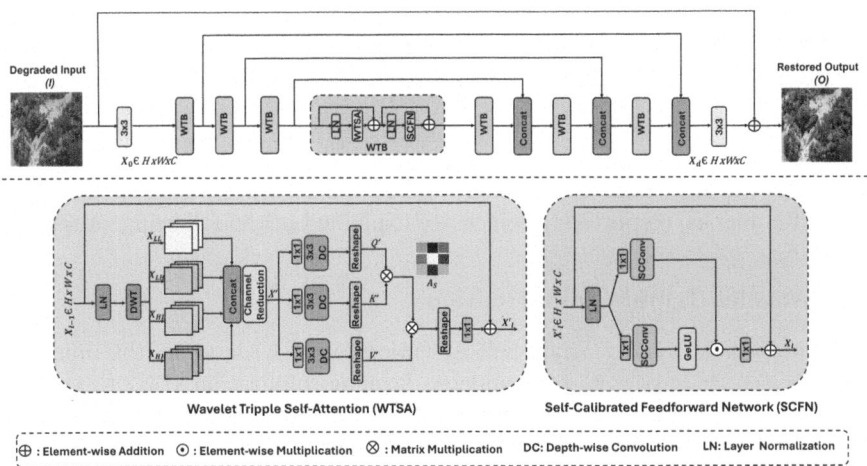

Fig. 1. Illustration of the proposed U-ENHANCE. It consists of a Wavelet Transformer Block (WTB) with a Wavelet Tripple Self-Attention (WTSA) and a Self-Calibrated Feed-forward Network (SCFN).

encoding and decoding, the deep features, $X_d \in H \times W \times C$, are passed through an Output Projection module, another 3×3 standard convolution, which restores the feature map to its original dimensions of $H \times W \times 3$. The final restored image O is produced by combining this output with the original input via a residual connection, ensuring enhanced image restoration.

3.2   Wavelet Transformer Block

Existing underwater image enhancement methods based on Transformers emphasize spatial domain attention, often neglecting the critical role of frequency information [19,45]. While some techniques incorporate frequency domain features, they frequently struggle to effectively capture both frequency and spatial characteristics due to a lack of a unified integration strategy. To overcome this limitation, we introduce the Wavelet Triple Self-Attention (WTSA) mechanism, which replaces the traditional Multi-head Self-Attention (MSA) in conventional transformer blocks. Our approach effectively fuses wavelet and spatial domain information, enabling superior feature extraction for restoring underwater-degraded effects like color distortion and blurriness while preserving structural details and edges. Wavelets offer a key advantage by preserving localized spatial information even in the frequency domain, providing a multiscale representation that captures both fine-grained frequency details and spatial localization, crucial for maintaining structural integrity across different scales [38,41]. Additionally, we introduce a Self-Calibrated Feedforward Network (SCFN) to ensure efficient feature refinement. This network optimizes the processing of both spatial and frequency domain features, further enhancing the model's ability to accurately restore underwater images. The computational workflow of the Wavelet Transformer Block (WTB) for underwater image enhancement can be described as follows:

$$X'_\ell = X_{\ell-1} + \mathrm{WTSA}(\mathrm{LN}(X_{\ell-1})) \tag{1}$$
$$X_\ell = X'_\ell + \mathrm{SCFN}(\mathrm{LN}(X'_\ell)) \tag{2}$$

where $\mathrm{LN}(\cdot)$ is layer normalization, X'_l and X_l denote the outputs of the WTSA and SCFN blocks, respectively, which are explained in the following subsection.

3.3   Wavelet Tripple Self-attention

In the proposed Wavelet Triple Self-Attention (WTSA) module, the input features $X_{l-1} \in H \times W \times C$ first undergo layer normalization (LN) to stabilize the input distribution. Following normalization, the features are transformed into the frequency domain using the Discrete Wavelet Transform (DWT), which decomposes them into four sub-bands: $(X_{LL}, X_{LH}, X_{HL}, X_{HH})$ as shown below in Eq. 3 :

$$(X_{LL}, X_{LH}, X_{HL}, X_{HH}) = \mathrm{DWT}(\mathrm{LN}(X_{\ell-1})) \tag{3}$$

Decomposing input features into these four sub-bands offers several advantages. Firstly, it allows for the preservation of both low-frequency and high-frequency components of the signal, which is essential for accurately capturing the overall structure and intricate details of the input data. Secondly, working in the frequency domain enables better noise reduction, as certain frequency components can be attenuated while retaining important features. These sub-bands retain the same number of channels as the input features but have spatial dimensions downsampled by a factor of 2. After the DWT, the four sub-bands are concatenated, and a channel reduction operation is applied to reduce the number of channels to match the input feature dimensions as described in Eq. 4 :

$$X' = \mathrm{CR}(\mathrm{Concat}(LL, LH, HL, HH)) \tag{4}$$

where, CR is channel reduction convolution.

To restore the spatial resolution, the concatenated features are upsampled using bilinear interpolation to return them to their original spatial dimensions. The use of bilinear interpolation is advantageous because it provides a smooth upsampling process, preserving the visual quality of the images by minimizing artifacts. Additionally, bilinear interpolation is computationally efficient, allowing for quick reconstruction of spatial dimensions without significantly increasing processing time. This is followed by a 1×1 pointwise convolution to mix the channels effectively and a 3×3 depth-wise convolution to refine the spatial features and capture local dependencies. Subsequently, the features are processed through the Triple Attention mechanism. Here, the self-attention computation is decomposed into three separate components: horizontal self-attention, vertical self-attention, and channel self-attention, based on the concept of Triple Multi-Dconv Head Transposed Attention (TMDTA) [53]. This approach divides the feature correlation into three distinct directions as shown in Fig. 2, with the query (Q'), key (K'), and value (V') tensors being reshaped accordingly. Specifically, the query, key and value tensors are projected in three dimensions: horizontally (Q_H, K_H, V_H), vertically (Q_W, K_W, V_W), and along the channel dimension (Q_C, K_C, V_C), resulting in three attention matrices: $A_H \in R^{H \times H}$, $A_W \in R^{W \times W}$, and $A_C \in R^{C \times C}$, instead of the conventional full-pixel attention matrix $R^{HW \times HW}$. Eventually these three different attention matrices are concatenated to generate A_s matrix as defined by the Eq. 5 and 6 where, A_x represents the self-attention along the horizontal, vertical and channel dimesions.

$$A_S = \mathrm{Concat}(A_C, A_H, A_W) \tag{5}$$

$$A_x = \mathrm{Softmax}\left(\frac{Q_x K_x^T}{\alpha}\right) \times V_x \tag{6}$$

where, $x \in \{C, H, W\}$.

By performing matrix multiplications in each of these three directions separately, the WTSA module preservs essential spatial and frequency information. The result is a more efficient feature representation that captures long-range dependencies across spatial and channel dimensions, leading to improved restoration of underwater-degraded images. This combination of wavelet-based frequency decomposition, spatial refinement, and directional attention ensures that the WTSA module efficiently captures both local and global features. The final output of WTSA block is X_l' which is shown in Eq. 1.

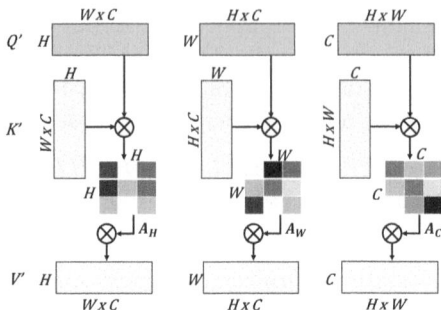

Fig. 2. Illustration of the Triple attention concept [53]. Characteristic pixel attention is decomposed into horizontal, vertical, and channel self-attention.

3.4   Self-calibrated Feedforward Network

Previous studies on transformer architectures primarily employ standard feed-forward networks (FFNs) in the transformer block to handle the token-wise transformations. However, these implementations typically overlook long-range spatial dependencies and inter-channel correlations, which are crucial for more discriminative feature representation. To address this, we propose a novel Self-Calibrated Feedforward Network (SCFN) that integrates self-calibrated convolutions [27] within the feedforward layers, enhancing both spatial and inter-channel information processing.

In particular, the SCFN builds upon the idea of self-calibrated convolution [27], which adapts the receptive field dynamically to consider richer contextual information. The result is a feedforward network that not only encodes local information but also expands its scope to adaptively capture long-range dependencies around each spatial position, thus improving the overall representation capability. The full process of the SCFN is formulated as follows (see Eq. 7):

$$X_l = X'_l + \psi_1 \Big[\text{GELU} \left(\text{SCConv} \left(\psi_1 \left(\text{LN}(X'_l) \right) \right) \right) \odot \text{SCConv} \left(\psi_1 \left(\text{LN}(X'_l) \right) \right) \Big]$$
(7)

where, $\psi_1(.)$ and SCConv denotes the 1×1 convolution and self-calibrated convolution, respectively.

This novel SCFN introduces an adaptive mechanism that dynamically adjusts the spatial and channel-wise dependencies, resulting in more robust and discriminative feature extraction. Moreover, by incorporating self-calibration, the SCFN avoids unnecessary complexity or parameter overhead while significantly improving the model's ability to capture multi-scale information, essential for underwater image enhancement.

4 Experimental Details

4.1 Training Losses

In training our proposed architecture, we utilized a comprehensive total loss function, denoted as L_T, which combines several distinct loss components. These components include perceptual loss (L_1) [18], Charbonnier loss (L_2) [3], multi-scale structural similarity index (MS-SSIM) loss (L_3) [48], and gradient loss (L_4) [36]. Together, these losses are integrated to ensure a robust and effective training process, addressing both perceptual quality and structural accuracy. The total loss function is formulated as follows:

$$L_T = \lambda_1 L_1 + \lambda_2 L_2 + \lambda_3 L_3 + \lambda_4 L_4 \tag{8}$$

where, $\lambda_{1,2,3,4} \in \{2, 3, 1, 2.5\}$ are empirically determined weighting factors. This combination of loss functions is crucial for optimizing our model, enabling it to capture various aspects of intrinsic image attributes and produce visually appealing, high-quality output images. The losses are explained individually as follows:

Perceptual Loss (L_1): Perceptual loss measures the perceptual similarity between generated and target images by utilizing feature representations from a pre-trained neural network. This approach has been shown to improve the quality of generated images across various image-generation tasks. Let O represent the target image and G_t represent the generated image. Using a pre-trained VGG19 [40] network (ϕ_i) we extract feature maps at different layers. The perceptual loss, L_1, is calculated as the difference between the feature maps of the target and generated images:

$$L_1 = \sum_{i=1}^{N=4} \|\phi_i(O) - \phi_i(G_t)\|_2^2 \tag{9}$$

Here, ϕ_i represents the feature extraction function at layer i of the CNN, and ($N = 4$) is the total number of layers considered for perceptual loss calculation.

Charbonnier loss (L_2): Training the network with MSE loss often results in blurry reconstructions because it maximizes the log-likelihood of a Gaussian distribution. To address this issue, we chose the Charbonnier loss, a differentiable version of the L_1 norm. The Charbonnier loss is computed between the restored images (O) and their corresponding ground-truth images (G_t), and it is defined as follows:

$$L_2 = \mathbb{E}_{O \sim Q(O), G_t \sim Q(G_t)} \sqrt{(O - G_t)^2 + \epsilon} \tag{10}$$

where, $Q(O)$ and $Q(G_t)$ are the distributions of the restored image (O) and the ground-truth image (G_t), respectively. Additionally, the value of ϵ is empirically set to 1×10^{-3}.

MS-SSIM loss (L_3): The Structural Similarity (SSIM) loss primarily addresses a single input resolution. In contrast, the Multi-Scale SSIM (MS-SSIM) loss provides greater flexibility by taking into account different input resolutions.

$$L_3 = 1 - (MSSSIM(O, G_t)) \tag{11}$$

Gradient loss (L_4): Generally, the Charbonnier loss prioritizes low-frequency components. However, when training the network to incorporate high-frequency details, the gradient loss becomes crucial. This second-order loss function enhances the sharpness of edges in the output [29]. Here, $\hat{G}_O$ and $\hat{G}_{G_t}$ represent the distributions of $Q(O)$ and $Q(G_t)$ respectively.

$$L_4 = \mathbb{E}_{\hat{G}_O \sim Q(O), \hat{G}_O \sim Q(G_t)} \left\| \hat{G}_{G_t} - \hat{G}_O \right\|_1 \tag{12}$$

4.2　Datasets

To perform a comparative analysis, we utilized the synthetic Underwater Image Enhancement Benchmark (UIEB) [21] along with the real-world underwater datasets U45 [22] and C60 [21]. The Underwater Image Enhancement Benchmark (UIEB) dataset contains 890 pairs of underwater images, including both degraded and clean versions, representing various scenes to capture the diversity of underwater environments. The dataset is divided into 800 image pairs for training, selected at random, while the remaining 90 pairs are designated for testing. Additionally, the dataset provides 60 real-world degraded underwater images for evaluation purposes. U45 consists of 45 real-world images that exhibit features like color casts, low contrast, and degradation effects similar to haze in underwater environments.

4.3　Training Details

To mitigate the limited size of the UIEB dataset for training, we employed several data augmentation strategies to expand the dataset's diversity. These augmentations included horizontal and vertical flips, noise addition, and contrast adjustments. This approach significantly increased the variability in the training data, improving the robustness and performance of the model. In total, 4800 augmented image pairs from the UIEB dataset were used for training, while 90 images were reserved for testing. All input images were uniformly resized

Table 1. Quantitative comparison of different UIE methods on the synthetic UIEB dataset (↑: higher is better, ↓: lower is better, red and blue indicate **best** and second best values, respectively).

Method	PSNR ↑	SSIM ↑	UIQM ↑	Parameters (M) ↓
UDCP [11]	13.81	0.692	1.825	-
UIBLA [34]	15.78	0.731	2.014	-
RGHS [15]	14.57	0.791	2.410	-
WaterNet [21]	19.81	0.864	2.818	193.70
FUnIE-GAN [17]	21.03	0.775	3.092	7.02
CLUIE-Net [24]	20.37	0.890	2.674	31.00
Ours	22.07	0.911	2.701	5.52

to 256×256 pixels for consistency. The model was trained using the ADAM optimizer with an initial learning rate of 3×10^{-4}, which was progressively adjusted through a cosine annealing schedule. The implementation was done in PyTorch, and the training was conducted on an NVIDIA GeForce RTX 2080 GPU.

4.4 Results on Synthetic Dataset

The effectiveness of the proposed approach is evaluated through a quantitative comparison with current state-of-the-art methods, using metrics like PSNR and SSIM. We also compared the parameters (in millions, M) of learning-based methods. Table 1 presents the quantitative outcomes on the popular UIEB dataset, while Fig. 3 illustrates qualitative results. The proposed method exhibits superior performance in comparison to the state-of-the-art techniques.

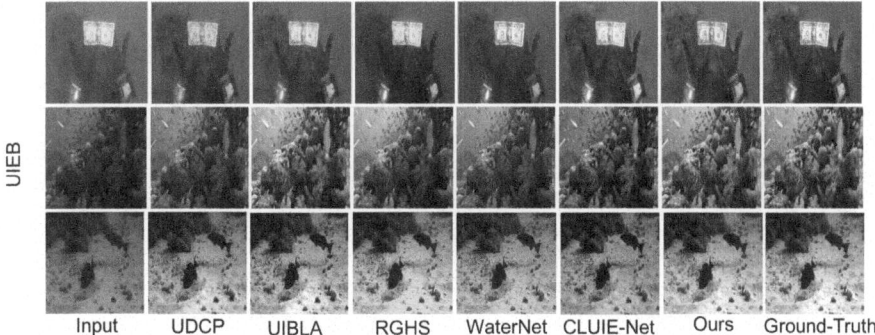

Input   UDCP   UIBLA   RGHS   WaterNet   CLUIE-Net   Ours   Ground-Truth

Fig. 3. Visual comparison of our method and state-of-the-art techniques on the full-reference UIEB dataset [21].

4.5   Results on Real-World Dataset

To evaluate the effectiveness of our proposed method in real-world conditions, we presented quantitative results on the C60 dataset [21]. Our analysis includes a comprehensive evaluation using metrics such as the Underwater Image Quality Measure (UIQM), Underwater Image Sharpness Measure (UISM), Naturalness Image Quality Evaluator (NIQE), and Blind/Referenceless Image Spatial Quality Evaluator (BRISQUE). The quantitative outcomes are summarized in Table 2. In addition, the qualitative comparison of the C60 and U45 datasets is illustrated in Fig. 4. The results highlight a substantial improvement in color correction and overall visibility in the enhanced images, which can be attributed to the novel modules incorporated in our proposed approach.

Table 2. Quantitative comparison of different UIE methods on the real-world C60 dataset.

Method	UIQM ↑	UISM ↑	NIQE ↓	BRISQUE ↓
UDCP [11]	4.371	6.849	6.246	29.412
UIBLA [34]	4.654	7.084	6.230	27.968
RGHS [15]	3.942	7.347	6.141	25.485
WaterNet [21]	4.348	7.322	5.310	23.115
CLUIE-Net [24]	4.194	7.418	5.848	26.369
Ours	**4.713**	**7.531**	**5.269**	25.870

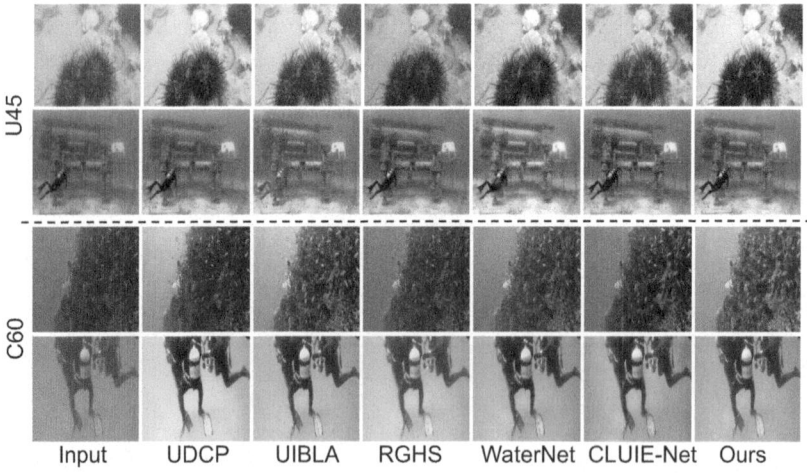

Input   UDCP   UIBLA   RGHS   WaterNet   CLUIE-Net   Ours

Fig. 4. Visual comparison of our method and state-of-the-art techniques on real-world U45 [22] and C60 [21] datasets.

5 Ablation Studies

To validate the effectiveness of the proposed components, we conduct a series of ablation experiments using the UIEB dataset.

5.1 Impact of the Proposed Wavelet Triple Self-Attention

To evaluate the effectiveness of the proposed Wavelet Triple Self-Attention (WTSA) module, we performed ablation experiments by comparing the model's performance with and without this module, provided in Table 3. Ablation experiments revealed significant improvements in underwater image enhancement when the Wavelet Triple Self-Attention (WTSA) module was included. Without WTSA, the model struggled with noise reduction and detail preservation, resulting in lower image quality. Incorporating WTSA led to higher performance across all metrics, including PSNR and SSIM, due to the wavelet-based frequency decomposition and triple attention mechanism, which efficiently captured long-range dependencies. These results underscore the importance of WTSA in enhancing image clarity while maintaining computational efficiency.

Table 3. Ablation studies conducted on various network configurations using the UIEB benchmark.

Network Setting	PSNR ↑	SSIM ↑
Baseline	20.82	0.651
Baseline + WTSA	21.53	0.787
Baseline + SCFN	21.74	0.834
Ours (Baseline + WTSA + SCFN)	**22.07**	**0.911**

5.2 Impact of the Proposed Self-Calibrated Feedforward Network

To validate the effectiveness of the proposed Self-Calibrated Feedforward Network (SCFN) in U-ENHANCE, we conducted ablation experiments comparing performance with and without this module, highlighted in Table 3. The baseline model without SCFN showed a noticeable drop in image restoration quality, particularly in color accuracy and structural clarity. In contrast, incorporating SCFN led to superior results across all metrics. SCFN improved feature refinement by dynamically adjusting the receptive field, enhancing the integration of spatial and frequency domain features. These findings confirm SCFN's critical role in boosting U-ENHANCE's overall performance.

6   Applicability to Higher Level Computer Vision Task

In underwater environments, reduced visibility often impairs the performance of computer vision tasks. To address this, enhancing underwater images can serve as a crucial pre-processing step, improving the accuracy of downstream applications. To validate this, we performed an experiment centred on underwater depth estimation. We first applied a depth estimation algorithm to the degraded images and then compared the results with those obtained from images enhanced through our method as well as other existing techniques. As illustrated in Fig. 5, our enhanced images outperformed all other methods in depth estimation, demonstrating superior accuracy. This outcome underscores the effectiveness of our approach for depth estimation. The enhanced image quality ensures more reliable data for subsequent computational tasks, which is particularly important in the challenging underwater domain, where clarity and detail can significantly impact the success of vision-based algorithms.

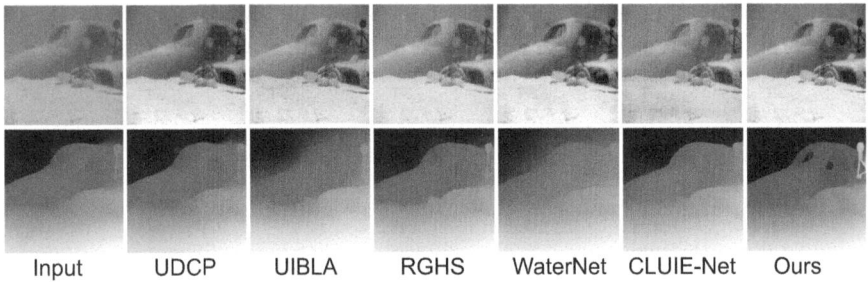

| Input | UDCP | UIBLA | RGHS | WaterNet | CLUIE-Net | Ours |

Fig. 5. Application of our proposed method and existing techniques as a pre-processing step for depth estimation on the underwater U45 dataset [22].

7   Conclusion

In this paper, we proposed U-ENHANCE, a novel framework for underwater image enhancement that effectively integrates wavelet-based frequency decomposition with spatial domain attention to address the complex challenges of underwater-degraded images. By introducing the Wavelet Triple Self-Attention (WTSA) mechanism, we demonstrated that self-attention across three dimensions—horizontal, vertical, and channel-wise—enables superior capture of multi-scale features, essential for preserving fine details and structural integrity. Additionally, the Self-Calibrated Feedforward Network (SCFN) refined feature representation by dynamically adjusting the receptive field, enhancing the model's ability to process spatial and frequency domain information. Through extensive experiments on benchmark datasets, U-ENHANCE consistently outperformed state-of-the-art methods, delivering superior restoration of color accuracy, clarity, and structural details. These results confirm the effectiveness of our approach and its potential to advance the field of underwater image enhancement.

Acknowledgement. This work was supported by Project MoES/PAMC/DOM/ 04/2022 (E-12710), Project TIHIITG202204 and Project CRG/2022/006876. Also, I would like to thank all the CVPR Lab members for their support.

References

1. Bailey, G.N., Flemming, N.C.: Archaeology of the continental shelf: marine resources, submerged landscapes and underwater archaeology. Quatern. Sci. Rev. **27**(23–24), 2153–2165 (2008)
2. Blidberg, D.R.: The development of autonomous underwater vehicles (AUV); a brief summary. In: IEEE ICRA, vol. 4, pp. 122–129 (2001)
3. Bruhn, A., Weickert, J., Schnörr, C.: Lucas/kanade meets horn/schunck: combining local and global optic flow methods. Int. J. Comput. Vision **61**, 211–231 (2005)
4. Chen, X., Li, H., Li, M., Pan, J.: Learning a sparse transformer network for effective image deraining. In: Proceedings of the IEEE/CVF Conference on Computer Vision and Pattern Recognition, pp. 5896–5905 (2023)
5. Chen, Z., Zhang, Y., Gu, J., Kong, L., Yuan, X., et al.: Cross aggregation transformer for image restoration. Adv. Neural. Inf. Process. Syst. **35**, 25478–25490 (2022)
6. Chi, Z., Shu, X., Wu, X.: Joint demosaicking and blind deblurring using deep convolutional neural network. In: 2019 IEEE International Conference on Image Processing (ICIP), pp. 2169–2173. IEEE (2019)
7. Chi, Z., Wang, Y., Yu, Y., Tang, J.: Test-time fast adaptation for dynamic scene deblurring via meta-auxiliary learning. In: Proceedings of the IEEE/CVF Conference on Computer Vision and Pattern Recognition, pp. 9137–9146 (2021)
8. Chiang, J.Y., Chen, Y.C.: Underwater image enhancement by wavelength compensation and dehazing. IEEE Trans. Image Process. **21**(4), 1756–1769 (2011)
9. Coleman, D.F., Newman, J.B., Ballard, R.D.: Design and implementation of advanced underwater imaging systems for deep sea marine archaeological surveys. In: OCEANS 2000 MTS/IEEE Conference and Exhibition. Conference Proceedings (Cat. No. 00CH37158), vol. 1, pp. 661–665. IEEE (2000)
10. Drews, P., Nascimento, E., Moraes, F., Botelho, S., Campos, M.: Transmission estimation in underwater single images. In: Proceedings of the IEEE International Conference on Computer Vision Workshops, pp. 825–830 (2013)
11. Drews, P.L., Nascimento, E.R., Botelho, S.S., Campos, M.F.M.: Underwater depth estimation and image restoration based on single images. IEEE Comput. Graphics Appl. **36**(2), 24–35 (2016)
12. Fabbri, C., Islam, M.J., Sattar, J.: Enhancing underwater imagery using generative adversarial networks. In: 2018 IEEE International Conference on Robotics and Automation (ICRA), pp. 7159–7165. IEEE (2018)
13. Fu, M., Liu, H., Yu, Y., Chen, J., Wang, K.: DW-GAN: a discrete wavelet transform GAN for nonhomogeneous dehazing. In: Proceedings of the IEEE/CVF Conference on Computer Vision and Pattern Recognition, pp. 203–212 (2021)
14. Han, J., et al.: Underwater image restoration via contrastive learning and a real-world dataset. Remote Sens. **14**(17), 4297 (2022)

15. Huang, D., Wang, Y., Song, W., Sequeira, J., Mavromatis, S.: Shallow-water image enhancement using relative global histogram stretching based on adaptive parameter acquisition. In: Schoeffmann, K., et al. (eds.) MMM 2018. LNCS, vol. 10704, pp. 453–465. Springer, Cham (2018). https://doi.org/10.1007/978-3-319-73603-7_37
16. Huang, S., Wang, K., Liu, H., Chen, J., Li, Y.: Contrastive semi-supervised learning for underwater image restoration via reliable bank. In: Proceedings of the IEEE/CVF Conference on Computer Vision and Pattern Recognition, pp. 18145–18155 (2023)
17. Islam, M.J., Xia, Y., Sattar, J.: Fast underwater image enhancement for improved visual perception. IEEE Robot. Autom. Lett. **5**(2), 3227–3234 (2020)
18. Johnson, J., Alahi, A., Fei-Fei, L.: Perceptual losses for real-time style transfer and super-resolution. In: Leibe, B., Matas, J., Sebe, N., Welling, M. (eds.) ECCV 2016. LNCS, vol. 9906, pp. 694–711. Springer, Cham (2016). https://doi.org/10.1007/978-3-319-46475-6_43
19. Khan, R., et al.: SpectroFormer: multi-domain query cascaded transformer network for underwater image enhancement. In: Proceedings of the IEEE/CVF Winter Conference on Applications of Computer Vision, pp. 1454–1463 (2024)
20. Li, C., Anwar, S., Hou, J., Cong, R., Guo, C., Ren, W.: Underwater image enhancement via medium transmission-guided multi-color space embedding. IEEE Trans. Image Process. **30**, 4985–5000 (2021)
21. Li, C., et al.: An underwater image enhancement benchmark dataset and beyond. IEEE Trans. Image Process. **29**, 4376–4389 (2019)
22. Li, H., Li, J., Wang, W.: A fusion adversarial underwater image enhancement network with a public test dataset. arXiv preprint arXiv:1906.06819 (2019)
23. Li, J., Skinner, K.A., Eustice, R.M., Johnson-Roberson, M.: WaterGAN: unsupervised generative network to enable real-time color correction of monocular underwater images. IEEE Robot. Autom. Lett. **3**(1), 387–394 (2017)
24. Li, K., et al.: Beyond single reference for training: underwater image enhancement via comparative learning. IEEE Trans. Circuits Syst. Video Technol. **33**(6), 2561–2576 (2022)
25. Liang, J., Cao, J., Sun, G., Zhang, K., Van Gool, L., Timofte, R.: SwinIR: image restoration using Swin transformer. In: Proceedings of the IEEE/CVF International Conference on Computer Vision, pp. 1833–1844 (2021)
26. Liu, H., Wu, Z., Li, L., Salehkalaibar, S., Chen, J., Wang, K.: Towards multi-domain single image dehazing via test-time training. In: Proceedings of the IEEE/CVF Conference on Computer Vision and Pattern Recognition, pp. 5831–5840 (2022)
27. Liu, J.J., Hou, Q., Cheng, M.M., Wang, C., Feng, J.: Improving convolutional networks with self-calibrated convolutions. In: Proceedings of the IEEE/CVF Conference on Computer Vision and Pattern Recognition, pp. 10096–10105 (2020)
28. Liu, S., Fan, H., Wang, Q., Han, Z., Guan, Y., Tang, Y.: Wavelet-pixel domain progressive fusion network for underwater image enhancement. Knowl. Based Syst. **299**, 112049 (2024)
29. Mathieu, M., Couprie, C., LeCun, Y.: Deep multi-scale video prediction beyond mean square error. arXiv preprint arXiv:1511.05440 (2015)
30. Naveen, P.: Advancements in underwater imaging through machine learning: Techniques, challenges, and applications. Multimedia Tools and Appl., 1–20 (2024). https://doi.org/10.1007/s11042-024-20091-4
31. Paull, L., Saeedi, S., Seto, M., Li, H.: AUV navigation and localization: a review. IEEE J. Oceanic Eng. **39**(1), 131–149 (2013)
32. Peng, L., Zhu, C., Bian, L.: U-shape transformer for underwater image enhancement. IEEE Trans. Image Process. **32**, 3066–3079 (2023)

33. Peng, Y.T., Cao, K., Cosman, P.C.: Generalization of the dark channel prior for single image restoration. IEEE Trans. Image Process. **27**(6), 2856–2868 (2018)

34. Peng, Y.T., Cosman, P.C.: Underwater image restoration based on image blurriness and light absorption. IEEE Trans. Image Process. **26**(4), 1579–1594 (2017)

35. Ren, T., Xu, H., Jiang, G., Yu, M., Luo, T.: Reinforced Swin-Convs transformer for underwater image enhancement. arXiv preprint arXiv:2205.00434 (2022)

36. Ribeiro, J., Elsayed, E.: A case study on process optimization using the gradient loss function. Int. J. Prod. Res. **33**(12), 3233–3248 (1995)

37. Rout, D.K., Kapoor, M., Subudhi, B.N., Thangaraj, V., Jakhetiya, V., Bansal, A.: Underwater visual surveillance: a comprehensive survey. Ocean Eng. **309**, 118367 (2024)

38. Scholl, S.: Fourier, gabor, morlet or wigner: comparison of time-frequency transforms. arXiv preprint arXiv:2101.06707 (2021)

39. Shortis, M., Abdo, E.H.D.: A review of underwater stereo-image measurement for marine biology and ecology applications. In: Oceanography and Marine Biology, pp. 269–304 (2016)

40. Simonyan, K., Zisserman, A.: Very deep convolutional networks for large-scale image recognition. arXiv preprint arXiv:1409.1556 (2014)

41. Sun, Q., Ren, Y., Jiao, L., Li, X., Shang, F., Liu, F.: MWQ: multiscale wavelet quantized neural networks. arXiv preprint arXiv:2103.05363 (2021)

42. Tang, Y., Kawasaki, H., Iwaguchi, T.: Underwater image enhancement by transformer-based diffusion model with non-uniform sampling for skip strategy. In: Proceedings of the 31st ACM International Conference on Multimedia, pp. 5419–5427 (2023)

43. Uplavikar, P.M., Wu, Z., Wang, Z.: All-in-one underwater image enhancement using domain-adversarial learning. In: CVPR Workshops, pp. 1–8 (2019)

44. Vaswani, A.: Attention is all you need. In: Advances in Neural Information Processing Systems (2017)

45. Wang, D., Sun, Z.: Frequency domain based learning with transformer for underwater image restoration. In: Khanna, S., Cao, J., Bai, Q., Xu, G. (eds.) Pacific Rim International Conference on Artificial Intelligence, pp. 218–232. Springer (2022). https://doi.org/10.1007/978-3-031-20862-1_16

46. Wang, Y., Liu, H., Chau, L.P.: Single underwater image restoration using adaptive attenuation-curve prior. IEEE Trans. Circuits Syst. I Regul. Pap. **65**(3), 992–1002 (2017)

47. Wang, Z., Cun, X., Bao, J., Zhou, W., Liu, J., Li, H.: UFormer: a general u-shaped transformer for image restoration. In: Proceedings of the IEEE/CVF Conference on Computer Vision and Pattern Recognition, pp. 17683–17693 (2022)

48. Wang, Z., Simoncelli, E.P., Bovik, A.C.: Multiscale structural similarity for image quality assessment. In: The Thrity-Seventh Asilomar Conference on Signals, Systems & Computers, 2003, vol. 2, pp. 1398–1402. IEEE (2003)

49. Zamir, S.W., Arora, A., Khan, S., Hayat, M., Khan, F.S., Yang, M.H.: Restormer: efficient transformer for high-resolution image restoration. In: Proceedings of the IEEE/CVF Conference on Computer Vision and Pattern Recognition, pp. 5728–5739 (2022)

50. Zhao, C., Cai, W., Dong, C., Hu, C.: Wavelet-based fourier information interaction with frequency diffusion adjustment for underwater image restoration. In: Proceedings of the IEEE/CVF Conference on Computer Vision and Pattern Recognition, pp. 8281–8291 (2024)

51. Zhou, J., et al.: UGIF-Net: an efficient fully guided information flow network for underwater image enhancement. IEEE Trans. Geosci. Remote Sens. **61**, 4206117 (2023)
52. Zhou, J., Zhang, D., Ren, W., Zhang, W.: Auto color correction of underwater images utilizing depth information. IEEE Geosci. Remote Sens. Lett. **19**, 1–5 (2022)
53. Zhou, Y., Lin, J., Ye, F., Qu, Y., Xie, Y.: Efficient lightweight image denoising with triple attention transformer. In: Proceedings of the AAAI Conference on Artificial Intelligence, vol. 38, pp. 7704–7712 (2024)
54. Zhuang, P., Ding, X.: Underwater image enhancement using an edge-preserving filtering Retinex algorithm. Multimedia Tools Appl., 1–21 (2020). https://doi.org/10.1007/s11042-019-08404-4

WARMOS: Enhancing Weather-Affected Referred Moving Object Segmentation

Prafulla Saxena[1]([envelope]) [iD], Dinesh Kumar Tyagi[1] [iD], Santosh Kumar Vipparthi[2] [iD], and Subrahmanyam Murala[2,3] [iD]

[1] Malaviya National Institute of Technology Jaipur, Jaipur, India
2020rcp9580@mnit.ac.in
[2] Indian Institute of Technology Ropar, Bara Phool, India
[3] SCSS Trinity College Dublin, Dublin, Ireland

Abstract. Environmental noise, such as haze and rain, poses significant challenges in video surveillance, in tasks like referred video object segmentation. These weather-related disturbances introduce excessive pixel variance, making moving object segmentation more complex. In this work, we focus on addressing the issue of adverse weather conditions by simulating the effects of haze and rain in videos and employing a robust noise removal model. The model effectively reduces pixel variance caused by environmental factors. This enhanced framework is precious for referred moving object segmentation, where objects identification done based on text queries. By integrating our noise removal module, we ensure better alignment of features, which enhances the precision of referred moving segmentation. Our approach maintains temporal consistency, making object segmentation more reliable under challenging weather conditions while preserving the original video quality by removing weather noise. We have employed separate noise removal modules for haze and rain environmental noise. A ResNet based classifier model trained to identify the noise class on the fly. To demonstrate the effectiveness of our methodology, we selected an ROVS benchmark to assess segmentation performance. Experiments on the DAVIS 2017 dataset show that our proposed methodology performs well on weather-affected videos, significantly improving the benchmark metrics Jaccard (J) and F-measure (F) indices after removing weather noise. Using the benchmark SgMg model for referred segmentation, the mean J&F score is 63.64 without environmental noise. When haze is introduced to the dataset, the mean J&F score drops to 58.71. After applying WARMOS approach, the mean J&F improves to 60.50. A similar pattern is observed for rain: when rain is introduced, the mean J&F score is 61.00, and after applying WARMOS, it improves to 61.06. This highlights our approach's significance in mitigating the impact of environmental noise.

Keywords: Moving Object Segmentation · Referred segmentation · Weather noise · Haze and rain removal · Deep learning

1   Introduction

Weather disturbances such as haze, rain, and fog pose significant challenges to video surveillance tasks, including referred video object segmentation (RVOS) [1]. These weather-related noises degrade the quality of video frames, making it difficult to accurately capture the visual features needed for effective object segmentation. The visibility of objects is often compromised due to blurred or occluded regions, which can severely affect the performance of segmentation models. Figure1 demonstrates the effects of haze and rain in real world images with simulation. In tasks like moving object segmentation (MOS) [2–8], where temporal and spatial accuracy is crucial, the interference caused by weather noise can lead to substantial errors in identifying and tracking objects.

Fig. 1. Visual degradation of a video frame after environmental noise like haze and rain. Due to haze and rain quality of an image degrades and compromised the object visibility. Images from DAVIS 2017 [9] dataset

Referred video object segmentation (RVOS) [1, 10–13] is an emerging field at the intersection of computer vision and natural language processing. It focuses on segmenting an object based on a language expression, combining both visual and linguistic information. While standard object segmentation primarily focuses on the spatial domain within a single frame, RVOS introduces a new layer of complexity by adding natural language as a reference. However, some of approaches often ignore temporal information [14], which is essential in video-based tasks. Standard moving object segmentation (MOS) [2,15], methods, in contrast, excel in capturing temporal changes across frames to identify moving objects but fail to incorporate semantic understanding or contextual information.

When weather disturbances like rain or haze are present in captures videos and images, these challenges are exacerbated in video segmentation applications [16]. The combination of low visibility, dynamic backgrounds, and weather noise hampers the model's ability to accurately target specific objects described through text. In scenarios like Referred video object segmentation, where precise spatiotemporal features are required for identifying referred objects, weather disturbances can severely degrade performance. Objects may become partially or fully obscured, leading to incorrect or incomplete segmentation. This further complicates the task, as spatiotemporal information and text-based embedding must work in conjunction, while weather noise interferes with both.

Furthermore, weather noise like haze and rain significantly impacts the effectiveness of moving object segmentation, adding complexity to Referred moving segmentation as well. Techniques for removing such weather-related noise, like dehazing [17–20] and rain removal [21–23], are essential to improving the quality of video frames. By enhancing visibility and removing these disturbances, the model can more accurately focus on the relevant object characteristics, leading to improved performance in referred video segmentation.

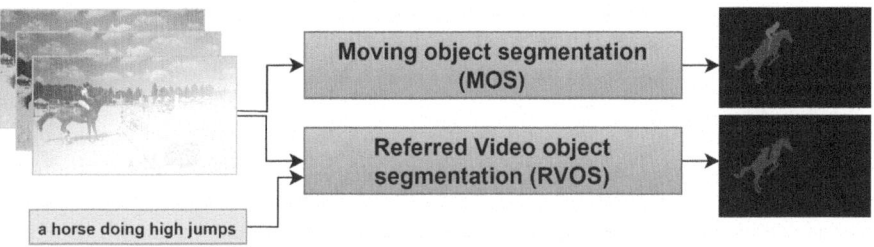

Fig. 2. In RVOS, the segmentation of moving objects depends on a text query. Whether in MOS task, all moving objects are segmented irrespective of the text query.

Though several learning-based approaches have been proposed to improve the MOS task by designing various deep learning models, including dealing with challenges like illumination changes, low visibility, and dynamic backgrounds, addressing weather noise remains a critical area of focus. When referring to moving objects through language expressions, the segmentation task becomes even more complex, requiring advanced handling of cross-modal sources like vision and language. So in this work we have proposed an approach to enhance the Referred moving object segmentation in weather-affected scenarios. Figure2 shows the referred object segmentation in weather affected scenario. We propose the following contribution to this paper.

- Proposed WARMOS, a weather noise removal framework for improving the existing referred video object segmentation performance.
- Shows the adverse effect of weather disturbance in referred segmentation task by simulating haze and rain in real world dataset.
- Utilises a benchmark SgMg, referred video segmentation model and improves its performance in weather affected scenarios.
- Evaluate on Ref-DAVIS17 benchmark dataset and verify the proposed methodology's effectiveness with quantitative and qualitative analysis.

2 Related Work

RVOS focuses on segmenting specific objects in a video based on a text reference. In this task, the representation of objects is influenced by the natural language query used to refer to them. Various approaches, such as those in [10–12], handle the segmentation of referred objects frame by frame as the video progresses. However, when referring to moving objects, a deeper understanding of their movement is necessary. Segmenting a moving object referred to by a text query requires the integration of both temporal and visual features [13,24]. Therefore, the combined use of spatio-temporal and linguistic features is crucial for the RVOS task. Additionally, language embedding must be incorporated into the dataset for RVOS, which results in a limited number of available databases in the literature. The RVOS benchmark [25] introduces a dataset and evaluation metrics where segmentation is guided by a natural language query.

Environmental disturbances like rain or haze removal in video frames is a crucial task for improving visual quality in video surveillance tasks, especially in outdoor environments. These weather disturbances degrade visibility, making it challenging for models to accurately detect and segment objects in various general computer vision tasks like anomaly detection [26–30], video segmentation etc. Numerous statistical and deep learning models have been developed to address these issues. Traditional statistical approaches focus on physical models that account for atmospheric scattering effects. For example, the dark channel prior (DCP) [16] is a widely-used statistical method for haze removal, based on the observation that haze-free images contain pixels with low intensity in at least one color channel. DCP has been extended and refined in various works [18] to improve dehazing performance.

In recent years, deep learning-based models have shown remarkable success in both haze and rain removal. Methods like DehazeNet [19] leverage convolutional neural networks (CNNs) to automatically learn haze-relevant features from training data. This method bypasses the need for handcrafted features, making it more flexible in diverse conditions. Similarly, rain removal techniques have also evolved, with early methods focusing on low-level image processing [21], and more recent approaches utilizing deep networks to separate rain streaks from background scenes. The deep detail network (DDN) [22] is a notable method that splits the image into two layers: a rain-free background and a rain-streak layer, using deep CNNs to refine the background.

These deep learning methods significantly outperform traditional techniques, especially in complex scenarios with heavy haze or rain. Hybrid approaches that combine physical models with deep learning [23] have also shown promise by leveraging the strengths of both worlds. The ongoing development of larger and more diverse datasets, such as RESIDE [20] for haze removal and Rain800 [31] for rain removal, has also contributed to the rapid advancement of weather noise removal technologies. These databases and methods provide improved benchmarks and evaluation metrics, helping researchers better understand and tackle the challenges posed by weather-induced noise in videos.

3   Proposed Work

In this work we have utilised a state-of-the-art SgMg [1] referred video segmentation model for evaluation. SgMg model is a deep learning based end-to-end framework which is trained on ref-DAVIS 2017 dataset. We have simulated haze and rain environmental noise to the dataset and evaluate the performance on weather-affected dataset first. Degradation in the performance motivates us to develop towards weather noise removal framework. In the later sections we have elaborated the weather simulation method from which environmental noise has been introduced in the dataset.

3.1   Simulation Method

We have utilised procedural image generation techniques that use randomness and geometric principles to simulate haze and rain effects.

Haze Generation Methodology. In the context of haze generation, we rely on the standard haze formation model, where the observed hazy image $I(x)$ is a combination of the scene radiance $J(x)$ and the atmospheric light A. The transmission map $t(x)$ controls the blending of these two components. The mathematical representation is as follows:

Hazy Image Model

$$I(x) = J(x) \cdot t(x) + A \cdot (1 - t(x)) \tag{1}$$

where:

- $I(x)$ is the observed hazy image,
- $J(x)$ is the scene radiance (haze-free image),
- $t(x)$ is the transmission map,
- A is the global atmospheric light.

Transmission Map (Haze Map) Generation. The transmission map $H(x)$ is generated using a random gradient based on image coordinates, modeled as:

$$H(x) = \alpha_1 \cdot xv + \alpha_2 \cdot yv \tag{2}$$

where:

- xv and yv are 2D gradient fields based on the image coordinates,
- α_1 and α_2 are random weights controlling the gradient's direction and intensity.

Final Hazy Image. The final hazy image is computed by blending the original image with the haze layer as follows:

$$I_{hazy}(x) = I(x) \cdot (1 - H(x)) + 255 \cdot H(x) \tag{3}$$

where:

- 255 represents the white haze layer.

Rain Generation Method. For rain generation, we simulate stick-like raindrops by calculating their positions and orientations. Gaussian blurring is applied to enhance the raindrop streaks.

Raindrop Endpoint Calculation. To generate a raindrop, we compute its endpoint based on its length and orientation angle:

$$x_1 = x_0 + L \cdot \cos(\theta) \tag{4}$$

$$y_1 = y_0 + L \cdot \sin(\theta) \tag{5}$$

where:

- (x_1, y_1) is the endpoint of the raindrop,
- L is the length of the raindrop,
- θ is the angle of orientation.

Gaussian Blur for Raindrop Streaks. To simulate the appearance of raindrop streaks, we apply Gaussian blur, which is mathematically represented as:

$$G(x) = \frac{1}{\sqrt{2\pi\sigma^2}}e^{-\frac{x^2}{2\sigma^2}} \tag{6}$$

where σ is the standard deviation that controls the spread of the raindrop streaks.

Final Rainy Image. The final rainy image is computed by blending the original image with the raindrop layer:

$$I_{rainy}(x) = I(x) \cdot (1 - \beta) + R(x) \cdot \beta \tag{7}$$

where:

- $I(x)$ is the original image,
- $R(x)$ is the generated raindrop layer,
- β is the blending factor that controls the intensity of the rain effect.

3.2 Enhancement of Weather-Affected Referred Video Object Segmentation (WARMOS) Framework

Referred Video Object Segmentation Architecture. In this work, we utilize the Spectrum-guided Multigranularity (SgMg) [1] approach, which performs direct segmentation on encoded features and leverages visual details to refine the segmentation masks. SgMg employs spectrum-guided intra-frame global interactions in the spectral domain to enhance multimodal representation, allowing for more accurate segmentation across diverse input types practicality in complex video analysis tasks. The SgMg performance has been shown in the Table 1 on Ref-DAVIS 2017 dataset.

Table 1. Quantitative results of SgMg [1] on Original Ref-DAVIS-2017. **J**: Jaccard, **F**: F-measure, **J&F** is average of J and F

Method	Backbone	J&F	J	F
SgMg[[1]])	Video-Swin-B	63.64	61.22	66.06

Once the weather noise like haze and rain have been introduced to the dataset we directly evaluate them with Benchmark SgMg framework. The noise degraded the video quality hence we got sub-optimal results. Table 2 and Table 3 show the weather-affected results on the Ref-DAVIS 2017 dataset.

Table 2. Quantitative results of SgMg [1] on Ref-DAVIS-2017 affected by weather noise **Haze**. **J**: Jaccard, **F**: F-measure, **J&F** is average of J and F

Method	Backbone	J&F	J	F
SgMg[[1]])	Video-Swin-B	58.71	56.48	60.95

Table 3. Quantitative results of SgMg [1] on Ref-DAVIS-2017 affected by weather noise **Rain**. **J**: Jaccard, **F**: F-measure, **J&F** is average of J and F

Method	Backbone	J&F	J	F
SgMg[[1]])	Video-Swin-B	61.00	58.03	63.69

Weather Noise Removal Framework. To reduce the adverse effects of environmental noise, we used a convolutional neural network-based method Light-DehazeNet (LD-Net) [17]. LD-Net works by estimating both the transmission map and atmospheric light through a modified atmospheric scattering model. Additionally, a color visibility restoration technique is applied to avoid color distortion in the dehazed image. We fine-tuned the LD-Net model separately for haze and rain removal, training two different models with the same architecture. The improved results highlight the importance of this noise removal framework, which boosts the performance of RVOS in weather-affected situations. A ResNet [32] based classifier is trained to know the environmental noise class on the fly. Once the noise class is recognized, the corresponding noise removal model is enabled.

4   Experimental Setup

We have evaluated WARMOS on DAVIS-2017 [9] datasets which are widely used benchmarks in video object segmentation tasks. To assess the performance of the proposed method, quantitative results of the Jaccard and the F-measure similarity index have been compared with state-of-the-art methods. The Jaccard index (J), also known as Intersection over union (IoU), measures the similarity between predicted output and ground truth mask by computing the ratio of Intersection over union among two classes. Jaccard Index can be represented in Eq. 8

$$J = \frac{\text{Area of Overlap}}{\text{Area of Union}} = \frac{|A \cap B|}{|A \cup B|} \tag{8}$$

where A and B are the sets that represent the predicted output and group truth, and the F-measure index is evaluated as a harmonic mean of precision and recall. It provides a balanced evaluation of the model performance. F-measure can represented as in Eq. 9

$$F - measure = 2 \times \frac{precision \times recall}{precision + recall} \tag{9}$$

where precision is the ratio of predicted true positives (TP) to the total predicted positives (TP + FP), and recall is the ratio of true positives predicted to the total actual positives.

$$precision = \frac{TP}{TP + FP} \ , \ recall = \frac{TP}{TP + FN}$$

4.1   Dataset

The DAVIS-2017 [9] dataset consists of 90 videos, widely used as benchmarks for Video Object Segmentation (VOS). Of these 90 videos, 60 are designated for training, while the remaining 30 are set aside for testing. To adapt the dataset for Referred Video Object Segmentation (RVOS) tasks, it is augmented with

Regular Expressions (REs), which act as text queries to identify specific objects in the videos. Ground-truth maps are then dynamically generated for each frame based on the input text query. If a video contains multiple objects, the relevant objects are masked according to the corresponding text query.

4.2   Training Details

SgMg Framework. We have leverages SgMg [1] framework for RVOS task. This is a PyTorch based implementation. We utilised trained weights available to us by the authors for the inference purpose.

Weather Noise Removal Framework. We have fine-tuned ResNet model for classification purpose. The classifier accuracy is 99%. We employed Light-DehazeNet (LD-Net) and utilizes torch-1.13 framework over NVIDIA A100 GPU with torch-1.13 and Cuda 11.7 to train and fine-tune our haze and rain removal network. The objective of the weather noise removal with RVOS task is to generate accurate binary mask that highlights relevant objects in video frames despite weather disturbance in original video frames. To enhance the robustness of our model, we applied various data augmentations during training, such as random flips, rotations, random crops, etc. Our training process spanned 60 epochs.

5   Results and Discussion

In this section we have discussed the performance of our proposed method and shows its effectiveness with quantitative and qualitative results. We have reported the mean Intersection over Union also known as Jaccard (J) and average F-measure (F) to evaluate the segmentation performance of our model.

5.1   Quantitative Analysis

We evaluate our method on the Ref-Davis 2017 dataset and shown the results in Table 4 and Table 5. The quantitative results of WARMOS demonstrate satisfying performance and improved the results when compared with Table 2 and 3 that evaluated in weather affected scenarios. These results highlight the robustness and efficacy of our approach in handling complex video object segmentation guided by text query.

Table 4. Quantitative results of SgMg [1] on Ref-DAVIS-2017 after Haze noise removal **J**: Jaccard, **F**: F-measure, **J&F** is average of J and F

Method	Backbone	J&F	J	F
SgMg[[1]])	Video-Swin-B	60.50	58.00	63.00

5.2   Qualitative Analysis

In Figure-3 we have shown qualitative results of our WARMOS approach. Results indicates that our approach performs well in segmenting objects referred by text query. In Figure 3 two set of examples are considered from DAVIS-2017 dataset from dog-jumps and black-scooter category. Label-A shows actual image, label-B shows ground truth, label-C shows Weather simulated frame, haze effect in left dog-jump category and rain effect in right scooter-black category. Label-D shows the output of our approach which segments referred object well in challenging weather scenario.

Table 5. Quantitative results of SgMg [1] on Ref-DAVIS-2017 after Rain noise removal. **J**: Jaccard, **F**: F-measure, **J&F** is average of J and F

Method	Backbone	J&F	J	F
SgMg[[1]])	Video-Swin-B	61.05	58.58	63.52

5.3   Discussion

Environmental noise compromises visibility in video frames, thereby negatively affecting the segmentation task. This work proposes the WARMOS approach, which removes weather noise such as haze and rain to enhance segmentation output. We utilized state-of-the-art deep learning-based haze and rain removal methods to mitigate the negative effects of noise. We simulated haze and rain conditions on the DAVIS-2017 dataset to train and fine-tune the models. The addition of weather noise removal techniques improved the performance of the state-of-the-art SgMg segmentation model. Our experiments on the DAVIS-2017 dataset showed that while the SgMg model's performance was hindered by environmental noise, WARMOS successfully restored video clarity, improving the

Fig. 3. A: DAVIS-2017 Video frame, B: Ground truth, C: Weather noise simulated image, Haze in left and Rain in right, D: Output of WARMOS approach based on text query highlighted in red color. (Color figure online)

mean J&F scores from 58.71 to 60.50 in haze conditions and from 61.00 to 61.05 in rain-affected scenarios. This improvement highlights the importance of noise removal in maintaining segmentation accuracy in adverse weather conditions.

6   Conclusion

In this work, we have demonstrated the challenges posed by environmental noise such as haze and rain in the task of Referred Video Object Segmentation (RVOS). Weather conditions introduce pixel variance, which hampers the segmentation performance by obscuring object appearances and disrupting temporal and spatial consistency. To address this, we simulated haze and rain effects in video sequences and proposed a robust noise removal framework to mitigate these disturbances, thereby improving segmentation performance. Our approach leverages the state-of-the-art Spectrum-guided Multigranularity (SgMg) model for video object segmentation, which is guided by text-based queries. We observed that the introduction of environmental noise in the DAVIS-2017 dataset resulted in significant degradation in segmentation accuracy. To combat this, we employed Light-DehazeNet (LD-Net) method that effectively removes weather noise. Fine-tuning LD-Net for haze and rain removal significantly improved the segmentation performance on weather-affected videos. The effectiveness of our WARMOS framework was demonstrated through both quantitative and qualitative results. In conclusion, our approach effectively mitigates the adverse effects of environmental noise on RVOS tasks by ensuring that the performance remains optimal even under challenging real-world conditions.

References

1. Miao, B., Bennamoun, M., Gao, Y., Mian, A.: Spectrum-guided multi-granularity referring video object segmentation. In: Proceedings of the IEEE/CVF International Conference on Computer Vision, pp. 920–930 (2023)
2. Mandal, M., Saxena, P., Vipparthi, S.K., Murala, S.: Candid: robust change dynamics and deterministic update policy for dynamic background subtraction. In: IEEE/ICPR, pp. 2468–2473 (2018)
3. Saxena, P., Biradar, K., Tyagi, D.K., Vipparthi, S.K:. Richex: A robust inter-frame change exposure for segmenting moving objects. In: IEEE CVF/ICIP, pp. 2172–2176. IEEE (2022)
4. Patil, P.W., Biradar, K.M., Dudhane, A., Murala, S.: An end-to-end edge aggregation network for moving object segmentation. In: Proceedings of the IEEE/CVF CVPR, pp. 8149–8158 (2020)
5. Patil, P.W., Murala, S.: MSFgNet: a novel compact end-to-end deep network for moving object detection. IEEE Trans. Intell. Transport. Syst. 20, 4066–4077 (2018)
6. Akilan, T., Wu, Q.J., Safaei, A., Huo, J., Yang, Y.: A 3D CNN-LSTM-based image-to-image foreground segmentation. IEEE Trans. Intell. Transport. Syst. 21, 959–971 (2019)
7. Cheng, J., Tsai, Y.H., Hung, W.C., Wang, S., Yang, M.H.: Fast and accurate online video object segmentation via tracking parts. In: Proceedings of the IEEE CVPR, pp. 7415–7424 (2018)

8. Hu, P., Liu, J., Wang, G., Ablavsky, V., Saenko, K., Sclaroff, S.: Dipnet: dynamic identity propagation network for video object segmentation. In: Proceedings of the IEEE/CVF WACV, pp. 1904–1913 (2020)

9. Pont-Tuset, J., Perazzi, F., Caelles, S., Arbeláez, P., Sorkine-Hornung, A., Van Gool, L.: The 2017 davis challenge on video object segmentation. arXiv preprint arXiv:1704.00675 (2017)

10. Bellver, M., Ventura, C., Silberer, C., Kazakos, I., Torres, J., Giro-i-Nieto, X.: A closer look at referring expressions for video object segmentation. Multimed. Tools Appl. **82**(3), 4419–4438 (2023)

11. Wu, J., Jiang, Y., Sun, P., Yuan, Z., Luo, P.: Language as queries for referring video object segmentation. In: Proceedings of the IEEE/CVF CVPR, pp. 4974–4984 (2022)

12. Botach, A., Zheltonozhskii, E., Baskin, C.: End-to-end referring video object segmentation with multimodal transformers. In: Proceedings of the IEEE/CVF CVPR, pp. 4985–4995 (2022)

13. Saxena, P., Roy, S.M., Tyagi, D.K., Vipparthi, S.K., Murala, S., Balasubramanian, R.: Refmos: A robust referred moving object segmentation framework based on text query. In: 2024 IEEE International Conference on Advanced Video and Signal Based Surveillance (AVSS), pp. 1–7. IEEE, (2024)

14. Ye, L., Rochan, M., Liu, Z., Wang, Y.: Cross-modal self-attention network for referring image segmentation. In: IEEE/CVF Conference on Computer Vision and Pattern Recognition, pp. 10502–10511 (2019)

15. St-Charles, P.-L., Bilodeau, G.-A., Bergevin, R.: Subsense: a universal change detection method with local adaptive sensitivity. IEEE Trans. Image Process. **24**(1), 359–373 (2014)

16. He, K., Sun, J., Tang, X.: Single image haze removal using dark channel prior. IEEE Trans. Pattern Anal. Mach. Intell. **33**(12), 2341–2353 (2010)

17. Ullah, H., et al.: Light-dehazenet: a novel lightweight CNN architecture for single image dehazing. IEEE Trans. Image processing **30**, 8968–8982 (2021)

18. Meng, G., Wang, Y., Duan, J., Xiang, S., Pan, C.: Efficient image dehazing with boundary constraint and contextual regularization. In: IEEE International Conference on Computer Vision, pp. 617–624 (2013)

19. Cai, B., Xiangmin, X., Jia, K., Qing, C., Tao, D.: Dehazenet: an end-to-end system for single image haze removal. IEEE Trans. Image Process. **25**(11), 5187–5198 (2016)

20. Li, B., et al.: Benchmarking single-image dehazing and beyond. IEEE Trans. Image Process. **28**(1), 492–505 (2018)

21. Garg, K., Nayar, S.K.: Vision and Rain. Int. J. Comput. Vision **75**(1), 3–27 (2007). https://doi.org/10.1007/s11263-006-0028-6

22. Fu, X., Huang, J., Zeng, D., Huang, Y., Ding, X., Paisley, J.: Removing rain from single images via a deep detail network. In: IEEE Conference on Computer Vision and Pattern Recognition, pp. 3855–3863 (2017)

23. Zhang, H., Patel, V.M.: Density-aware single image de-raining using a multi-stream dense network. In: IEEE Conference on Computer Vision and Pattern Recognition, pp. 695–704 (2018)

24. Li, D., et al.: You only infer once: cross-modal meta-transfer for referring video object segmentation. Proc. AAAI Conf. Artif. Intell.D **36**, 1297–1305 (2022)

25. Xu, N.: Youtube-vos: A large-scale video object segmentation benchmark. arXiv preprint arXiv:1809.03327 (2018)

26. Biradar, K., Dube, S., Vipparthi, S.K.: Dearest: deep convolutional aberrant behavior detection in real-world scenarios. In: 2018 IEEE 13th International Conference on Industrial and Information Systems (ICIIS), pp. 163–167. IEEE (2018)
27. Biradar, K.M., Gupta, A., Mandal, M., Vipparthi, S.K.: Challenges in time-stamp aware anomaly detection in traffic videos. arXiv preprint arXiv:1906.04574 (2019)
28. Biradar, K.M., Mandal, M., Dube, S., Vipparthi, S.K., Tyagi, D.K.: Triplet-set feature proximity learning for video anomaly detection. Image Vision Comput. **150**, 105205 (2024). https://doi.org/10.1016/j.imavis.2024.105205
29. Biradar, K.M., Gupta, A., Mandal, M., Vipparthi, S.K.: Challenges in time-stamp aware anomaly detection in traffic videos. In: Proceedings of the IEEE/CVF Conference on Computer Vision and Pattern Recognition (CVPR) Workshops (2019)
30. Dube, S., Biradar, K., Vipparthi, S.K., Tyagi, D.K.: MAG-Net: a memory augmented generative framework for video anomaly detection using extrapolation. In: Raman, B., Murala, S., Chowdhury, A., Dhall, A., Goyal, P. (eds.) Computer Vision and Image Processing: 6th International Conference, CVIP 2021, Rupnagar, India, December 3–5, 2021, Revised Selected Papers, Part II, pp. 426–437. Springer International Publishing, Cham (2022). https://doi.org/10.1007/978-3-031-11349-9_37
31. Zhang, H., Sindagi, V., Patel, V.M.: Image de-raining using a conditional generative adversarial network. IEEE Transactions on Circuits and Systems for Video Technology **30**(11), 3943–3956 (2020). https://doi.org/10.1109/TCSVT.2019.2920407
32. Koonce, B.: Resnet 50. In: Convolutional Neural Networks with Swift For Tensorflow: Image Recognition and Dataset Categorization, pp. 63–72 (2021)

Robust Anomaly Detection Through Transformer-Encoded Feature Diversity Learning

Kuldeep Biradar[1]([envelope]), Dinesh Kumar Tyagi[1], Ramesh Babu Battula[1], and P. Subbarao[2]

[1] Malaviya National Institute of Technology Jaipur, Jaipur, India
{2018rcp9503,dktyagi.cse,rbbattula.cse}@mnit.ac.in
[2] Vignan University, Guntur, India
drpsr_it@vignan.ac.in

Abstract. Detecting irregularities in weather data serves multiple practical applications. For example, nowcasting focuses on predicting atmospheric conditions for the next 0 to 4 h, which is essential for effective emergency response and disaster management. Anomaly detection also plays a crucial role in forecasting extreme weather events. However, the complexity increases when considering both anomalies and varying weather conditions. A common approach to anomaly detection is weakly supervised video-level labeling, which aims to identify frames containing abnormal events and is typically framed as a multiple instance learning (MIL) problem. While existing methods perform well, the prevalence of negative instances significantly hinders their ability to detect positive instances, particularly rare abnormal segments. To address this, we aim to extract distinctive features by enhancing the observable differences between various classes using a single branch. We propose a novel method, Transformer Encoded Feature Video Anomaly Detection (TEF-VAD), which exclusively utilizes attention mechanisms, specifically Multi-Head Attention Learning. This approach combines feature magnitude learning loss, class-specific loss, and a TEF-VAD-enhanced MIL classifier training loss, thereby training a model to effectively identify positive examples and improve the MIL method's robustness for detecting positive instances in abnormal videos. Our extensive experiments demonstrate that the MIL model enhanced by our Transformer method significantly improves sample efficiency and the detection of subtle anomalies, outperforming several state-of-the-art techniques on benchmark datasets like UCF-Crime and ShanghaiTech.

Keywords: Anomaly Detection · Transformer · UCF Crime

1 Introduction

Identifying irregularities in weather data serves various practical purposes. For instance, nowcasting aims to forecast atmospheric conditions within the next

M. Cho et al. (Eds.): ACCV 2024 Workshops, LNCS 15482, pp. 118–131, 2025.
https://doi.org/10.1007/978-981-96-2641-0_8

0 to 4 h, which is vital for emergency response and disaster mitigation. Similarly, anomaly detection is instrumental in predicting and analyzing extreme weather phenomena over time. Integrating weather factors with applications such as anomaly detection [1], object detection [2], and segmentation poses significant challenges for researchers. However, when we take into account both anomalies and varying weather conditions, the situation becomes quite complex. To address this, we have various types of datasets, such as UCF-Crime and XD, which encompass multiple weather scenarios along with the associated anomalies. Given that abnormal events are infrequent in videos, recent research has predominantly been conducted within the weakly supervised learning framework [1,3–9], which involves solely video-level annotations. Video anomaly detection aims to predict frame-level anomaly scores at which an abnormal event takes place. Abuse, stealing, aggression and other similar behaviors are examples of anomalies in the context of surveillance.

Despite years of research in video anomaly detection (VAD) with different types of weather, creating a model capable of accurately detecting anomalies in videos remains a challenging task. This difficulty stems from the requirement for the model to comprehend the inherent distinctions between normal and abnormal events, particularly rare and widely varying anomalous events. Earlier studies have addressed VAD as an unsupervised learning task [10–14]. One commonly utilized approach is the concept of reconstructing features from normal training data. Based on the features employed, all available methods typically involve training an autoencoder framework or generative reconstruction-based methods (GAN) with a deep neural network. These methods aim to ensure the reconstruction of normal frame events with minimal reconstruction mistake. Nevertheless, the likelihood of failure increases significantly under diverse weather conditions. For instance, if a model is trained exclusively on data from normal weather conditions and is then tested in hazy or rainy conditions, there is a considerable risk of encountering high reconstruction errors, leading to inaccurate results. Consequently, it becomes evident that nearly all methods based on training set reconstruction of frames cannot ensure the detection of unusual occurrences.

Here, we tackle the subject of weakly supervised video anomaly detection (WS-VAD), where acquiring video-level labels are often more practical and can yield more robust outcomes compared to unsupervised techniques. By giving different types of data at the time of training and equalizing the amount of abnormal-normal snippets evenly across the training set, WSVAD techniques trained using multiple-instance learning (MIL) algorithms have recently addressed the aforementioned issues [1,3–5]. The normal videos are used to select the normal snippets at random, while the abnormal videos are used to select the snippets with the greatest anomaly scores.

MIL partially resolves the previously mentioned issues but introduces following problems: Selecting normal snippets at random from normal videos might lead to relatively simple modeling, potentially hindering convergence of training; if a video contains multiple abnormal snippets, the opportunity for a more effective training session that includes multiple abnormal snippets per video is lost; relying on classification scores provides a feeble training signal that may not facilitate a clear distinction among normal and abnormal snippets. Model trained with single weather with same anomaly. All of these problems are made worse by techniques that ignore important temporal dependencies.

In order to tackle the MIL challenges mentioned earlier, we introduce a new method called Transformer-Encoded Feature-Based Video Anomaly Detection(TEF-VAD). In this approach, we utilize features encoded using a transformer and select the top-k snippets inspired by RTFM [6] of anomaly to train MIL which works on different types of weather it may be rain, fogg, day time, or it may be night time etc. Also we learned class-specific loss to ensure that the features of each class are patterned in a similar manner. Using k instances of the abnormal and normal films with the greatest classification scores, the MIL method trains a classifier. Proposed method addresses the challenges associated with MIL in the following ways: Robust Features encoded by using a transformer encoded feature; it increases the likelihood of selecting truly anomalous frames from abnormal videos; by choosing hard negative normal snippets from normal videos, which are more difficult to model, it promotes better training convergence; it allows the inclusion of a greater number of anomalous frames for each abnormal video in the training process.

We assess the efficacy of proposed TEF-VAD method on two benchmark anomaly detection datasets, ShanghaiTech [11] and UCF-Crime [1]. Our results demonstrate the significant outperformance of our method compared to the current state-of-the-art techniques across all benchmarks. We also establish that our method achieves markedly improved sample efficiency and subtle anomaly discriminability compared to widely used MIL methods. Furthermore, our proposed method showcases effective anomaly snippet detection as indicated by the AUC metric.

Our contributions can be outlined as:

- The Transformer encoded feature VAD (TEF-VAD) architecture for anomaly detection, which aligns with the concept of predicting frame-level anomaly scores during anomalous events, utilizing video-level annotation.
- We achieve the extraction of robust features using a transformer, facilitated by a multi-head attention mechanism.
- The learning of TEF-VAD involves the utilization of three different types of losses: the Transformer encoded feature magnitude learning loss; class-specific loss; and TEF-VAD-enabled MIL classifier training loss.
- The robustness of our method is demonstrated through experiments conducted on two benchmark datasets which improves 2.21% in ShanghaiTech and 1.44% in UCF-Crime anomaly dataset.

2   Related Work

The video anomaly detection research landscape can be categorized into two main classes: unsupervised and weakly-supervised VAD.

Unsupervised VAD: Techniques that only use unlabeled training data or that carry out direct training and testing on testing data are referred to be unsupervised methods. A method to identify changes in a video sequence by identifying unique frames was proposed by Del et al. [15], while Tudor et al. [16] introduced unmasking technology [17] to train a binary classifier iteratively in order to identify the most discriminant features. A recent method by Zaheer et al. [18] establishes cross-supervision between a generator and a discriminator to take advantage of the low frequency of anomalies. Moreover, One-Class Classification (OCC) approaches treat the problem in an unsupervised way and assume the availability of just normal training data. Usually, when building a model, researchers use only normal data and identify events that deviate from the model; this allows them to identify anomalies. Previous works made use of hand-crafted look and motion aspects [19–23]. Recent techniques are using features from pre-trained deep neural networks to construct an anomaly classifier, thanks to developments in deep learning [10,24]. Moreover, self-supervised feature learning techniques exist ? [25–27]. One well-liked strategy uses temporal prediction [11,28,29]. Unsupervised approaches, however, can raise false alarms for normal patterns that are not seen because it is not feasible to include every possible kind of normalcy in a single dataset.

Weakly Supervised VAD: Video-level labels are more useful in weakly supervised learning approaches [1,3–5,30] for distinguishing between abnormal and normal events. There are two types of existing weakly supervised VAD approaches: encoder-independent methods and encoding-based methods. **Encoder-independent** techniques only help the classifier get trained. For example, a deep multiple instance learning model that treats a video as a "snippet" and its numerous parts as discrete "instances" was first developed by Sultani et al. [1]. [31] proposed supervised anomaly detection. Wan et al. [4] presented dynamic MIL loss and regularization led by the center, while Zhang et al. [5] introduced inner-bag score gap regularization. Both a feature encoder and a classifier are trained using encoding-based approaches. To encode motion-aware information, Zhu et al. [30] presented an MIL model with attention, integrated with an autoencoder based on optical flow. Weakly supervised VAD was approached as a label noise learning task by Zhong et al. [3], who also used graph convolutional networks (GCNs) to filter label noise for iterative model training. Nevertheless, the method proved to be inefficient and progressed slowly.

Multiple Instance Learning. Weakly supervised learning is frequently accomplished through the use of Multiple Instance Learning (MIL). MIL handles a video as a "bag" and its component clips as distinct "instances" in video-related applications [1,32,33]. Indirect supervision for instance-level learning can be obtained from video-level labels by using a certain feature/score aggregation function. Many aggregation functions are used, including attention pooling [32,33] and max pooling [1,5,30].

We also use an encoder-based methodology and an online, fine-grained app-roach in our work. On the other hand, we encode features using a transformer, and then use the top-k snippets for Multiple Instance Learning (MIL) to optimize our solution.

3   Proposed Method

In the context of Weakly Supervised Video Anomaly Detection (WSVAD), the training data for detecting anomaly events at the frame-level typically consists of normal video V_n and abnormal video V_a, each having a label at the video level $Y = [0(Normal), 1(Anomaly)]$. During training, each input video, whether normal or abnormal is divided into T segments, which are then placed in the negative bag (Normal) and positive bag (Abnormal), as illustrated in Fig. 1. Additionally, these snippets provide features of dimension D, obtained from the pre-trained I3D model [34], denoted as f_m^D. Here, f represents the feature of D dimensions, and m equals 1 if the snippet is abnormal and 0 if the snippet is normal. The proposed architecture utilizes encoded features of the transformer, denoted by $E_\alpha(f_m^D)$ where α is a learning parameter denoted in Fig. 1. Fur-ther the model employed by Transformer Encoded Feature VAD (TEF-VAD) is denoted as $F_\beta(E_\alpha(f_m^D))$, and results in a 2-dimensional feature [0 (Normal), 1 (Anomaly)], indicating the classification of the T video segments as normal or anomalous, with the learning parameters α and β specified below which visual-ize in Fig. 2. The learning of this model encompaces a collective optimization involving end-to-end Transformer encoded feature magnitude learning, class-specific feature learning, and TEF-VAD-enabled MIL classifier training, utilizing the specified loss represented in Eq. 1.

$$\min_{\alpha,\beta} \sum_{m,n=1}^{D} l_t(E_\alpha(f_m), E_\alpha(f_n), y_m, y_n) + l_{mil}(F_\beta(E_\alpha(f_m)), y_m) + l_{cs}(E_\alpha(f_m), y_m)$$

(1)

where l_t represents Transformer encoded feature magnitude learning loss, l_{mil} represents TEF-VAD-enabled MIL classifier training loss and l_{cs} represents class specific loss

Next, we will elaborate on the rationale behind our suggested TEF-VAD, accompanied by a comprehensive description of the units.

3.1   Transformer Encoder Feature

In standard NLP models, the input sequence is typically processed sequentially, either word by word or character by character. In contrast, a transformer model processes the input sequence in parallel, enabling the model to analyze the entire input sequence at once [35]. A common formulation of the weakly super-vised video-level label anomaly detection method is a multiple instance learning

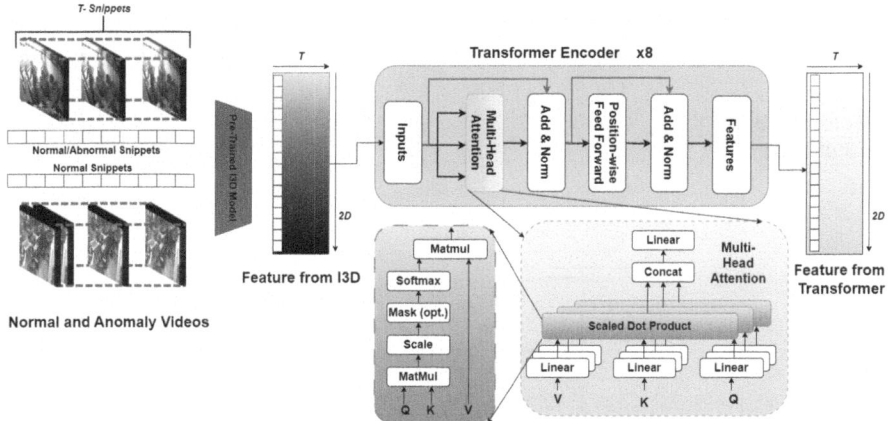

Fig. 1. Videos with weak labels are divided into T snippets and then fed into the Pre-Trained Model for Encoding 2D-Dimentions Features(Anomaly(D)-Normal(D)) using Transformer.

(MIL) issue. Finding video segments featuring anomalous incidents is the primary objective. In our scenario, instead of processing individual characters, we pass temporal features, and leverage the capabilities of the transformer to extract robust features. A series of feedforward neural network layers and multi-head self-attention layers make up the transformer model. In the multi-head self-attention mechanism, the input sequence undergoes processing through three linear layers to produce three sets of projections: values(V), queries(Q) and keys(K). The dot product of the query and key projections is calculated and then scaled by the square root of the dimension of the key vectors. Subsequently, a softmax function is employed to generate a set of attention weights for each token in the sequence. These attention weights are utilized to compute a weighted sum of the value vectors, resulting in a context vector for each token in the sequence shown in Fig. 1.

3.2 Transformer Encoded Feature Magnitude Learning and Class-Specific Learning

Transformer encoded features were further harnessed for VAD using MIL, as depicted in Fig. 2. In this framework, we calculated Transformer encoded feature magnitude learning and Class-Specific Learning losses, illustrated using Eq. 2 and Eq. 3 respectively. In this method, the largest snippet feature magnitudes from normal videos are reduced, and those from abnormal videos are increased (i.e. using top k samples and top k batch samples), inspired by [6,36]. The class-specific loss l_{cs} is represented in equation to ensure that the characteristics of each class are patterned in a similar way.

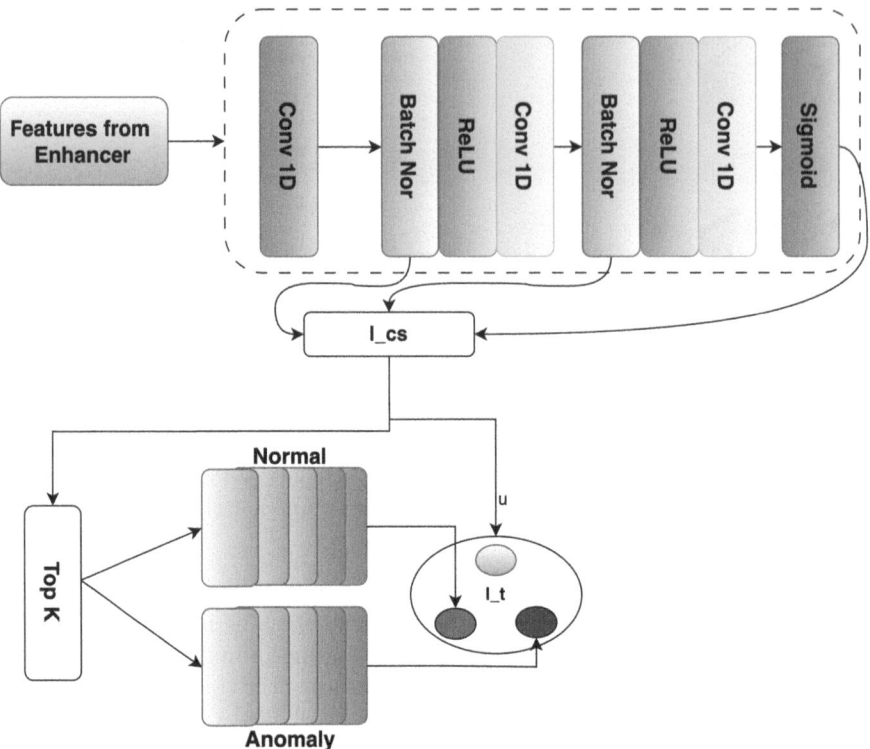

Fig. 2. Learning of proposed Transformer based encoded feature MIL using different type of losses(l_t, l_{cs}).

Building on our understanding of statistical normality modeling in Batch-Norm within VAD, the mean vector μ serves as a statistical indicator of normality, allowing us to differentiate between normal snippets and potential anomalies.

To promote divergence between potential abnormal features and the mean vector $\hat{\mu}$, while simultaneously clustering the normal features, Specifically, following our l_t criterion, we select K potential abnormal features magnitude from abnormal videos and K normal features magnitude from normal videos. further those substracted with normalised mean $\hat{\mu}$ that exhibit the highest K l_t scores shown in Eq. 2

$$l_t(E_\alpha(f_a), E_\alpha(f_n), y_m, y_n) = \begin{cases} \max(0, m - d_{\alpha,k}(X_n - \hat{\mu}, X_a - \hat{\mu})) & \text{if } y_m = 1 \text{ and } y_n = 0 \\ 0 & \text{otherwise} \end{cases}$$

$$(2)$$

As we have a feature vectors of $E_\alpha(f_n)_k$ and $E_\alpha(f_a)_k$ for normal and anomaly respectively. We took magnitude of

$$l_{cs}(E_\alpha(f_m), y_m)) = L_1(E_\alpha(f_n)_k, y_n) + L_1(E_\alpha(f_a)_k, y_a) \qquad (3)$$

where L_1 represents l1-loss , $E_\alpha(f_n)_k$ represents normal top k snipet features from transformer similarly $E_\alpha(f_a)_k$ for abnormal.

In order to train the segment classifier (MIL), We employ a binary cross-entropy classification loss function with a specified set of top k snippets.

4   Datasets and Experiment Setup

We assess our model across two varied benchmark datasets, established for anomaly detection in videos labeling: ShanghaiTech, UCF-Crime

4.1   Dataset

The **UCF-Crime** dataset [1] is a collection of 1900 uncut films for anomaly detection, amounting to 128 h in total. These videos are taken from actual security cameras on the streets and inside buildings presenting dynamic and varied backgrounds as opposed to the still backgrounds observed in the samples present ShanghaiTech dataset. There are an equal number of normal and anomalous videos in the training and testing sets. Together with labels at the video level, the dataset includes 1,610 training videos covering 13 anomaly classes, and it also incorporates 290 test clips with labels at the frame level.

ShanghaiTech is relatively a medium sized dataset with 437 videos altogether and 13 distinct background scenes that were obtained from street video footage captured from fixed angles. These, 307 videos are categorized as normal, while 130 are classified as anomaly videos. The dataset cited in [11] has been commonly adopted as a benchmark for detecting anomolies. Zhong et al. [3] modified the dataset via utilizing the inclusion of anomalous inferencing videos within the training data, generating a weakly supervised training dataset. This assures that the train and test samples contain representations of all 13 background scenarios. We have adopted the same methodology as described in [3] to adapt ShanghaiTech for the weakly supervised scenario.

The proposed anomaly detection method is structured in two distinct phases: feature extraction and classification. For feature extraction, we employ the I3D (Inflated 3D ConvNet), which excels at capturing spatiotemporal dynamics from video data. However, it is important to note that this multi-step approach may limit its effectiveness for real-time anomaly detection due to inherent processing latencies. To validate our method, we have utilized two widely recognized datasets: UCF-Crime and ShanghaiTech. UCF-Crime, in particular, provides diverse real-time data drawn from CCTV footage, social media, and YouTube, encompassing a variety of anomaly types. While these datasets are popular in the field, we recognize the necessity for broader experimentation. Future work will involve expanding our analysis to additional datasets to strengthen the generalizability of our findings across various anomaly detection contexts. Although our results demonstrate superior performance within these specific datasets.

4.2   Assessment Metrics

Following previous research, we select the evaluation metric to measure the effectiveness of our proposed technique on the UCF-Crime and ShanghaiTech datasets: the area under the curve (AUC) of the receiver operating characteristic (ROC) curve at the frame level.

4.3   Implementation Details

According to [1], the features of the I3D pre-trained network [34] were taken from the 'mix 5c' layers correspond to 2048D. Following feature extraction, there are 32 film clips in each video denoted as $T = 32$. Throughout all investigations, the margin remains fixed at $m = 100$, while $k = 5$ in Eq. (2). The model delineated in Sect. 3 Fig. 2 includes three dense layers, having 512, 128, and 1 nodes. Each node succeeded by a dropout function and a ReLU activation function having the dropout rate of 0.3. Before fully connected we used transformer encoded layer to extract features from I3D features. We use an exhaustive training strategy with our TEF-VAD technique, employing the optimizer Adam [37] with a batch size of 32 over 50 epochs and a weight decay of 0.0001. For the ShanghaiTech and UCF-Crime datasets, a learning rate of 0.001 is set. Samples chosen at random from 32 normal and anomalous videos make up each mini-batch. The method's implementation is completed in PyTorch [38]. We use the same standard setup

Table 1. Comparative analysis of frame-level AUC performance with other state-of-the-art (SOTA) methods on the ShanghaiTech dataset. AUC_o represent the AUC calculated on the complete test set.

Category	Methods	Features	$AUC_o(\%)$
Unsupervised VAD	Conv-AE [10]	-	60.85
	Frame-Pred [11]	-	73.40
	Stacked-RNN [12]	-	68.00
	VEC [39]	-	74.80
	Mem-AE [40]	-	71.20
	MNAD [41]	-	70.50
Weakly-Supervised VAD	GCN-Anomaly [3]	TSN-Flow	84.13
	GCN-Anomaly [3]	C3D-RGB	76.44
	GCN-Anomaly [3]	TSN-RGB	84.44
	Zhang et al. [5]	I3D-RGB	82.50
	Sultani et al. [1]	I3D-RGB	85.33
	AR-Net [4]	I3D-Flow	82.32
	AR-Net [4]	I3D-RGB & I3D-Flow	91.24
	AR-Net [4]	I3D-RGB	85.38
	Proposed	I3D-RGB	93.45

as in [1,3–5,30] to ensure fair comparison, and we use the published results with the same foundation as ours for all baselines.

4.4 Results

Table 1 provides a breakdown of the AUC results at the frame level for the Shang-haiTech dataset. Our proposed method demonstrates superior performance as measured against earlier state-of-the-art (SOTA) unsupervised learning techniques [10–12,39,41], as well as weakly supervised approaches [1,4,5]. Using I3D-RGB features, proposed TEFVAD model achieves the greatest AUC score on this dataset, reaching 93.45%. Furthermore, with the same I3D-RGB features, our Transformer-encoded feature learning MIL approach surpasses existing SOTA MIL-based methods [56, 62, 74] by a margin ranging from 2% to 9%. These results highlight the advancements facilitated by our proposed transformer-encoded feature learning technique.

Table 2. Comparative analysis of frame-level AUC performance with other state-of-the-art (SOTA) methods on the UCF-Crime dataset. AUC_o and AUC_a represent the AUC calculated on the complete test set and solely on abnormal test videos, respectively

Category	Methods	Features	$AUC_o(\%)$	$AUC_a(\%)$
Unsupervised VAD	SVM Baseline		50	50
	Sohrab et al. [42]	-	58.50	-
	Conv-AE [10]	-	50.60	-
	Lu et al. [21]	-	65.51	-
	GODS [43]	-	70.46	-
	BODS [43]	-	68.26	-
Weakly-Supervised VAD	Zhang et al. [5]	I3D-RGB	78.66	-
	Sultani et al. [1]	C3D-RGB	75.41	54.25
	Zhang et al. [5]	I3D-RGB	78.66	-
	Motion-Aware [30]	TSN-Flow	79.10	62.18
	GCN-Anomaly [3]	TSN-RGB	82.12	59.02
	Wu et al. [44]	I3D-RGB	82.44	-
	WSAL [7]	I3D-Flow	85.38	67.38
	RTFM [6]	I3D-RGB	84.30	-
	UMIL [8]	I3D-Flow	86.75	68.68
	Proposed	I3D-RGB	83.71	70.12

Table 2 compares our Proposed Method (TEFVAD) with other well-known state-of-the-art (SOTA) techniques, including Unsupervised VAD (UVAD) and WSVAD. When compared to all other approaches, the Proposed Method earns the greatest AUC_A on the UCF-Crime dataset., marking an enhancement of +1.44%. Similarly, for AUC_O, our method demonstrates improvements relative to all competing methods except for those specified (RTFM [6], WSAL [7], UMIL [8]).

In particular, our improvement in the AUC_A metric demonstrates the exceptional performance of the Proposed TEFVAD Method in AUC_O is not only attributed to the existence of clear-cut normal videos, but also arises from its enhanced capability to detect anomalous segments within abnormal videos. Furthermore, across both datasets, Weakly Supervised VAD exhibits substantial enhancements over Unsupervised VAD in terms of AUC_O, empirically confirming the inherent challenges in detecting open-set anomalies in UVAD. Conversely, the improvements in AUC_A are relatively minor, for instance, 54.25% in comparison to 50.00% on UCF-Crime. This highlights the existing bias in current WSVAD methods towards the explicit differentiation between normal and abnormal, leading to a high incidence of false positives and negatives in uncertain segments from abnormal videos.

5   Conclusion

In this research, we presented a Multiple Instance Learning (MIL) approach that utilizes Transformer-encoded features to develop an impartial anomaly classifier and a customized representation for WSVAD. This MIL training scheme necessitates a confident set comprising visually apparent normal/abnormal video snippets, derived from Transformer-encoded features with a substantial margin in positive and negative snippets. The application of MIL leads to notable enhancements in classification performance. The combination of these distinctive features and classifier in our approach yields a substantial An enhancement over present state-of-the-art techniques. Our approach is empirically substantiated by achieving top-tier performance on two benchmark datasets for WSVAD.

References

1. Sultani, W., Chen, C., Shah, M.: Real-world anomaly detection in surveillance videos. In: Proceedings of the IEEE Conference on Computer Vision and Pattern Recognition, pp. 6479–6488 (2018)
2. Saxena, P., Biradar, K., Tyagi, D.K., Vipparthi, S.K.: Richex: A robust interframe change exposure for segmenting moving objects. In: 2022 IEEE International Conference on Image Processing (ICIP), pp. 2172–2176. IEEE (2022)
3. Zhong, J.X., Li, N., Kong, W., Liu, S., Li, T.H., Li, G.: Graph convolutional label noise cleaner: Train a plug-and-play action classifier for anomaly detection. In: Proceedings of the IEEE/CVF Conference on Computer Vision and Pattern Recognition, pp. 1237–1246 (2019)
4. Wan, B., Fang, Y., Xia, X., Mei, J.: Weakly supervised video anomaly detection via center-guided discriminative learning. In: 2020 IEEE International Conference on Multimedia and Expo (ICME), pp. 1–6. IEEE (2020)

5. Zhang, J., Qing, L., Miao, J.: Temporal convolutional network with complementary inner bag loss for weakly supervised anomaly detection. In: 2019 IEEE International Conference on Image Processing (ICIP), pp. 4030–4034. IEEE (2019)

6. Tian, Y., Pang, G., Chen, Y., Singh, R., Verjans, J.W., Carneiro, G.: Weakly-supervised video anomaly detection with robust temporal feature magnitude learning. In: Proceedings of the IEEE/CVF International Conference on Computer Vision, pp. 4975–4986 (2021)

7. Lv, H., Zhou, C., Cui, Z., Chunyan, X., Li, Y., Yang, J.: Localizing anomalies from weakly-labeled videos. IEEE Trans. Image Process. **30**, 4505–4515 (2021)

8. Lv, H., Yue, Z., Sun, Q., Luo, B., Cui, Z., Zhang, H.: Unbiased multiple instance learning for weakly supervised video anomaly detection. In: Proceedings of the IEEE/CVF Conference on Computer Vision and Pattern Recognition, pp. 8022–8031 (2023)

9. Peng, W., et al.: Vadclip: adapting vision-language models for weakly supervised video anomaly detection. Proc. AAAI Conf. Artif. Intell. **38**, 6074–6082 (2024)

10. Hasan, M., Choi, J., Neumann, J., Roy-Chowdhury, A.K., Davis, L.S.: Learning temporal regularity in video sequences. In: Proceedings of the IEEE Conference on Computer Vision and Pattern Recognition, pp. 733–742 (2016)

11. Liu, W., Luo, W., Lian, D., Gao, S.: Future frame prediction for anomaly detection–a new baseline. In: Proceedings of the IEEE Conference on Computer Vision and Pattern Recognition, pp. 6536–6545 (2018)

12. Luo, W., Liu, W., Gao, S.: A revisit of sparse coding based anomaly detection in stacked rnn framework. In: Proceedings of the IEEE International Conference on Computer Vision, pp. 341–349 (2017)

13. Dube, S., Biradar, K., Vipparthi, S.K., Tyagi, D.K.: MAG-Net: a memory augmented generative framework for video anomaly detection using extrapolation. In: Raman, B., Murala, S., Chowdhury, A., Dhall, A., Goyal, P. (eds.) Computer Vision and Image Processing: 6th International Conference, CVIP 2021, Rupnagar, India, December 3–5, 2021, Revised Selected Papers, Part II, pp. 426–437. Springer International Publishing, Cham (2022). https://doi.org/10.1007/978-3-031-11349-9_37

14. Biradar, K.M., Gupta, A., Mandal, M., Vipparthi, S.K.: Challenges in time-stamp aware anomaly detection in traffic videos. *arXiv preprint*arXiv:1906.04574, 2019

15. Del Giorno, A., Bagnell, J.A., Hebert, M.: A discriminative framework for anomaly detection in large videos. In: Leibe, B., Matas, J., Sebe, N., Welling, M. (eds.) Computer Vision – ECCV 2016: 14th European Conference, Amsterdam, The Netherlands, October 11-14, 2016, Proceedings, Part V, pp. 334–349. Springer International Publishing, Cham (2016). https://doi.org/10.1007/978-3-319-46454-1_21

16. Tudor Ionescu, R., Smeureanu, S., Alexe, B., Popescu, M.: Unmasking the abnormal events in video. In: Proceedings of the IEEE International Conference on Computer Vision, pp. 2895–2903 (2017)

17. Koppel, M., Schler, J., Bonchek-Dokow, E.: Measuring differentiability: Unmasking pseudonymous authors. J. Mach. Learn. Res. **8**(6) 2007

18. Zaheer, M.Z., Mahmood, A., Khan, M.H., Segu, M., Yu, F., Lee, S.I.: Generative cooperative learning for unsupervised video anomaly detection. In: Proceedings of the IEEE/CVF Conference on Computer Vision and Pattern Recognition, pp. 14744–14754 (2022)

19. Adam, A., Rivlin, E., Shimshoni, I., Reinitz, D.: Robust real-time unusual event detection using multiple fixed-location monitors. IEEE Trans. Pattern Anal. Mach. Intell. **30**(3), 555–560 (2008)

20. Antić, B., Ommer, B.: Video parsing for abnormality detection. In: 2011 International Conference on Computer Vision, pp. 2415–2422. IEEE, (2011)
21. Lu, C., Shi, J., Jia, J.: Abnormal event detection at 150 fps in matlab. In: Proceedings of the IEEE International Conference On Computer Vision, pp. 2720–2727 (2013)
22. Wang, S., Miao, Z.: Anomaly detection in crowd scene. In: IEEE 10th International Conference on Signal Processing Proceedings, pp. 1220–1223. IEEE (2010)
23. Mehran, R., Oyama, A., Shah, M.: Abnormal crowd behavior detection using social force model. In: 2009 IEEE Conference on Computer Vision and Pattern Recognition, pp. 935–942. IEEE (2009)
24. Ravanbakhsh, M., Nabi, M., Mousavi, H., Sangineto, E., Sebe, N.: Plug-and-play cnn for crowd motion analysis: An application in abnormal event detection. In: 2018 IEEE Winter Conference on Applications of Computer Vision (WACV), pp. 1689–1698. IEEE (2018)
25. Sabokrou, M., Fathy, M., Hoseini, M., Klette, R.: Real-time anomaly detection and localization in crowded scenes. In: Proceedings of the IEEE Conference on Computer Vision and Pattern Recognition Workshops, pp. 56–62 (2015)
26. Xu, D., Ricci, E., Yan, Y., Song, J., Sebe, N.: Learning deep representations of appearance and motion for anomalous event detection. arXiv preprint arXiv:1510.01553 (2015)
27. Biradar, K.M., Mandal, M., Dube, S., Vipparthi, S.K., Tyagi, D.K.: Triplet-set feature proximity learning for video anomaly detection. Image Vision Comput. **150**, 105205 (2024). https://doi.org/10.1016/j.imavis.2024.105205
28. Lv, H., Chen, C., Cui, Z., Xu, C., Li, Y., Yang, J.: Learning normal dynamics in videos with meta prototype network. In: Proceedings of the IEEE/CVF Conference on Computer Vision and Pattern Recognition, pp. 15425–15434 (2021)
29. Shi, X., Chen, Z., Wang, H., Yeung, D.Y., Wong, W.K., Woo, W.C.: Convolutional lstm network: a machine learning approach for precipitation nowcasting. In: Advances in Neural Information Processing Systems, vol. 28 (2015)
30. Zhu, Y., Newsam, S.: Motion-aware feature for improved video anomaly detection. *arXiv preprint*arXiv:1907.10211 (2019)
31. Biradar, K., Dube, S., Vipparthi, S.K.: Dearest: deep convolutional aberrant behavior detection in real-world scenarios. In: 2018 IEEE 13th International Conference on Industrial and Information Systems (ICIIS), pp. 163–167. IEEE, (2018)
32. Nguyen, P., Liu, T., Prasad, G., Han, B.: Weakly supervised action localization by sparse temporal pooling network. In: Proceedings of the IEEE Conference on Computer Vision and Pattern Recognition, pp. 6752–6761 (2018)
33. Hong, F.T., Huang, X., Li, W.H., Zheng, W.S.: Mini-net: multiple instance ranking network for video highlight detection. In: Computer Vision–ECCV 2020: 16th European Conference, Glasgow, UK, August 23–28, 2020, Proceedings, Part XIII 16, pp. 345–360. Springer (2020)
34. Kay, W., et al.: The kinetics human action video dataset. *arXiv preprint*arXiv:1705.06950 (2017)
35. Vaswani, A.: Attention is all you need. In: Advances in Neural Information Processing Systems, vol. 30 (2017)
36. Zhou, Y., Qu, Y., Xu, X., Shen, F., Song, J., Tao Shen, H.: BatchNorm-based weakly supervised video anomaly detection. IEEE Transact. Circ. Syst. Video Technol. **34**(12), 13642–13654 (2024). https://doi.org/10.1109/TCSVT.2024.3450734
37. Diederik, P.K.: Adam, A method for stochastic optimization. *arXiv preprint* arXiv:1412.6980 (2014)

38. Paszke, A., et al.: Pytorch: An imperative style, high-performance deep learning library. In: Advances in Neural Information Processing Systems, vol. 32 (2019)
39. Yu, G.: Cloze test helps: effective video anomaly detection via learning to complete video events. In: Proceedings of the 28th ACM International Conference on Multimedia, pp. 583–591 (2020)
40. Gong, D., et al.: Memorizing normality to detect anomaly: Memory-augmented deep autoencoder for unsupervised anomaly detection. In: Proceedings of the IEEE/CVF International Conference on Computer Vision, pp. 1705–1714 (2019)
41. Park, H., Noh, J., Ham, B.: Learning memory-guided normality for anomaly detection. In: Proceedings of the IEEE/CVF Conference on Computer Vision and Pattern Recognition, pp. 14372–14381 (2020)
42. Sohrab, F., Raitoharju, J., Gabbouj, M., Iosifidis, A.: Subspace support vector data description. In: 2018 24th International Conference on Pattern Recognition (ICPR), pp. 722–727. IEEE, (2018)
43. Wang, J., Cherian, A.: Gods: generalized one-class discriminative subspaces for anomaly detection. In: Proceedings of the IEEE/CVF International Conference on Computer Vision, pp. 8201–8211 (2019)
44. Wu, P., et al.: Not only look, but also listen: Learning multimodal violence detection under weak supervision. In: Computer Vision–ECCV 2020: 16th European Conference, Glasgow, UK, August 23–28, 2020, Proceedings, Part XXX 16, pp. 322–339. Springer (2020)

Adversarial Weather-Resilient Image Retrieval: Enhancing Restoration Using Captioning for Robust Visual Search

Prem Shanker Yadav[(⊠)] [ID], Kushall Singh [ID], Dinesh Kumar Tyagi, and Ramesh Babu Battula

Malaviya National Institute of Technology, Jaipur 302017, Rajasthan, India
2018rcp9157@mnit.ac.in

Abstract. Accurate image retrieval in real-world scenarios is often hampered by degraded or noisy images, particularly those affected by adverse weather conditions such as rain, fog, or snow. Traditional retrieval methods that rely solely on feature extraction struggle to handle these degraded inputs and image captioning models are similarly limited in their ability to interpret distorted images. To address these challenges, we propose a novel framework that integrates image restoration with image captioning to create a robust image retrieval system capable of handling images degraded by adverse weather. Additionally, we introduce an integrated loss function to optimize restoration and captioning processes for degraded images. Our system enhances retrieval performance in challenging weather conditions by leveraging improved visual content alongside semantic context. Evaluations on Flicker8k dataset demonstrate that our approach significantly outperforms traditional image retrieval systems, particularly in scenarios where weather-induced degradation presents a challenge.

Keywords: Feature Differences loss · Unseen Changes · Textual Significance

1 Introduction

In recent years, advancements in image retrieval systems have significantly improved the ability to search and retrieve images based on their content. However, these systems often face challenges such as adverse weather conditions, which can severely impact image quality and the effectiveness of traditional image retrieval methods. Uncertain conditions like rain, blur, fog, noise, snow, etc. can obscure key features, making it difficult for conventional models to accurately interpret and retrieve relevant images.

Image restoration techniques have become a potential solution to the problem of improving the visual quality of degraded images and enhancing the performance of image retrieval systems. Recent studies have shown that image restoration is effective in mitigating the effects of adverse weather conditions, allowing

M. Cho et al. (Eds.): ACCV 2024 Workshops, LNCS 15482, pp. 132–145, 2025.
https://doi.org/10.1007/978-981-96-2641-0_9

for clearer and more accurate image analysis [30]. In particular, image restoration methods that use deep learning models have shown promising results in reconstructing high-quality images from degraded inputs [13,17].

In parallel, image captioning has proven to be a valuable tool for understanding and describing the content of images. By generating textual descriptions of images, image captioning systems facilitate more nuanced image retrieval by enabling semantic search based on the content of the images [15,27]. The combination of image restoration and captioning offers a novel approach to enhance image retrieval systems, particularly in challenging conditions. Figure 1 demonstrates how textual descriptions of images can facilitate the retrieval of scenes under adverse conditions. Captioning significantly improves the accuracy of this retrieval process.

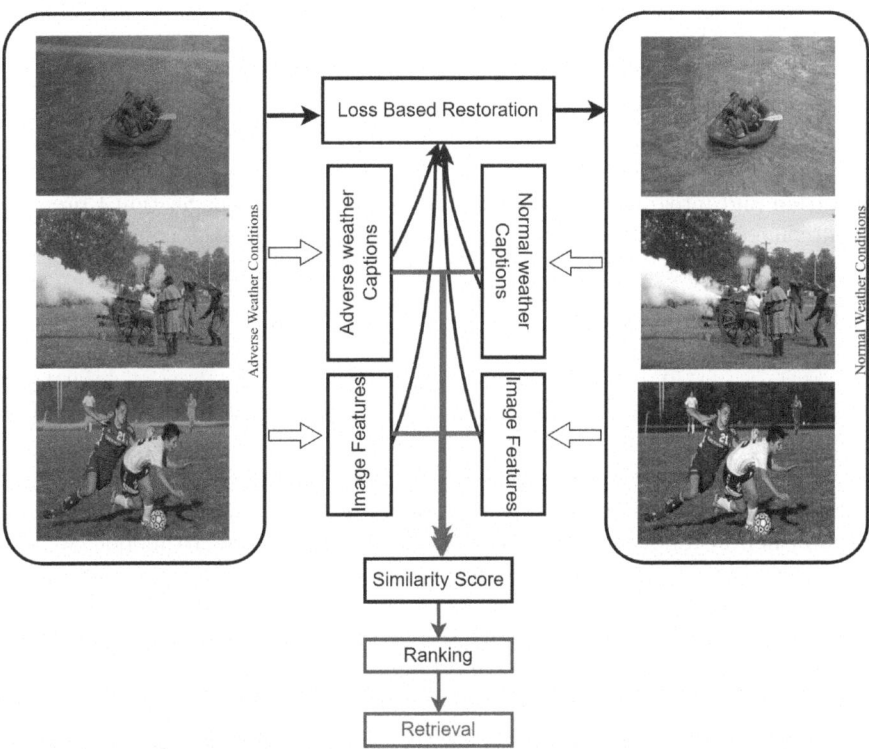

Fig. 1. Image retrieval using restoration by captioning and using Loss Based Restoration on weather-degraded images

This work proposes a novel framework that integrates image restoration with image captioning to improve image retrieval performance under adverse weather conditions.

Our approach involves the following key contributions:

- **Integrated Loss Functions for Image Restoration:** We utilize the integrated loss function to restore images degraded by adverse weather conditions, thereby improving image quality, clarity, and feature visibility.
- **Proposed Captioning-based Restoration:** We integrate deep learning-based image captioning to retrieve natural images with minimal synthetic image generation.
- **Integrated Retrieval-Restoration Framework:** We propose a framework that merges the restored images with their generated captions, thereby improving retrieval accuracy, particularly in challenging scenarios where conventional methods are less effective.

By integrating these components, our framework aims to provide a robust solution for image retrieval in adverse weather conditions, offering significant improvements over existing methods. Our proposed approach addresses the limitations of current systems and provides a more effective means of retrieving relevant images in challenging environments.

We organize the remaining sections in the following order: Section 2 describes a previously devised literature review. In Sect. 3, The designed framework for the restoration and integration with image retrieval is presented. Section 4 presents the outcome of the developed technique. Section 5 presents the conclusion.

2   Literature Review

The development of image retrieval systems has been an important field of study in computer vision. Traditional image retrieval systems have used feature extraction techniques such as (Scale-Invariant Feature Transform [34]) and HOG (Histogram of Oriented Gradients) to represent visual content. The study in [26] introduced a Bayesian encoder for metric learning in image retrieval, replacing neural amortization with Laplace Approximation to model network weight distributions and improve uncertainty estimates. However, these algorithms are constrained when photos are damaged by external variables such as poor weather, resulting in erroneous or partial feature extraction. The paper [21] presented a method for estimating hurricane rain rates using SAR images analyzed through an Artificial Neural Network (ANN). The ANN helps interpret complex radar data to provide accurate rain rate measurements, which are important for weather forecasting and understanding hurricane dynamics. The paper [23] presented a novel approach for retrieving nighttime videos using a Temporal Weighting Appearance-Aligned Network. This network improves the retrieval process by addressing temporal variations and appearance changes in low-light conditions. By incorporating these aspects, the proposed method aims to enhance the accuracy and relevance of video retrieval in challenging nighttime scenarios. Extreme noise or low-quality nighttime videos could diminish the network's effectiveness, and it may struggle to adapt to novel scenarios not covered in the training data. Deep learning-based image retrieval models have recently evolved, capable of learning improved feature representations using Convolutional Neural Networks (CNNs) and Vision Transformers. Despite their gains, these approaches still struggle in difficult situations such as rain, fog, and snow, when critical visual features are frequently concealed, resulting in poor retrieval performance [8].

Adverse weather conditions, such as rain, fog, and snow, can significantly degrade the visual quality of images by introducing distortions like occlusions, motion blur, and low contrast, which can obscure important features. Traditional image retrieval systems often fail to properly interpret images affected by these factors, leading to misclassifications or the failure to retrieve relevant results. Studies have shown that weather-induced degradation reduces the performance of many feature extraction and classification techniques in image retrieval. As a result, there is an increasing demand for more robust retrieval systems capable of restoring image clarity and mitigating weather-induced distortions to enhance the accuracy of retrieval [12,33].

Image restoration has been extensively studied as a solution to address image degradation caused by adverse weather conditions. Techniques such as dehazing [3], rain removal [7], and super-resolution have shown their effectiveness in restoring images to a clearer state, making them more suitable for subsequent image analysis tasks, including image retrieval. Early works, like He et al.'s dark channel prior method [5,11] for image dehazing, paved the way for more advanced restoration techniques. This method introduced the dark channel prior, a simple yet effective image prior for single-image haze removal. By leveraging the observation that haze-free images tend to have very low-intensity pixels in at least one color channel, the method estimates haze thickness and recovers high-quality, haze-free images, with the added benefit of producing a useful depth map as a byproduct.

The field of image restoration has significantly evolved from CNN-based methods [25,29,32] to transformer-based approaches [6,20], which excel in capturing long-range dependencies and improving restoration performance. Techniques like window-based approaches [18] and transposed attention [28] have been employed to manage computational efficiency while maintaining restoration quality. Unified models, such as IPT [4] and AirNet [16], aim to handle multiple degradations, though their effectiveness remains limited across diverse tasks [22,31]. On the other hand, prior-based methods, that utilize external information like high-resolution images or pre-trained models have been instrumental in improving restoration results [14,19]. More recent approaches, such as text-based priors combined with image-based priors, have introduced a new paradigm, enabling models to learn degradation information at a textual level, thereby providing clean guidance for enhanced restoration performance [2].

Parallel to image restoration, image captioning has become an invaluable tool for describing the content of images in natural language. Image captioning models generate textual descriptions that capture the objects, scenes, and activities depicted in an image, facilitating the semantic search for image retrieval tasks. Traditional image retrieval systems rely heavily on low-level features, while captioning introduces a higher-level semantic understanding of image content. Karpathy and Fei-Fei [15] introduced a method for aligning image regions with natural language descriptions, which greatly improved the accuracy of image caption matching. This work laid the foundation for more advanced models like the Show and Tell model [24] and the attention-based Show, Attend and Tell model [27], which further enhanced the capability of captioning systems to provide meaningful, context-aware descriptions of images. The integration of such captioning techniques into image retrieval systems enables more robust search capabilities, even under challenging conditions like those posed by adverse weather.

The combination of image restoration and captioning represents a novel approach to image retrieval. By first restoring the visual quality of images affected by weather conditions and then generating detailed captions describing the content, such systems can significantly improve retrieval accuracy in scenarios where traditional methods fall short. Anderson et al. [1] introduced a bottom-up and top-down attention mechanism for image captioning, which has the potential to be combined with restoration techniques to enhance the retrieval process. Wang et al. [25] explored the integration of restoration techniques with image recognition systems, suggesting that restoring image quality can improve both recognition and retrieval outcomes. However, this combined approach has yet to be fully explored in the context of adverse weather conditions, presenting a gap in the current literature. While considerable progress has been made in both image restoration and captioning individually, the integration of these two approaches for robust image retrieval under adverse weather conditions has received limited attention. Existing image retrieval models primarily focus on improving feature extraction but often fail to address image degradation caused by environmental factors. Similarly, while image captioning models can provide semantic descriptions, their performance is hindered when operating on degraded images.

Our work focuses on addressing these limitations by proposing a novel framework that integrates image restoration with deep learning-based image captioning to enhance image retrieval in adverse weather conditions.

3   Methodology

Our framework combines advanced image restoration techniques with deep learning-based image captioning to improve image retrieval in challenging weather conditions. This approach enhances both image clarity and semantic understanding, resulting in more precise and meaningful retrieval outcomes.

The complete architecture of our framework is depicted in Fig. 2.

3.1   Dataset Preparation

We use a dataset comprising normal ($\mathcal{I}_n$) and weather-degraded ($\mathcal{I}_w$) images, along with their respective captions ($\mathcal{C}$). Each pair ($\mathcal{I}_n, \mathcal{I}_w$) forms the basis for learning the difference in feature distributions caused by weather degradation. The dataset is represented as follows:

$$\mathcal{D} = (\mathcal{I}_n, \mathcal{I}_w, \mathcal{C}) \tag{1}$$

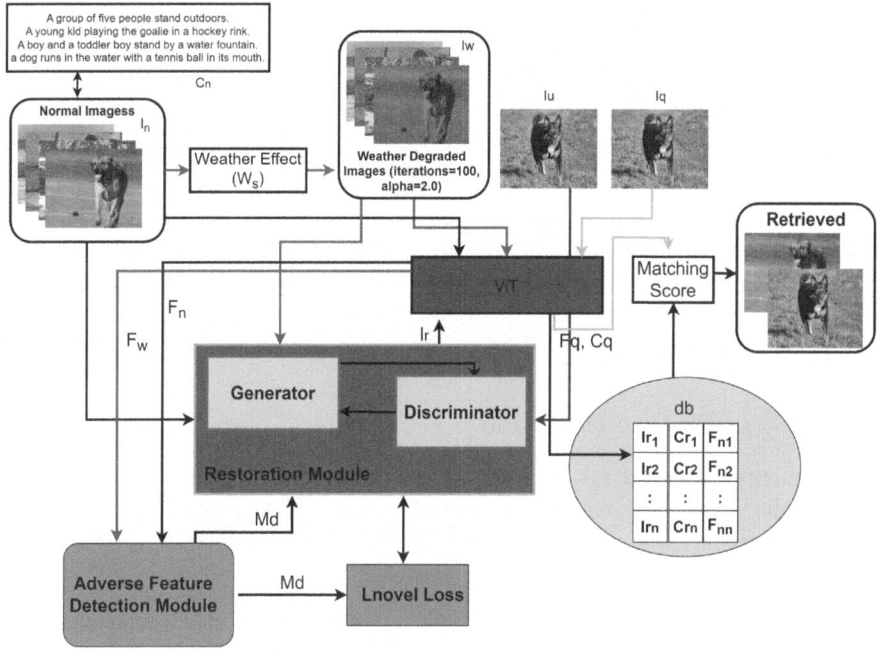

Fig. 2. Proposed Framework for image retrieval using restoration and captioning

To get adverse weather conditions, We apply various weather simulation effects [9] $\mathcal{W}_s$ to normal images. This generates synthetic weather-degraded images:

$$\mathcal{I}_w = \mathcal{W}_s(\mathcal{I}_n), \quad \mathcal{W}_s \in \text{rain, snow, fog} \tag{2}$$

where $\mathcal{I}_n$ and $\mathcal{I}_w$ are the images, and $\mathcal{C}$ is the corresponding caption. Figure 3 shows normal and degraded images.

Fig. 3. Normal and degraded images due to fog, rain, and snow

3.2   Feature Extraction

We utilize a dual-input feature extraction module that processes $\mathcal{I}_n$ and $\mathcal{I}_w$. A Vision Transformer (ViT) extracts feature representations for each image:

$$\mathcal{F}_n = \text{ViT}(\mathcal{I}_n), \quad \mathcal{F}_w = \text{ViT}(\mathcal{I}_w) \tag{3}$$

where $\mathcal{F}_n$ and $\mathcal{F}_w$ are the feature vectors of the normal and weather-degraded images, respectively. The Vision Transformer captures high-level features crucial for semantic understanding and restoration. It significantly contributes to the embedding process and enhances the quality of caption generation. We are using BLIP (Bootstrapping Language-Image Pre-training) to generate captions.

3.3   Adverse Feature Detection

Using the extracted features $\mathcal{F}_w$, the Adverse Feature Detection Module identifies specific regions or distortions related to adverse weather effects. Adverse Feature Detection Module is a layered structure of convolution shown in the Fig. 4. This produces an adverse degraded map $\mathcal{M}_d$:

$$\mathcal{M}_d = \text{AdverseNet}(\mathcal{F}_w) \tag{4}$$

where $\mathcal{M}_d$ highlights the areas of $\mathcal{I}_w$ most affected by weather conditions, guiding the restoration process.

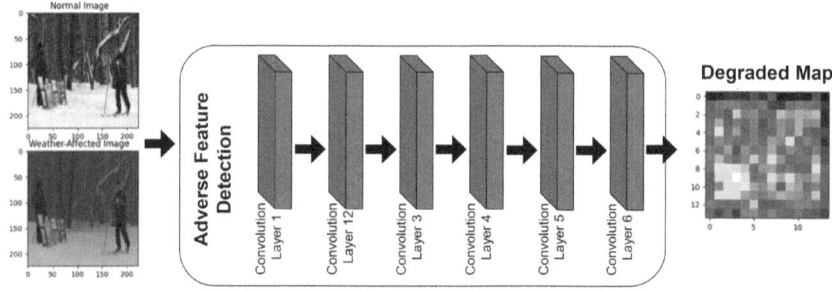

Fig. 4. Layered structure of AdverseNet

3.4   Image Restoration

The restoration process is driven by a dynamic restoration network $\mathcal{R}$ Generative Adversarial Networks (GAN-UNet), which restores $\mathcal{I}_w$ using the degraded map $\mathcal{M}_d$ and outputs a restored image $\hat{\mathcal{I}}w$. The network optimizes a novel composite loss function $\mathcal{L}$novel composed of the following key components:

- **Feature Loss** ($\mathcal{L}_f$): Ensures the restored image $\hat{\mathcal{I}}_w$ retains the visual characteristics of the original image $\mathcal{I}_n$.

$$\mathcal{L}_f = |\mathcal{F}_n - \text{ViT}(\hat{\mathcal{I}}_w)|^2 \tag{5}$$

- **Captioning Loss** ($\mathcal{L}_c$): Aligns the generated caption $\hat{\mathcal{C}}_w$ of the restored image with its ground truth caption $\mathcal{C}$.

$$\mathcal{L}_c = \text{CrossEntropy}(\mathcal{C}, \hat{\mathcal{C}}_w) \tag{6}$$

- **Adverse Feature Loss** ($\mathcal{L}_a$): Minimizes weather-induced distortions in the restored image based on the degraded map $\mathcal{M}_d$.

$$\mathcal{L}_a = \|\mathcal{M}_d\|^2 \tag{7}$$

- **Semantic Consistency Loss** ($\mathcal{L}_{sc}$): A novel loss component ensuring that the restored image retains not only visual but also semantic consistency with the original caption. It measures the similarity between captions generated for both $\mathcal{I}_n$ and $\hat{\mathcal{I}}_w$:

$$\mathcal{L}_{sc} = 1 - \text{CosineSimilarity}(\hat{\mathcal{C}}_n, \hat{\mathcal{C}}_w) \tag{8}$$

- **Adaptive Restoration Loss** ($\mathcal{L}_{ar}$): An adaptive loss that dynamically adjusts restoration based on the severity of weather degradation. This loss weights the restoration error inversely to the clarity level of $\mathcal{I}_w$:

$$\mathcal{L}_{ar} = \frac{\|\hat{\mathcal{I}}_w - \mathcal{I}_n\|^2}{1 + \|\mathcal{M}_d\|} \tag{9}$$

The total novel loss function used to train the restoration model is:

$$\mathcal{L}_{\text{novel}} = \lambda_f \mathcal{L}_f + \lambda_c \mathcal{L}_c + \lambda_a \mathcal{L}_a + \lambda_{sc} \mathcal{L}_{sc} + \lambda_{ar} \mathcal{L}_{ar} \tag{10}$$

where λ_f, λ_c, λ_a, λ_{sc}, and λ_{ar} are hyperparameters controlling the weight of each loss component.

3.5   Image Captioning and Feedback Mechanism

After restoration, a transformer-based image captioning model generates a detailed caption $\hat{\mathcal{C}}_w$ for the restored image $\hat{\mathcal{I}}_w$. The caption feedback loop ensures that the generated caption aligns with the original context, helping to further refine the restoration process. The caption feedback is incorporated into the total loss function:

$$\mathcal{L}_{\text{novel}} \leftarrow \mathcal{L}_{\text{novel}} + \lambda_c \cdot \text{CaptionFeedback}(\mathcal{C}, \hat{\mathcal{C}}_w) \tag{11}$$

This feedback loop dynamically adjusts the restoration process, improving semantic consistency and contextual accuracy.

3.6   Feature-Driven Adverse Detection for Unseen Images

For unseen images $\mathcal{I}_u$, the model predicts the presence of adverse weather effects by comparing the features of the unseen image $\mathcal{F}_u$ with learned normal image features $\mathcal{F}_n$:

$$\mathcal{F}_u = \mathrm{ViT}(\mathcal{I}_u) \tag{12}$$

The model generates a prediction of adverse features $\hat{\mathcal{M}}_d$ if weather degradation is detected. The restoration network then restores the image accordingly:

$$\mathcal{I}_r u = \mathcal{R}(\mathcal{I}_u, \mathcal{F}_u, \hat{\mathcal{M}}_d) \tag{13}$$

3.7   Image Retrieval

Once the restoration is complete, the restored image I_r and its caption C_r are stored in an index (db). During the retrieval phase, a query image $\mathcal{I}_q$ is matched with stored images based on the similarity of their features $\mathcal{F}_q$ and captions C_q. The Matching-Score is calculated as:

$$\mathrm{Similarity}(\mathcal{I}_q, \mathcal{I}_{\mathrm{db}}) = \alpha \cdot \mathrm{cosine\_similarity}(\mathcal{F}_q, \mathcal{F}_{\mathrm{db}}) + \beta \cdot \mathrm{Similarity\_Score}(C_q, C_{\mathrm{db}}) \tag{14}$$

where α and β are weighting parameters, ensuring a balanced emphasis on feature and caption similarity. Similarity-Score is calculated using Term Frequency-Inverse Document Frequency (TF-IDF) and Bag of Words.

4   Experiments and Result Analysis

This section describes the experiments conducted to evaluate the performance of the proposed image retrieval framework under adverse weather conditions. The experiments are designed to assess the effectiveness of image restoration, caption generation, and retrieval quality using both quantitative metrics and qualitative analysis.

Dataset

The experiments were conducted on the Flicker8k dataset, which includes a diverse collection of images with corresponding captions. The dataset features images captured in various contexts, with captions provided to evaluate the semantic accuracy of the generated descriptions. There are 8,092 images and five captions describing each image. Since weather conditions are not explicitly labeled; we applied approaches [9] and implemented them to generate adverse weather conditions in Flicker8k images.

Evaluation Metrics. We employ a range of metrics to evaluate the effectiveness of the image retrieval framework by BLUE, Precision, Recall, Peak Signal-to-Noise Ratio (PSNR), and Structural Similarity Index (SSIM).

4.1   Results and Analysis

Image Captioning Effect in Retrieval. Table 1 presents a comparative analysis of the image captioning effect in image retrieval in adverse conditions. We used BLUE, Precision, Recall and F1-score to present the accuracy and relevance of the image retrieval system.

Table 1. Image Captioning Affected Retrieval.

Method	BLEU	Precision	Recall	F1-score
CNN+LSTM	20.02	0.743	0.703	0.717
CNN+Bi-LSTM	21.20	0.764	0.744	0.754
Deep CNN+LSTM [10]	22.3	-	-	-
Show and Tell [24]	27.1	-	-	-
ViT (our)	40.10	0.842	0.807	0.824

Image Restoration Performance. Table 2 shows the quantitative performance of our proposed framework on image restoration tasks. The results are measured using PSNR and SSIM.

Table 2. Effect of various losses on Restoration Performance.

Method	PSNR	SSIM
Proposed Framework	27.94	0.91
$(\mathcal{L}_f)$	23.5	0.81
$(\mathcal{L}_f)+(\mathcal{L}_c)+ (\mathcal{L}_{sc})$	25.9	0.85

Ablation Study. We conduct an ablation study to examine the contributions of individual components in our loss function. Table 3 presents the results of removing specific components and their impact on captioning performance.

The ablation results confirm that each loss component plays a crucial role in improving the overall captioning performance. The Captioning Loss $(\mathcal{L}_c)$ and Feature Loss $(\mathcal{L}_f)$ contribute significantly to the quality of the generated captions.

Qualitative Analysis. Figure 5 showcases a qualitative comparison of retrieved images under various weather conditions. The proposed framework retrieves images with high visual and semantic similarity, even in challenging weather conditions, compared to the baseline methods.

Table 3. Ablation Study on loss Captioning Performance (BLEU).

Configuration	BLEU
Without Adaptive Restoration Loss ($\mathcal{L}_{ar}$)	37
Without Captioning Loss ($\mathcal{L}_c$)	34
Without Semantic Consistency ($\mathcal{L}_{sc}$)	33
Without Feature Loss ($\mathcal{L}_f$)	32

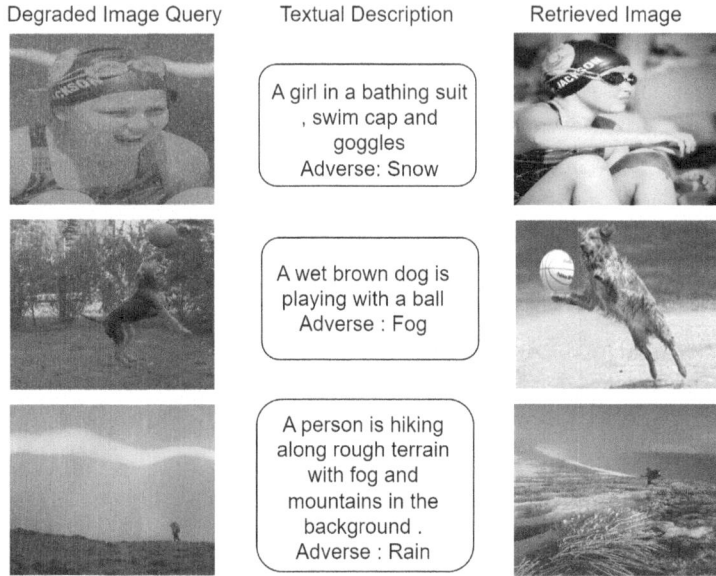

Fig. 5. The proposed model retrieves images that better match both the visual content and the captions under adverse weather conditions.

4.2   Discussion

The experimental results demonstrate the effectiveness of our network. The proposed framework significantly improves image quality by effectively mitigating weather-induced distortions. The high PSNR and SSIM values reflect this. Including the Caption Feedback Mechanism and maintaining semantic consistency in the loss function improves image retrieval quality. The model successfully retrieves semantically relevant images even in challenging weather scenarios. The ablation study confirms the importance of the key components of our loss function, highlighting the importance of feature and caption loss in achieving optimal performance.

Our framework integrates image restoration and deep learning-based image captioning to enhance image retrieval in adverse weather conditions. This methodology improves both image clarity and semantic comprehension, which leads to more accurate and meaningful retrieval results.

5   Conclusion

We proposed a novel framework for robust image retrieval in adverse weather conditions, integrating image restoration and caption generation. By utilizing weather-degraded and normal image pairs, our approach restores images affected by conditions such as rain, snow, and fog while simultaneously generating accurate captions that preserve semantic consistency. Through extensive experiments on the Flickr8k dataset, we demonstrated that our model significantly enhances image retrieval performance in weather-affected images. The restoration component of the framework effectively mitigates weather distortions, while the transformer-based captioning model ensures that the generated captions align with the visual content. The combination of these two processes enables more accurate and semantically relevant image retrieval.

Future work will focus on fine-tuning the model for specific weather conditions and expanding the framework to larger and sequential event-based datasets for better generalization. Additionally, exploring weakly supervised learning could enhance the system's efficiency and applicability across various domains.

References

1. Anderson, P., et al.: Bottom-up and top-down attention for image captioning and visual question answering. In: Proceedings of the IEEE conference on computer vision and pattern recognition, pp. 6077–6086 (2018)
2. Bai, Y., Wang, C., Xie, S., Dong, C., Yuan, C., Wang, Z.: TextIR: a simple framework for text-based editable image restoration. arXiv preprint arXiv:2302.14736 (2023)
3. Cai, B., Xu, X., Jia, K., Qing, C., Tao, D.: DehazeNet: an end-to-end system for single image haze removal. IEEE Trans. Image Process. **25**(11), 5187–5198 (2016)
4. Chen, H., et al.: Pre-trained image processing transformer. In: Proceedings of the IEEE/CVF Conference on Computer Vision and Pattern Recognition, pp. 12299–12310 (2021)
5. Cui, G., Ma, Q., Zhao, J., Yang, S., Chen, Z.: Image dehazing algorithm based on optimized dark channel and haze-line priors of adaptive sky segmentation. JOSA A **40**(6), 1165–1182 (2023)
6. Dosovitskiy, A.: An image is worth 16×16 words: transformers for image recognition at scale. arXiv preprint arXiv:2010.11929 (2020)
7. Fu, X., Huang, J., Zeng, D., Huang, Y., Ding, X., Paisley, J.: Removing rain from single images via a deep detail network. In: Proceedings of the IEEE Conference on Computer Vision and Pattern Recognition, pp. 3855–3863 (2017)
8. Gkelios, S., Boutalis, Y., Chatzichristofis, S.A.: Investigating the vision transformer model for image retrieval tasks. In: 2021 17th International Conference on Distributed Computing in Sensor Systems (DCOSS), pp. 367–373. IEEE (2021)
9. Gupta, H., Kotlyar, O., Andreasson, H., Lilienthal, A.J.: Robust object detection in challenging weather conditions. In: Proceedings of the IEEE/CVF Winter Conference on Applications of Computer Vision, pp. 7523–7532 (2024)
10. Gupta, N., Jalal, A.S.: Integration of textual cues for fine-grained image captioning using deep CNN and LSTM. Neural Comput. Appl. **32**(24), 17899–17908 (2020)

11. He, K., Sun, J., Tang, X.: Single image haze removal using dark channel prior. IEEE Trans. Pattern Anal. Mach. Intell. **33**(12), 2341–2353 (2010)
12. He, X., Jia, T., Li, J.: Learning degradation-aware visual prompt for maritime image restoration under adverse weather conditions. Front. Mar. Sci. **11**, 1382147 (2024)
13. Hodges, C., Bennamoun, M., Rahmani, H.: Single image dehazing using deep neural networks. Pattern Recogn. Lett. **128**, 70–77 (2019)
14. Jiang, Y., Chan, K.C., Wang, X., Loy, C.C., Liu, Z.: Robust reference-based super-resolution via C2-matching. In: Proceedings of the IEEE/CVF Conference on Computer Vision and Pattern Recognition, pp. 2103–2112 (2021)
15. Karpathy, A., Fei-Fei, L.: Deep visual-semantic alignments for generating image descriptions. In: Proceedings of the IEEE Conference on Computer Vision and Pattern Recognition, pp. 3128–3137 (2015)
16. Li, B., Liu, X., Hu, P., Wu, Z., Lv, J., Peng, X.: All-in-one image restoration for unknown corruption. In: Proceedings of the IEEE/CVF Conference on Computer Vision and Pattern Recognition, pp. 17452–17462 (2022)
17. Li, R., Cheong, L.F., Tan, R.T.: Heavy rain image restoration: Integrating physics model and conditional adversarial learning. In: Proceedings of the IEEE/CVF Conference on Computer Vision and Pattern Recognition, pp. 1633–1642 (2019)
18. Liang, J., Cao, J., Sun, G., Zhang, K., Van Gool, L., Timofte, R.: SwinIR: image restoration using Swin transformer. In: Proceedings of the IEEE/CVF International Conference on Computer Vision, pp. 1833–1844 (2021)
19. Lin, X., et al.: DiffbIR: towards blind image restoration with generative diffusion prior. arXiv preprint arXiv:2308.15070 (2023)
20. Liu, Z., et al.: Swin transformer: hierarchical vision transformer using shifted windows. In: Proceedings of the IEEE/CVF International Conference on Computer Vision, pp. 10012–10022 (2021)
21. Liu, Z., et al.: Retrieval of hurricane rain rate from SAR images based on artificial neural network. IEEE J. Sel. Top. Appl. Earth Obser. Remote Sen. **17**, 15067–15076 (2024)
22. Potlapalli, V., Zamir, S.W., Khan, S.H., Shahbaz Khan, F.: PromptIR: prompting for all-in-one image restoration. In: Advances in Neural Information Processing Systems, vol. 36 (2024)
23. Ruan, W., Tao, Y., Ruan, L., Shu, X., Qiao, Y.: Temporal weighting appearance-aligned network for nighttime video retrieval. IEEE Signal Process. Lett. **29**, 2008–2012 (2022)
24. Vinyals, O., Toshev, A., Bengio, S., Erhan, D.: Show and tell: a neural image caption generator. In: Proceedings of the IEEE Conference on Computer Vision and Pattern Recognition, pp. 3156–3164 (2015)
25. Wang, Y., Li, Y., Wang, G., Liu, X.: Multi-scale attention network for single image super-resolution. In: Proceedings of the IEEE/CVF Conference on Computer Vision and Pattern Recognition, pp. 5950–5960 (2024)
26. Warburg, F., Miani, M., Brack, S., Hauberg, S.: Bayesian metric learning for uncertainty quantification in image retrieval. In: Advances in Neural Information Processing Systems, vol. 36 (2024)
27. Xu, K., et al.: Show, attend and tell: Neural image caption generation with visual attention. In: International Conference on Machine Learning, pp. 2048–2057. PMLR (2015)
28. Zamir, S.W., Arora, A., Khan, S., Hayat, M., Khan, F.S., Yang, M.H.: Restormer: efficient transformer for high-resolution image restoration. In: Proceedings of the

IEEE/CVF Conference on Computer Vision and Pattern Recognition, pp. 5728–5739 (2022)

29. Zamir, S.W., et al.: Learning enriched features for real image restoration and enhancement. In: Vedaldi, A., Bischof, H., Brox, T., Frahm, J.-M. (eds.) ECCV 2020. LNCS, vol. 12370, pp. 492–511. Springer, Cham (2020). https://doi.org/10.1007/978-3-030-58595-2_30

30. Zhang, H., Patel, V.M.: Densely connected pyramid dehazing network. In: Proceedings of the IEEE Conference on Computer Vision and Pattern Recognition, pp. 3194–3203 (2018)

31. Zhang, J., et al.: Ingredient-oriented multi-degradation learning for image restoration. In: Proceedings of the IEEE/CVF Conference on Computer Vision and Pattern Recognition, pp. 5825–5835 (2023)

32. Zhang, K., Zuo, W., Chen, Y., Meng, D., Zhang, L.: Beyond a gaussian denoiser: residual learning of deep CNN for image denoising. IEEE Trans. Image Process. **26**(7), 3142–3155 (2017)

33. Zhang, Z., et al.: NTIRE 2024 challenge on bracketing image restoration and enhancement: Datasets methods and results. In: Proceedings of the IEEE/CVF Conference on Computer Vision and Pattern Recognition, pp. 6153–6166 (2024)

34. Zheng, L., Yang, Y., Tian, Q.: Sift meets CNN: a decade survey of instance retrieval. IEEE Trans. Pattern Anal. Mach. Intell. **40**(5), 1224–1244 (2017)

RW-SVD: A Surround View Rough Weather Video Anomaly Dataset and a Brief Overview of Existing Datasets

Sachin Dube$^{(\boxtimes)}$, Dinesh Kumar Tyagi, and Ramesh Babu Battula

Malaviya National Institute of Technology Jaipur, Jaipur 302017, India
sachin.rdubey.2011@gmail.com, {dktyagi.cse,rbbattula.cse}@mnit.ac.in

Abstract. Modern surveillance societies constantly face bottlenecks due to manual monitoring of huge amounts of data generated by surveillance infrastructure. The limitation of manual monitoring is further aggravated by challenging weather conditions such as fog, rain, mist, etc. This gave rise to automated surveillance making Video anomaly detection (VAD) one of the most sought-after domains in computer vision. The availability of data that contain weather-induced variations is a key factor in the effectiveness of Data-driven approaches that rely on data for precise modeling. To this end, we have presented a brief review of previous datasets and their limitations on parameters such as size, scene variations, activities covered, effect of weather phenomena, etc. To leverage the intricate relationship between data and model we present a novel human-centric surround view dataset where each scripted activity is recorded simultaneously by 4 strategically placed cameras to capture effects of varying distance, angle, height, and illumination on the same scene. The proposed dataset is arranged into 4 abnormal classes namely fighting, snatching, panic running, and kidnapping. It contains complex backgrounds, real-life objects (cycle, motorbike, four-wheeler), both indoor and outdoor environments as well as illumination change. To tackle ambiguity during the transition from normal to abnormal or vice-versa we conducted voting (subjective evaluation) with 10 volunteers. We further augmented the dataset with two of the most common weather phenomena namely haze and rain to bridge the gap between real-world challenges and dataset.

Keywords: Anomaly Detection · Database · RW-SVD

1 Introduction

Video anomaly detection (VAD) refers to autonomous identification as well as localisation of abnormal events in a video stream. It reduces dependency on manpower thereby increasing efficiency and effectiveness. It has become a necessary tool to realise the true potential of huge amounts of data collected from various sources. It forms the basis of various vision-based applications such as crime and violence detection [1,2], traffic monitoring and management [3], Accident detection [4], intelligent surveillance systems [5,6], disaster management

© The Author(s), under exclusive license to Springer Nature Singapore Pte Ltd. 2025
M. Cho et al. (Eds.): ACCV 2024 Workshops, LNCS 15482, pp. 146–162, 2025.
https://doi.org/10.1007/978-981-96-2641-0_10

[7] etc. Data distribution happens to be core of various learning-based VAD approaches that can be further divided into 4 categories: *1.supervised learning:* based approaches [8,9] require well-defined labels and are often trained on small single scene datasets such as UCSD [10,11], UMN [12], Subway [13] etc. *2.unsupervised learning:* based approaches [14,15] are generally powered by large scale datasets such as UCF-Crime [16] where, frame level labels are not feasible. These approaches usually try to build a latent representation of normality only from normal data and completely ignore abnormal data. *3.weakly supervised learning* : based approaches [17,18] act as a bridge between the above two approaches and uses large-scale datasets such as XD-Violence [19]. Instead of frame; a label is assigned to videos. However, it may be possible for a video to contain multiple snippets that can be normal or anomalous, that anomaly decision criterion happens to be a field of active research.

Weather-affected data has seen constant interest in traffic management, road safety, and autonomous driving domains due to apparent reasons. many datasets such as X-MAN [20], AI city challenge [21], Dawn and Dusk [22] etc. have been proposed to cater to that need. However, this trend is being picked up in the anomaly detection domain but the availability of data, especially human-centric data remains a challenge. Apart from this very specific challenge issues of data imbalance, multi-view real-world data, appropriate labels, ambiguous labels, etc. remain. To this end, we propose a human-centric surround view dataset that is augmented to reflect weather-induced challenges. We summarise our contribution through the following points:

- A brief review of most sought-after benchmark datasets. The review includes a mix of single-view, multi-view, large and small scale as well as weather-specific datasets. Information like technical details, strengths, and weaknesses is also provided
- A human-centric surround-view video anomaly dataset that contains the effect of rough weather conditions like haze and rain is proposed. It is obtained by augmenting previously proposed Anovil [23] by using statistical methods after fine-tuning acc. to data.
- Some samples of neutral weather, haze, and rain-affected frames for multiple combinations like normal-abnormal, indoor-outdoor, and different activities are presented along with frame-wise quantitative data.

The rest of the paper is divided into 3 sections. Section 2 gives a brief review of various datasets while the proposed dataset and augmentation technique are described in Sect. 3. It also contains snippets for subjective evaluation. Section 4 concludes the paper.

2 Review of Existing Benchmark Datasets

Here we present a brief overview of the most commonly used benchmark datasets by dividing them into three categories namely: single-view datasets, multiple-view datasets, and weather-related datasets. We further discuss recently proposed weather-centric datasets.

2.1   Single Scene Datasets

Car   Biker   Wheel chair

Fig. 1. Abnormal snippets from UCSD- Dataset

UCSD pedestrian Dataset for Anomaly Detection [10,11]*:* consists of a grayscale video clip of a crowded walkway. It is shot by a static camera under two different scenes: Peds1 and Peds2 as shown in Fig. 1. Pedestrians walking on the path are considered to be normal and the movement of cyclists, skaters, bikers, carts, wheelchairs, etc. in the pathway is considered to be anomalous. Even the movement of pedestrians outside the designated pathway (lawn) is also considered to be anomalous. Peds 1 includes a total of 70 video clips ranging from 5 to 10 s (200–400 frames). The training set consists of 34 video clips while the test set consists of 36 video clips. Peds 2 dataset is even smaller and contains 28 video clips ranging from 2–4 seconds (120–180 frames). The training set consists of 16 video clips while the test set consists of 12 video samples. Both datasets contain annotations at both pixel and frame level. Hence, it facilitates the evaluation of the localization performance of SOTA methods.

UMN Dataset [12]*:* is primarily designed to capture crowd dynamics and includes panic-driven situations such as sudden transitions from walking to running. It contains about 8000 frames (320×240) from 3 scenarios namely: lawn, plaza, and indoor. It is developed around only two events: pedestrians walking and panic running while trying to escape. The normal scene is characterised by individuals casually walking in various directions, while the abnormal scene contains individuals running in disarray. These videos are captured in varying scenes and environments ensuring diversity in motion patterns.

PETS Dataset [24]: pets 2009 dataset is primarily focused on video surveillance applications such as human and vehicular motion tracking, crowd dynamics, etc. It features real-world video clips captured by multiple synchronised cameras at 576×768. this helps with 3D- tracking, occlusion, and re-identification problems. Captured events include; pedestrians walking, moving vehicles, crowd movement, etc. The dataset is meticulously annotated with bounding boxes around people, and vehicles along with corresponding labels. Major challenges presented by the dataset are illumination and environmental variations, occlusion, and crowd variability. The pets2010 dataset is an extension of the pets2009 datasets and includes more complex scenarios such as loitering, unattended objects, aggressive behaviour, etc.

Subway Dataset [13]: It is a real-world dataset shot at a relatively low resolution and longer duration. It contains two video sequences: *1) subway entrance:* focuses on people entering the subway station. It is 96 min long (144249 frames) and contains 66 abnormal activities such as individuals walking in the wrong direction, abrupt Running, and stopping. *2 subway exit:* captures people leaving the subway station. It is 43 min long (64900 frames) and encompasses 19 types of anomalous events, such as loitering near the exit. Major challenges posed by the dataset include occlusion, illumination variation, and abrupt motion.

CUHK Avenue Dataset [25]: is shot to capture events at the CUHK Campus to represent the crowded urban environment. It consists of 30 min of footage at 25 FPS with 640×360 resolution. The training set encompasses 16 training videos containing only normal events while the test set consists of 21 video clips that include both normal as well as anomalous events. Throwing bags, sudden running, stopping, or changing direction is considered to be an anomaly. To further increase challenges, camera shake is introduced deliberately in test video frames. The dataset is temporally and spatially annotated.

UBnormal Dataset [8]: is a software-generated synthetic dataset (30 FPS) that offers diverse real-world scenarios in a controlled environment to overcome issues with real-world videos such as privacy concerns, inconsistent annotations, etc. As it is computer generated it offers high-resolution videos with consistency. The dataset is split into training (268), validation (64), and testing (211) sets. It provides frame-level as well as video-level annotations especially abnormal events annotated at pixel-level making it suitable for fully supervised methods also. Disjoint sets of anomalies are maintained in both: training and test sets to facilitate open-set formulation.

UBI-Fights dataset [26]: is focused on fighting and contains diverse real-life scenarios in indoor-outdoor, coloured-grayscale, fixed-movable cameras with varied orientations. Irrelevant frames that could hinder the learning process are removed manually and the dataset is resized to 640×360 pixels at 30 FPS. It contains a total of 1000 videos amounting to 80 h of playtime. However major chunk of the dataset is made up of normal videos (784) while the number of abnormal videos is less than 1/4th of the entire dataset (216). More than half of videos are less than 2 min while 98 videos have lengths of more than 10 min

Fig. 2. Snippets from MSAD dataset showing diversity and challenges offered by dataset [27].

MSAD (Multi-Scenario Anomaly Detection) dataset [27]*:* is a multi-view, multi-scenario dataset that contains a total of 720 videos of real-life scenarios captured by surveillance cameras. It contains 55 anomaly types: 20 non-human related and 35 Human related. Variations such as indoor-outdoor, effect of weather, multiple camera views, illumination variation as well diverse motion patterns present major challenges as shown in Fig. 2. Frame-level annotations are provided.

NTU CCTV-Fights Dataset [28]*:* contains 1000 videos ranging from 5 to 720 s with a total runtime of well over 17 h. These videos were recorded from CCTV, mobile, or dashboard cameras and were collected from YouTube. There are a total of 2414 annotated fight instances where the video contains at least one fight instance from the following categories: Pushing, punching, kicking, and wrestling involving two or more individuals. Frame-level annotation along with the starting and end points of each fighting instance are provided.

DoTA dataset [29]*:* contains 4677 videos focused on traffic anomaly in terms of three aspects: 1) **When** represents temporal aspects as it includes videos recorded at different times of day, leading to diverse traffic conditions (ex. Peak traffic in Rush hours). 2) **Where** represents geographic locations such as intersections, highways, and urban areas. 3) **What** represents types of traffic anomalies, such as Accident, Traffic violations, Blockage, and Pedestrian-related incidents These anomalies are annotated according to category allowing frame and video level detection. It includes challenges presented by multiple viewpoints, and weather conditions such as rain, fog, low visibility, etc. It finds primary application in the area of autonomous driving and intelligence surveillance systems.

D^2-**City Dataset** [30]: is a collection of 10000 dashboard camera videos shot at 720 and 1080 P and is primarily dedicated to the development of ADAS and intelligent driving systems. It includes bounding boxes for 1000 videos along with tracking information. These annotations cover 12 types of objects ranging from pedestrians, bicycles, tricycles, cars, buses, etc.

QMUL Junction Dataset [31]*:* focuses on traffic scene analysis and is shot at a busy public road regulated by traffic lights. It includes two sets of videos shot at 25 FPS and 360×288 resolution. The first one is an hour long (90,000 frames) while the second one is 52 min long (78,000 frames). The major challenge comes from complex and diverse interactions between vehicles and pedestrians, variable flow of traffic along with illumination variations, shadows, and noise. Traffic violations such as Traffic interruption, illegal u-turns, stoppages, etc. constitute anomalous samples. However, it lacks formal partitioning into training and test sets.

Street Scene Dataset [32]*:* The Street Scene dataset presents urban environments, focusing on capturing scenes from city streets. The training set contains 46 (56,847 frames) videos and the test set contains 55 (146,410 frames) videos shot at 1280 × 720 at 15 fps. Videos typically include a range of elements such as vehicles, pedestrians, buildings, traffic signs, and other street furniture. It contains 205 anomalous events and presents challenges like varied lighting conditions, weather, and urban layouts, making it valuable for developing robust AI systems that can interpret complex street environments.

CAVIAR dataset (Context-aware vision using image-based active recognition) [33]: is a publicly available dataset that can be used for pedestrian detection and tracking, anomaly detection, and activity recognition. It encompasses two scenarios shot at 25 FPS and 384 × 288 resolution at INIRA lab France and a shopping complex in Lisbon. It contains several scenarios such as walking, group interactions, Entering and leaving shops, window shopping, etc., and unusual activities like panic running, fighting, loitering, abandoning a package, etc. It has frame-level annotations of the activities mentioned above.

2.2 Multi-scene Datasets

BEHAVE Dataset [34]: focuses on crime-oriented behaviour. It contains 4 video clips of staged scenarios recorded in a controlled outdoor environment at 25 FPS and 640 × 480 pixels by a static camera. Fighting, chasing, and running together form the major portion of anomalous instances. It includes detailed annotation by drawing a bounding box for each interacting individual. It includes individual as well as group activities. However, it also lacks formal portioning into train and test sets.

Web Dataset [35]: is a crowd oriented dataset collected from various websites. It contains videos of crowds in various urban scenarios. It includes a total of 20 high-quality documentary videos. 12 videos contain normal scenes like walking, and running, while the anomalous sample contains 8 videos with activities like panic escape, fighting and protesters clashing.

MIT Traffic Dataset [36]: is a 90-minute long video (20 clips) shot by a static camera at 30 FPS and 720 × 480 resolution to capture traffic activity. The scene contains multiple vehicles and pedestrians with a less dense but more erratic and unorganised flow of traffic. Ground truth annotations are provided from some frames focused on pedestrian detection.

UCF Crime [16]: dataset is a collection of 1900 variable length untrimmed surveillance footage of real-world incidents exceeding playtime of 128 h. It is collected from various social media platforms and is further divided into 13 types of anomalous (criminal) events namely: Arrest, Abuse, Assault, Arson, Burglary, Fighting, Explosion, Robbery, Stealing, Shoplifting, Vandalism, Shooting, Road Accidents apart from Normal videos. It has a balanced distribution (950 normal and 950 abnormal videos) and is weakly labeled, i.e. each video is tagged with one type of anomaly without precise temporal annotations, posing a significant challenge for models to detect and localise anomalies. It includes variations in lighting, camera angles, background, and occlusions that are typical in real-world scenarios. The training set consists of 1610 videos while the Test set is made of 290 videos (150 abnormal and 140 abnormal).

XD-Violence Dataset [19]: is a large-scale, multimodal (includes aural and visual information), multi-scene dataset collected from movies, news reports, sports streaming, surveillance videos, etc. It's a comparatively balanced dataset that consists of 4,754 (2349 normal, 2405 abnormal) untrimmed videos totaling more than 217 h of footage, and covers 6 types of violent activities namely: Fighting, Rioting, Arson, Explosions, Abuse, and Shooting. It is also weakly supervised; however, it contains video-level annotation for the training set but frame-level annotations for the test set. The training set is made up of 3954 videos (2049 normal, 1905 abnormal) while the test set is made of 800 videos (300 normal, 500 abnormal). Currently, it happens to be the most comprehensive and voluminous public dataset for real-world VAD detection and introduces a range of challenges, such as varying camera angles, lighting conditions, and background noise.

2.3   Weather Related Datasets

While datasets shot outdoors are prone to weather phenomena and have a few weather-induced variations e.g. UCF-Crime [19], street scene [32], CUHK-Avenue [25] etc. yet, but to cater to research in weather-sensitive application few dedicated datasets are proposed as follows.

X-MAN: (Extreme Weather Anomaly Dataset) [20]: is focused on video Anomalies under extreme weather conditions such as heavy rain, snow, fog, etc. It directly addresses how extreme weather affects the visibility and detection of anomalies.

AI City Challenge Dataset (Track 3: Anomaly Detection) [21]: focuses on Traffic and urban anomaly detection and includes weather conditions like rain, fog, and cloudy weather. It explicitly claims to include weather variability, with the goal of simulating real-world driving and traffic anomalies that occur under different weather conditions.

Dawn and Dusk Dataset [22]: focuses on Anomaly detection in driving scenarios for autonomous driving. It includes challenging lighting conditions; Under the influence of fog, and rain. It also includes low-visibility scenarios in addition to varying light conditions such as dawn and dusk.

While the above datasets cater to the need for intelligent transport systems and ADAS-based applications There is no dedicated dataset for human-centric anomaly detection in urban settings under the influence of weather. We aim to fill this gap with RW-SVD.

3   Proposed Dataset RW-SVD

Information conveyed by visual data i.e. images and videos is highly dependent on various environmental factors as well as on composition. Change of perspective/viewpoint can change context; which, in turn, influences the interpretation of events recorded in a scene. We leverage this property and propose a Human-Centric surround-view dataset that captures a particular event of interest simultaneously with 4 different cameras placed at varying angles, heights, and distances from the subject as shown in Fig. 4.

Feeds received from different cameras show drastic variations in visual information due to changes in viewpoint, illumination, size, and background as shown

Fig. 3. Abnormal snippets from normal weather scenarios showing diverse backgrounds, viewpoints, illumination conditions, and action classes

Fig. 4. Shows two subjects fighting are simultaneously being recorded by 4 different cameras C1, C2, C3, and C4. They are placed at different heights h1,h2,h3, and h4 from the ground (hence subjects) as well as different distances d1,d2,d3, and d4 from the subjects respectively.

in Fig. 5. The salient features of the proposed dataset can be summarized as follows:

Fig. 5. Demonstrates the significance of angle and viewpoint while capturing the same scene. To further highlight the cumulative effect of height, view-point, and distance; symbolic frames are given in ovals

- Shot on four strategically placed cameras simultaneously to capture multiple views of the same scene to cover viewpoint changes shown in Fig. 6
- Shot on six different locations like a park, indoor/outdoor courts, hallway, road, and lift at a different time of day.
- Shot from varied distances and heights to capture subjects with varying size
- Scenes involve instances of kidnapping, snatching, Fighting, and panic running along with normal scenes during activities like strolling, playing, cycling, etc.
- Scenes were shot indoors/outdoors, in natural light-artificial light, and low light as well to cover illumination changes.

Figure 6 presents snippets of the same incident taken from different cameras to highlight the difference in visual information. The difference is so drastic that an anomalous event may look normal from a certain angle or vice versa. The wrong perspective can be deceiving at times. This combined with the transition from normality to abnormality or vice versa may lead to ambiguity and degradation in the performance of the model. To this end, we conducted voting for subjective assessment and labeling with the help of 10 volunteers. Figure 3 presents snippets of abnormal scenarios from proposed RW-SVD under normal weather conditions. They highlight the diversity encompassed by the dataset in

Fig. 6. Snippet from surround view dataset in normal weather of the same scene. It clearly presents the effect of viewpoint, distance, height, and background variation while keeping the same core activity and subjects.

terms of Spatiotemporal information, the effect of viewpoint and distance, Illumination variation, Diverse motion patterns, subjects, etc. under normal weather conditions. Frame-level annotations as normal or abnormal are provided for each video. The dataset contains a total of 101 videos with a good balance between normal (49) and abnormal videos (52). The training set consists of 53 videos (24 normal and 29 abnormal) while the test set contains 47 videos (25 normal and 22 abnormal). Hence, we have maintained a balance between normal and abnormal data as well as training and testing data. In extending our dataset to incorporate weather noise we artificially add haze and rain effects in video frames. The addition of weather noise like haze and rain makes our dataset more practical for real-world scenarios where environmental noises are more frequent. We have thoroughly extended our previously proposed AnoVIL [23] dataset by incorporating the needs of recent computer vision applications that are sensitive to weather conditions like haze and rain. So, we have meticulously augmented AnoVIL [23] to depict frames in rainy and hazy weather [37–43]. We have introduced haze and rain with a procedural image generation method with randomness and geometric principles. Statistical Augmentation is performed to incorporate effects in visibility due to haze and rain as follows:

3.1   Haze Generation Model

For haze addition we utilize the given methods to incorporate haze.

$$I(x) = J(x) \cdot t(x) + A \cdot (1 - t(x))$$

where: $I(x)$ stands for the observed hazy image, $J(x)$ stands for the scene radiance (haze-free image), $t(x)$ stands for the transmission map, A stands for the global atmospheric light.

Transmission Map (Haze Map) Generation. The transmission map $H(x)$ is generated using a random gradient based on image coordinates, which is given as:

$$H(x) = \alpha_1 \cdot xv + \alpha_2 \cdot yv \tag{1}$$

where: xv and yv are 2D gradient fields based on the image coordinates, α_1 and α_2 are random weights controlling the gradient's direction and intensity.

Fig. 7. Heavy haze conditions contained in proposed dataset both in indoor and outdoor conditions in case of normal activities

Haze Output. The final hazy image is computed as follows:

$$I_{hazy}(x) = I(x) \cdot (1 - H(x)) + 255 \cdot H(x)$$

where: 255 represents the white haze layer.
    Haze output can be seen in Figs. 7,8

3.2   Rain Generation Model

For rain generation, we introduce rain sticks by calculating their positions and orientations. Gaussian blurring is applied to enhance the raindrop streaks.

Raindrop Statistics. To generate a raindrop, we compute its endpoint based on its length and orientation angle:

$$x_{i+1} = x_i + L \cdot \cos(\theta)$$

$$y_{i+1} = y_i + L \cdot \sin(\theta)$$

where: (x_{i+1}, y_{i+1}) is the endpoint of the raindrop, L is the length of the raindrop, θ is the angle of orientation.

Fig. 8. Heavy haze conditions contained in proposed dataset both in indoor and outdoor conditions in case of 4 different types of abnormal activities with different background, scene density, and rate of motion

Fig. 9. Effect rain of in case of normal outdoor videos

Raindrop Streaks Generation with Gaussian Blur. To simulate the appearance of raindrops, we apply Gaussian blur, which is mathematically represented as:

$$G(x) = \frac{1}{\sqrt{2\pi\sigma^2}} e^{-\frac{x^2}{2\sigma^2}}$$

where σ is the standard deviation that controls the spread of the raindrop streaks.

Rainy Output. The final rainy image is computed by blending the original image with the raindrop layer:

$$I_{rainy}(x) = I(x) \cdot (1 - \beta) + R(x) \cdot \beta$$

Fig. 10. Effect of rain in case of abnormal outdoor videos.

where: $I(x)$ stands for the original image, $R(x)$ stands for the generated raindrop layer, β stands for the blending factor that controls the intensity of the rain effect.

Rainy output can be seen in Figs. 9,10

Videos were shot at 1080P however resolution was lowered to 320X240 to meet computation constraints. RW-SVD covers various routine activities like walking, playing, running, and playing as normal scenarios. Whereas, fighting, panic running, and snatching events are recorded as abnormal events. To tackle ambiguity in similar-looking actions, we have recorded the same scenario from 4 different perspectives. Frequently used objects such as bags, motorbikes, cycles, and cars are also included. Detailed frame-wise description is given in Table 1.

Table 1. Details of AnoVIL: No. of frames for 3 categories: normal weather, haze, and rain with Training, Testing set for Anomaly and Normal video frames are given below

Weather	Action	Training Frames	TestingFrames	Total
Normal Weather	Fighting	7090	4927	12017
	Kidnapping	1339	1014	2353
	Snatching	595	938	1533
	Panic Running	746	831	1577
	Normal	32855	60270	93125
Fogg	Fighting	7090	4927	12017
	Kidnapping	1339	1014	2353
	Snatching	595	938	1533
	Panic Running	746	831	1577
	Normal	32855	60270	93125
Rain	Fighting	6914	4837	11751
	Kidnapping	1339	1014	2353
	Snatching	595	938	1533
	Panic Running	746	831	1577
	Normal	29798	58213	88011
	Total	124642	201793	326435

4   Conclusion

We have developed a dataset that fills existing gaps such as the effect of weather; towards advancing the field. The proposed dataset focuses on human-centric anomalies through routine activities like walking, playing, and running as normal scenarios, and abnormal events such as fighting, panic running, kidnapping, and snatching. To address the challenges posed by varying viewpoints, the same scenarios are recorded from four different angles, ensuring comprehensive coverage of different perspectives. The dataset includes 49 normal and 52 abnormal videos, divided into training and evaluation sets, providing a robust benchmark for human-centric anomaly detection in video data.

References

1. Ke, X., Sun, T., Jiang, X.: Video anomaly detection and localization based on an adaptive intra-frame classification network. IEEE Trans. Multimedia **22**(2), 394–406 (2019)
2. Chang, S., Li, Y., Shen, S., Feng, J., Zhou, Z.: Contrastive attention for video anomaly detection. IEEE Trans. Multimedia **24**, 4067–4076 (2021)
3. Zaheer, M.Z., Mahmood, A., Khan, M.H., Segu, M., Yu, F., Lee, S.I.: Generative cooperative learning for unsupervised video anomaly detection. In: Proceedings of the IEEE/CVF Conference on Computer Vision and Pattern Recognition, pp. 14744–14754 (2022)

4. Pramanik, A., Sarkar, S., Maiti, J.: A real-time video surveillance system for traffic pre-events detection. Accident Anal. Prev. **154**, 106019 (2021)
5. Xia, K., Wang, L., Shen, Y., Zhou, S., Hua, G., Tang, W.: Exploring action centers for temporal action localization. IEEE Trans. Multimedia **25**, 9425–9436 (2023)
6. Zhai, Y., Wang, L., Tang, W., Zhang, Q., Zheng, N., Hua, G.: Action coherence network for weakly-supervised temporal action localization. IEEE Trans. Multimedia **24**, 1857–1870 (2021)
7. Maza, I., Caballero, F., Capitán, J., Dios, J.R.M., Ollero, A.: Experimental results in multi-UAV coordination for disaster management and civil security applications. J. Intell. Robot. Syst. **61**, 563–585 (2011)
8. Acsintoae, A., et al.: UbNormal: new benchmark for supervised open-set video anomaly detection. In: Proceedings of the IEEE/CVF Conference on Computer Vision and Pattern Recognition, pp. 20143–20153 (2022)
9. Biradar, K., Dube, S., Vipparthi, S.K.: DEAREST: deep convolutional aberrant behavior detection in real-world scenarios. In: 2018 IEEE 13th International Conference on Industrial and Information Systems (ICIIS), pp. 163–167. IEEE (2018)
10. Li, W., Mahadevan, V., Vasconcelos, N.: Anomaly detection and localization in crowded scenes. IEEE Trans. Pattern Anal. Mach. Intell. **36**(1), 18–32 (2013)
11. Mahadevanv, L., et al.: Anomaly detection in crowded scenes. In: 2010 IEEE Computer Society Conference on Computer Vision and Pattern Recognition (2010)
12. Unusual crowd activity dataset of university of minnesota (2006). http://mha.cs.umn.edu/proj_events.shtml/. Accessed 14 Sept 2024
13. Adam, A., Rivlin, E., Shimshoni, I., Reinitz, D.: Robust real-time unusual event detection using multiple fixed-location monitors. IEEE Trans. Pattern Anal. Mach. Intell. **30**(3), 555–560 (2008)
14. Dube, S., Biradar, K., Vipparthi, S.K., Tyagi, D.K.: MAG-Net: a memory augmented generative framework for video anomaly detection using extrapolation. In: International Conference on Computer Vision and Image Processing, pp. 426–437. Springer (2021). https://doi.org/10.1007/978-3-031-11349-9_37
15. Andrews, S., Tsochantaridis, I., Hofmann, T.: Support vector machines for multiple-instance learning. In: Advances in Neural Information Processing Systems, vol. 15 (2002)
16. Sultani, W., Chen, C., Shah, M.: Real-world anomaly detection in surveillance videos. In: Proceedings of the IEEE Conference on Computer Vision and Pattern Recognition, pp. 6479–6488 (2018)
17. Fan, Y., Yu, Y., Lu, W., Han, Y.: Weakly-supervised video anomaly detection with snippet anomalous attention. IEEE Trans. Circ. Syst. Video Technol. **34**, 5480–5492 (2024)
18. Feng, J.C., Hong, F.T., Zheng, W.S.: Mist: multiple instance self-training framework for video anomaly detection. In: Proceedings of the IEEE/CVF Conference on Computer Vision and Pattern Recognition, pp. 14009–14018 (2021)
19. Wu, P., et al.: Not only look, but also listen: learning multimodal violence detection under weak supervision. In: Vedaldi, A., Bischof, H., Brox, T., Frahm, J.-M. (eds.) ECCV 2020. LNCS, vol. 12375, pp. 322–339. Springer, Cham (2020). https://doi.org/10.1007/978-3-030-58577-8_20
20. Zhang, Y., Li, J., Liu, X.: X-man: a dataset for anomaly detection in extreme weather conditions. In: Proceedings of the IEEE/CVF Conference on Computer Vision and Pattern Recognition (CVPR) Workshops, pp. 1234–1243 (2022)
21. Yu, H., Li, Y., Ma, Y., Todd, A.: The 4th AI city challenge. In: Proceedings of the IEEE/CVF Conference on Computer Vision and Pattern Recognition Workshops (CVPRW), pp. 229–239 (2020)

22. Harbawee, L.A.: Artificial intelligence tools for facial expression analysis. PhD thesis, University of Exeter (United Kingdom) (2019)
23. Biradar, K.M., et al.: Triplet-set feature proximity learning for video anomaly detection. Image Vis. Comput. **150**, 105205 (2024)
24. Ellis, A., Ferryman, J.: Pets2010: dataset and challenge. AVSS, 00 (undefined), pp. 143–150 (2010)
25. Lu, C., Shi, J. and Jia, J.: Abnormal event detection at 150 fps in MATLAB. In: Proceedings of the IEEE International Conference on Computer Vision, pp. 2720–2727 (2013)
26. Degardin, B., Proença, H.: Human activity analysis: iterative weak/self-supervised learning frameworks for detecting abnormal events. In: 2020 IEEE International Joint Conference on Biometrics (IJCB), pp. 1–7. IEEE
27. Zhu, L., Wang, L., Raj, A., Gedeon, T., Chen, C.: A concise review and a new dataset. In: Advancing Video Anomaly Detection (2024)
28. Perez, M., Kot, A.C., Rocha, A.: Detection of real-world fights in surveillance videos. In: ICASSP 2019-2019 IEEE International Conference on Acoustics, Speech and Signal Processing (ICASSP), pp. 2662–2666. IEEE (2019)
29. Yao, Y., et al.: DOTA: unsupervised detection of traffic anomaly in driving videos. IEEE Trans. Pattern Anal. Mach. Intell. **45**, 444–459 (2022)
30. Che, Z., et al.: D2-city: A large-scale dashcam video dataset of diverse traffic scenarios. arxiv arXiv:1904.01975. (2019)
31. Tang, Y., et al.: Coin: a large-scale dataset for comprehensive instructional video analysis. In: Proceedings of the IEEE/CVF Conference on Computer Vision and Pattern Recognition, pp. 1207–1216 (2019)
32. Ramachandra, B., Jones, M.: Street scene: a new dataset and evaluation protocol for video anomaly detection. In: Proceedings of the IEEE/CVF Winter Conference on Applications of Computer Vision, pp. 2569–2578 (2020)
33. Chavez, E.: Caviar (context aware vision using image-based active recognition) dataset. http://homepages.inf.ed.ac.uk/rbf/CAVIAR/. University of Edinburgh, School of Informatics
34. Blunsden, S., Fisher, B.: The behave video dataset: ground truthed video for multi-person behavior classification. Ann. BMVA **4**, 1–11 (2010)
35. Yao, X., Liu, Z.: WebDataset: a dataset for anomaly detection in web-based scenarios. In: Proceedings of the IEEE Conference on Computer Vision and Pattern Recognition (CVPR), pp. 5625–5634. IEEE (2021)
36. Savarese, S., Feris, R., Sivic, J.: MIT traffic dataset. Massachusetts Institute of Technology (2009). http://web.mit.edu/
37. He, K., Sun, J., Tang, X.: Single image haze removal using dark channel prior. IEEE Trans. Pattern Anal. Mach. Intell. **33**(12), 2341–2353 (2010)
38. Garg, K., Nayar, S.K.: Vision and rain. Int. J. Comput. Vis. **75**, 3–27 (2007). https://doi.org/10.1007/s11263-006-0028-6
39. Fu, X., Huang, J., Zeng, D., Huang, Y., Ding, X., Paisley, J.: Removing rain from single images via a deep detail network. In: Proceedings of the IEEE Conference on Computer Vision and Pattern Recognition, pp. 3855–3863 (2017)
40. Meng, G., Wang, Y., Duan, J., Xiang, S., Pan, C.: Efficient image dehazing with boundary constraint and contextual regularization. In: Proceedings of the IEEE International Conference on Computer Vision, pp. 617–624 (2013)
41. Zhang, H., Patel, V.M.: Density-aware single image de-raining using a multi-stream dense network. In: Proceedings of the IEEE Conference on Computer Vision and Pattern Recognition, pp. 695–704 (2018)

42. Li, B., et al.: Benchmarking single-image dehazing and beyond. IEEE Trans. Image Process. **28**(1), 492–505 (2018)
43. Zhang, H., Sindagi, V., Patel, V.M.: Image de-raining using a conditional generative adversarial network. IEEE Trans. Circ. Syst. Video Technol. **30**(11), 3943–3956 (2019)

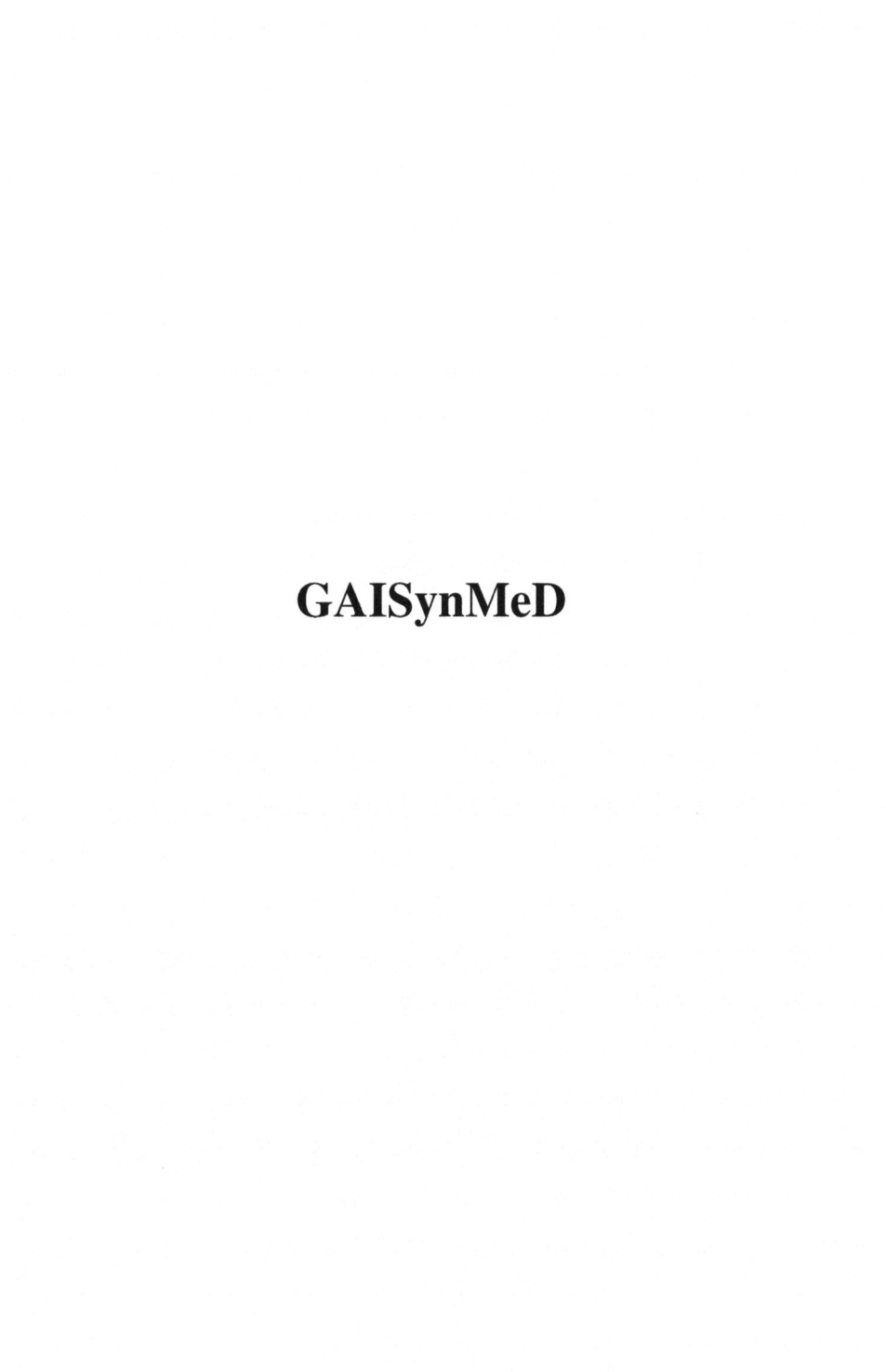

GAISynMeD

Unsupervised Skull Segmentation via Contrastive MR-to-CT Modality Translation

Kamil Kwarciak[1]([envelope]) [ID], Mateusz Daniol[1] [ID], Daria Hemmerling[1] [ID], and Marek Wodzinski[1,2] [ID]

[1] Department of Measurement and Electronics, AGH University of Krakow, Kraków, Poland
kkwarciak@student.agh.edu.pl
[2] Institute of Informatics, University of Applied Sciences Western Switzerland (HES-SO Valais), Sierre, Switzerland

Abstract. The skull segmentation from CT scans can be seen as an already solved problem. However, in MR this task has a significantly greater complexity due to the presence of soft tissues rather than bones. Capturing the bone structures from MR images of the head, where the main visualization objective is the brain, is very demanding. The attempts that make use of skull stripping seem to not be well suited for this task and fail to work in many cases. On the other hand, supervised approaches require costly and time-consuming skull annotations. To overcome the difficulties we propose a fully unsupervised approach, where we do not perform the segmentation directly on MR images, but we rather perform a synthetic CT data generation via MR-to-CT translation and perform the segmentation there. We claim that translating the process to the CT modality is essential, as it significantly simplifies the overall procedure by transforming the complex segmentation in MR into a more straightforward segmentation in CT. We address many issues associated with unsupervised skull segmentation including the unpaired nature of MR and CT datasets (contrastive learning), low resolution and poor quality (super-resolution), and generalization capabilities. We demonstrate the effectiveness of our methodology through a quantitative analysis using Dice and Surface Dice metrics on the validation dataset, as well as on the test set to highlight its adaptability to new datasets. The research has a significant value for downstream tasks requiring skull segmentation from MR volumes such as craniectomy or surgery planning and can be seen as an important step towards the utilization of synthetic data in medical imaging.

Keywords: Deep Learning · Generative AI · Modality Translation · Contrastive Learning · Super-Resolution · Synthetic Data · CT Synthesis

Supplementary Information The online version contains supplementary material available at https://doi.org/10.1007/978-981-96-2641-0_11.

1   Introduction

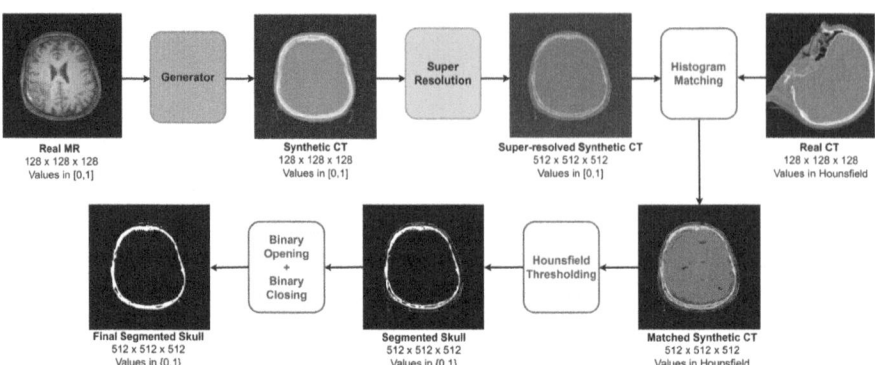

Fig. 1. Inference with the use of the proposed solution. Firstly, the generator from the CUT framework is applied. This is followed by a super-resolution module. Next, we perform histogram matching using a different CT sample from the dataset, which does not need to be correlated with the input MR image. After this, Hounsfield thresholding is applied, followed by binary operations, to achieve the final skull segmentation results. The entire process leverages synthetic CT for skull segmentation via MR-to-CT modality translation.

Skull segmentation is a common objective of many medical imaging studies, as it is usually the first step in processing pipelines and plays a crucial role in various diagnostic solutions. Among them, it is safe to mention: dental planning, patient positioning, or cranial implants design [39]. On one hand, this process in computed tomography (CT) may seem straightforward, as it commonly utilizes the thresholding in the Hounsfield scale [33], while on the other hand, when it comes to magnetic resonance (MR) modality, this procedure gets more complicated. The reason behind this is that the main strength of MR lies in providing detailed images of soft tissues rather than bone structures. Hence, the problem becomes highly complex, but also indispensable, as many challenges require working on the skull (or in general bone) data, while only MR images are available or the acquisition of CT is undesired. As an example, one can mention the case of pediatric imaging where MR is favored over CT to avoid of ionizing radiation [2].

Among the techniques used for the separation of skull and brain structures, skull stripping [7, 12, 35] seem to be standard solutions. Skull stripping is a technique that involves removing non-brain tissues, such as the skull and extracranial structures. This is typically achieved through a wide range of different methods such as classical morphological operations, or novel deep neural network architectures. Among the most widely used skull stripping methods, we find highlight Bet [35] which is the most classic approach and relies on deformable surface models, DeepBet [7] which can be seen as a deep learning-based extension of Bet and benefits from modifying the U-Net [32] and applying bias correction and normalization, and finally SynthStrip [12] which uses a 3-D U-Net and is trained on synthetic data. Then, to obtain a skull segmentation mask, one, can perform a subtraction between the input MR image and output of skull stripping and apply binary masking on top of it. However, it should be kept in mind that skull stripping methods are not directly dedicated to skull segmentation, and they may produce some artifacts. Various supervised segmentation models have been proposed in the literature to address different medical tasks. One notable solution is the SwinUNETR [10], suited for tasks such as brain tumor segmentation from MR images and lung tumor segmentation from CT images. However, the application of SwinUNETR for skull segmentation has not been explored, primarily due to the lack of annotated datasets that include MR images of the head along with corresponding skull segmentation masks. Additionally, supervised segmentation methods trained on specific tasks often exhibit poor generalization capabilities, struggling not only with different anatomical structures but also with the same anatomical structures represented in different datasets. Finally, an approach that can be seen as a potential go-to solution is the utilization of foundation models, specifically Segment Anything Model [18] and to be precise, its medical variant, MedSAM [21]. Due to the robust MedSAM training routine which also involves training on a large-scale dataset that can handle diverse segmentation tasks, segmenting the skull from an MR image should also be possible. Finally, the skull segmentation models could be trained in a supervised manner, however, such an approach would require a significant number of annotated volumes from various medical centers and acquisition protocols.

The concept of modality translation has already emerged in the field of generative AI, as a very wide range of models and architectures have been developed. The general concept of image-to-image translation is to find a mapping between images from the source and target domains. The popular techniques include GAN-based approaches [13,41], contrastive learning [25], or diffusion models [34]. Many of these solutions are applied in the medical image analysis to translate between modalities, where the most important one from the perspective of this work is MR-to-CT [24,40]. Existing solutions in this field yield satisfying outcomes. Nevertheless, their limitation lies in their inability to generalize effectively due to a fact of training on relatively homogeneous datasets. This causes a challenge for skull segmentation, especially in the context of clinical treatments, where data sources vary considerably. To take the matter further we find two major bottlenecks of contemporary strategies. Firstly, many approaches use solely 2-D data [20,40] rather than fully adopting a 3-D perspective, which is computationally cheaper, however limits inference performance in real-life scenarios with volumetric data, particularly when considering information between volumes. Secondly, various models are trained in the paired manner [24] which means that there is a direct label and alignment between images from both domains in the dataset, and in most of the clinical cases, the data is available only in one modality. One more concept that should also be mentioned in terms of working on MR-to-CT translation is the resolution of these imaging techniques. Most commonly, MR data has a lower resolution than CT data, and high-resolution CT is desired in surgical procedures, such as craniectomy. Hence, in translation works it is important to further target obtained synthetic CT images and upsample them into higher-resolution, with the use of classic interpolation techniques or more robust, super-resolution networks.

Contribution: In this study, we aim to explore the emerging field of synthetic data generation for medical imaging, specifically focusing on the application of modality translation in cranioplasty to achieve unsupervised skull segmentation in MR images. We solve the previously mentioned dilemmas by designing a pipeline that enables the training on highly diverse datasets of MR and CT images in a fully volumetric and unpaired manner with the use of contrastive learning. We additionally provide a super-resolution module to further address opportunities in clinical use cases, and finally, we experiment with downstream tasks, such as the generation of synthetic CT skulls of children, and modality translation on defective skulls. We compare the proposed approach to state-of-the-art skull stripping methods with postprocessing and medical segmentation foundation model, MedSAM. Importantly, our model does not require any annotations during training. We would also like to emphasize the difference between skull segmentation and skull stripping. The primary goal of skull stripping is to separate the brain from non-brain elements, including the skull. If the final output needs to include the skull, additional postprocessing steps will be required. In contrast, the goal of skull segmentation is to specifically extract the cranial bone, making non-bone structures unimportant. The tasks are entirely different and face distinct challenges.

2   Methods

2.1   Pipeline Overview

The two main components of our solution are the Contrastive Unpaired Translation (CUT) [25] module and the Laplacian Pyramid Super-Resolution Network (LapSRN) [19] module, both utilized for 3-D volumetric data processing. They work separately and use different techniques for sampling the data from the whole space of gathered datasets. CUT is responsible for the MR-to-CT translation in an unpaired manner, it takes unpaired samples and tries to generate a synthetic CT image for the corresponding MR image without a vision of a direct mapping. LapSRN is trained directly and only on high-resolution CT images to create a pyramidal structure of increasing resolutions, with the usage of extreme data augmentation (which we discuss further) to enhance the generalization capabilities of the model. The full architecture is presented in Fig. 2 and the inference process is presented in Fig. 1.

2.2   Preprocessing

For MR data we simply perform min-max normalization for every instance separately to obtain values in the range of [0,1] with preservation of the local dynamics. For CT data, we perform Hounsfield rescaling to remove the structures that

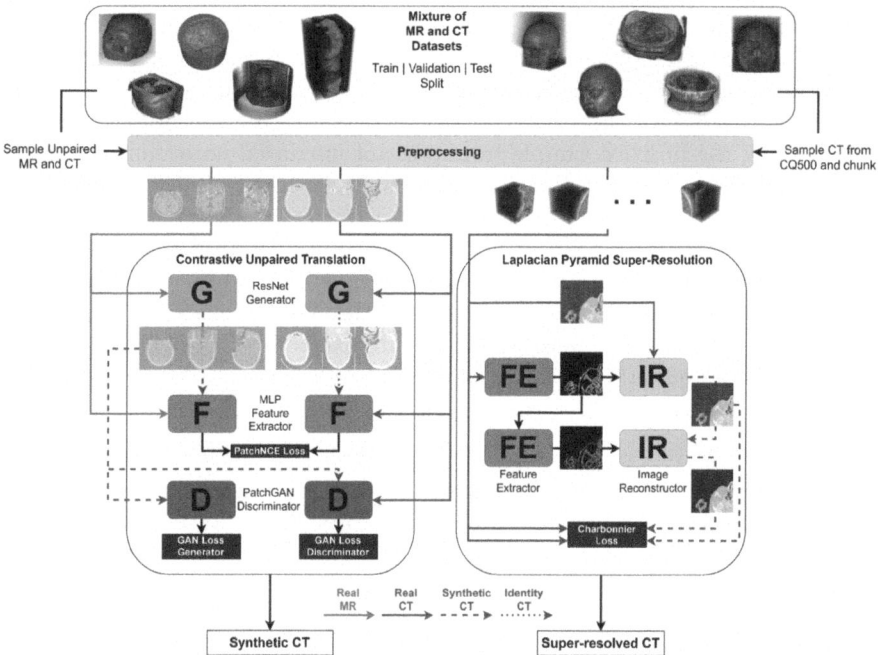

Fig. 2. The overview of the pipeline for the Contrastive Unpaired Translation and Laplacian Pyramid Super-Resolution Network. Note that we work on 3-D tensors, the 2-D representations are used only for visualization simplicity.

have values below $-500\ HU$ (air/background), as it enhances the training process, and has been found to be the most efficient for the whole dataset. After that, we also apply min-max scaling to values in the [0,1] range.

2.3   Contrastive Unpaired Translation

The CUT component of the pipeline is heavily inspired by the original work [25], however, we add several updates to address the domain specificity. The CUT framework consists of three networks, generator G, discriminator D, and feature extractor F, where each of them is implemented to operate on 3-D data representations. G in the context of this work is a ResNet-like encoder-decoder network, D is a PatchGAN discriminator [13], and F is a shallow multilayer perceptron. In general, G and D follow a classic GAN minimax game, where G aims to produce realistic CT data to fool a D network, while the D aims to distinguish between real and generated CT data:

$$\mathcal{L}_{\mathrm{GAN}}(D, G, \mathrm{MR}, \mathrm{CT}) = \mathbb{E}_{\mathbf{y}_{\mathrm{real}}^{\mathrm{CT}} \sim \mathrm{CT}} \left[\log D(\mathbf{y}_{\mathrm{real}}^{\mathrm{CT}}) \right] \tag{1}$$
$$+ \mathbb{E}_{\mathbf{x}_{\mathrm{real}}^{\mathrm{MR}} \sim \mathrm{MR}} \left[\log \left(1 - D(G(\mathbf{x}_{\mathrm{real}}^{\mathrm{MR}})) \right) \right],$$

where G as input takes a real MR image $\mathbf{x}_{\text{real}}^{\text{MR}}$ and produces synthetic CT image $\mathbf{y}_{\text{syn}}^{\text{CT}}$ by learning a mapping $G : \mathbf{x}_{\text{real}}^{\text{MR}} \rightarrow \mathbf{y}_{\text{syn}}^{\text{CT}}$. The second component of the CUT training objective is the use of noise contrastive estimation (NCE) [22]. Specifically, it employs the InfoNCE loss to maximize mutual information by distinguishing the positive sample from a set of unrelated noise samples. However, an important aspect of the CUT framework, in comparison to the original InfoNCE work, is that the comparison is not performed between a positive sample and noise samples. Instead, it is done by comparing patches sampled from different spatial locations. The source MR image $\mathbf{x}_{\text{real}}^{\text{MR}}$, together with its corresponding synthetic CT image $\mathbf{y}_{\text{syn}}^{\text{CT}}$, are passed through the encoding component of the generator G_{enc}, followed by the feature extractor F, to generate latent feature vector representations in a shared embedding space. Next, a reference patch $\mathbf{z}_{\text{syn}}^{\text{CT(ref)}}$ is extracted from the intermediate representation of $\mathbf{y}_{\text{syn}}^{\text{CT}}$ and compared to a positive patch $\mathbf{z}_{\text{real}}^{\text{MR}(+)}$ extracted from the intermediate representation of $\mathbf{x}_{\text{real}}^{\text{MR}}$ at the same spatial location, as well as to $N-1$ negative samples $\mathbf{z}_{\text{real}}^{\text{MR}(-)}$ also taken from the intermediate representation of $\mathbf{x}_{\text{real}}^{\text{MR}}$, but taken from different spatial locations. This setup yields the following formulation of InfoNCE loss, with the dot product used as a similarity measure:

$$\ell_{\text{MR}} = \mathcal{L}_{\text{InfoNCE}|\text{MR}_{\text{real}}, \text{CT}_{\text{syn}}} \left(\mathbf{z}_{\text{syn}}^{\text{CT(ref)}}, \mathbf{z}_{\text{real}}^{\text{MR}(+)}, \mathbf{z}_{\text{real}}^{\text{MR}(-)} \right) \tag{2}$$

$$= -\log \frac{\exp \left(\mathbf{z}_{\text{syn}}^{\text{CT(ref)}} \cdot \mathbf{z}_{\text{real}}^{\text{MR}(+)} \right)}{\exp \left(\mathbf{z}_{\text{syn}}^{\text{CT(ref)}} \cdot \mathbf{z}_{\text{real}}^{\text{MR}(+)} \right) + \sum_{j=1}^{N-1} \exp \left(\mathbf{z}_{\text{syn}}^{\text{CT(ref)}} \cdot \mathbf{z}_{\text{real}}^{\text{MR}(-)} \right)}.$$

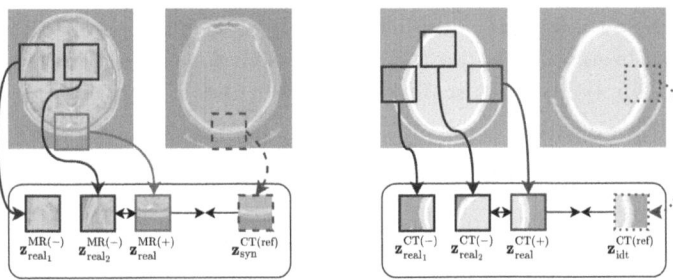

Fig. 3. Sampling procedure for patchwise contrastive estimation of real MR ↔ synthetic CT, and real CT ↔ identity CT. We show a 2-D view for better visualization.

The CUT methodology also incorporates the concept of identity mapping and utilizes InfoNCE in this setup to stabilize training, functioning similarly to an identity loss. These identity CT mappings are obtained via $G : \mathbf{y}_{\text{real}}^{\text{CT}} \rightarrow \mathbf{y}_{\text{idt}}^{\text{CT}} \approx \mathbf{y}_{\text{real}}^{\text{CT}}$. The latent representation generation process is applied in a similar manner: the data is passed through the encoding part of the generator G_{enc} and then through the feature extractor F. This process results in the following set of patches: $\mathbf{z}_{\text{idt}}^{\text{CT(ref)}}$, which is a reference patch from identity CT, and $\mathbf{z}_{\text{real}}^{\text{CT}(+)}$,

$z_{\text{real}}^{\text{CT}(-)}$ being positive and negative samples from real CT, this provides a second InfoNCE formulation as:

$$\ell_{\text{CT}} = \mathcal{L}_{\text{InfoNCE}|\text{CT}_{\text{real}},\text{CT}_{\text{idt}}}\left(z_{\text{idt}}^{\text{CT(ref)}}, z_{\text{real}}^{\text{CT}(+)}, z_{\text{real}}^{\text{CT}(-)}\right) \tag{3}$$

$$= -\log \frac{\exp\left(z_{\text{idt}}^{\text{CT(ref)}} \cdot z_{\text{real}}^{\text{CT}(+)}\right)}{\exp\left(z_{\text{idt}}^{\text{CT(ref)}} \cdot z_{\text{real}}^{\text{CT}(+)}\right) + \sum_{j=1}^{N-1}\exp\left(z_{\text{idt}}^{\text{CT(ref)}} \cdot z_{\text{real}}^{\text{CT}(-)}\right)}.$$

The idea of comparison behind Eq. 2 and 3 is presented in Fig. 3. Moreover, the CUT methodology broadens the mentioned InfoNCE loss components to the intermediate outputs of G_{enc} layers. This ensures that the features learned at multiple levels are aligned and collectively enhance the quality and consistency of the final generated output. If we denote the layers of the encoding part of G (G_{enc}) as L, then at the l-th layer, we obtain a set of features for the real MR and real CT images as follows: $\{z_{\text{real}(l)}^{\text{MR}}\}_L = \{F_l(G_{\text{enc}}^l(x_{\text{real}}^{\text{MR}}))\}_L$ and $\{z_{\text{real}(l)}^{\text{CT}}\}_L = \{F_l(G_{\text{enc}}^l(y_{\text{real}}^{\text{CT}}))\}_L$. If we also address the synthetic and identity representations we obtain the following sets: $\{z_{\text{syn}(l)}^{\text{CT}}\}_L = \{F_l(G_{\text{enc}}^l(G(x_{\text{real}}^{\text{MR}})))\}_L$ and $\{z_{\text{idt}(l)}^{\text{CT}}\}_L = \{F_l(G_{\text{enc}}^l(G(y_{\text{real}}^{\text{CT}})))\}_L$. Now, as we take S_l patches at layer l we have a set of S/s negative patches $(-)$ and s positive patch $(+)$, and we are able to formulate PatchNCE loss in two setups, the first one as the main training objective of synthesizing CT images from MR inputs:

$$\mathcal{L}_{\text{PatchNCE}}(G, F, \text{MR}) \tag{4}$$

$$= \mathbb{E}_{x_{\text{real}}^{\text{MR}} \sim \text{MR}} \sum_{l=1}^{L}\sum_{s=1}^{S_l} \ell_{\text{MR}}\left(z_{\text{syn}(l)}^{\text{CT(ref)}(s)}, z_{\text{real}(l)}^{\text{MR}(+)(s)}, z_{\text{real}(l)}^{\text{MR}(-)(S/s)}\right),$$

and the second being a stabilization term between real CT and its identity CT (real CT passed through the generator):

$$\mathcal{L}_{\text{PatchNCE}}(G, F, \text{CT}) \tag{5}$$

$$= \mathbb{E}_{y_{\text{real}}^{\text{CT}} \sim \text{CT}} \sum_{l=1}^{L}\sum_{s=1}^{S_l} \ell_{\text{CT}}\left(z_{\text{idt}(l)}^{\text{CT(ref)}(s)}, z_{\text{real}(l)}^{\text{CT}(+)(s)}, z_{\text{real}(l)}^{\text{CT}(-)(S/s)}\right).$$

To conclude, for CUT in terms of MR-to-CT translation, PatchNCE loss is computed in two manners, firstly by comparing $x_{\text{real}}^{\text{MR}}$ with $y_{\text{syn}}^{\text{CT}}$ as a main training objective, and secondly by comparing $y_{\text{real}}^{\text{CT}}$ with $y_{\text{idt}}^{\text{CT}}$ for training stabilization and regularization. Finally, we can formulate the final objective of CUT training as:

$$\mathcal{L}_{\text{CUT}} = \lambda_{\text{GAN}}\mathcal{L}_{\text{GAN}}(G, D, \text{MR}, \text{CT}) \tag{6}$$

$$+ \lambda_{\text{syn}}\mathcal{L}_{\text{PatchNCE}}(G, F, \text{MR}) + \lambda_{\text{idt}}\mathcal{L}_{\text{PatchNCE}}(G, F, \text{CT})$$

2.4   Laplacian Pyramid Super-Resolution

The second component of the designed pipeline is the modification of the original LapSRN [19] used for super-resolution of 3-D CT volumes. It uses two sub-networks: feature extractor and image reconstructor, which progressively generate images at higher resolutions from the low-resolution ones. The feature extractor is responsible for extracting the features at a coarse (lower) level and generating feature maps at a finer (higher) level. It enhances the representation of the input by capturing important details. The image reconstructor upsamples the lower-resolution input and then via element-wise summation with residuals obtained from the feature extractor creates a higher-resolution output with improved visual quality. By its nature, progressive reconstruction provides task-dependent flexibility and adjustability, as by bypassing the pyramid at certain levels we can obtain representations at the resolution required for a given task. Furthermore, to address the requirement of high generalization capabilities, our LapSRN uses not only a set of diverse data augmentations (flipping, affine transformations, motion artifacts, blurriness, contrast), but is also trained on overlapping chunks of input tensors (see Fig. 2). This configuration enhances the model's generalization capabilities, enabling it to effectively super-resolve single chunks and capture inconsistencies in various skulls, such as those with defects. Additionally, it significantly reduces memory requirements during training, as the chunks can be processed individually or combined into small micro-batches within the mini-batch. LapSRN uses Charbonnier loss, where for M chunks of the input tensor, and pyramid of L levels, the Charbonnier loss is defined as:

$$\mathcal{L}_{\text{Charbonnier}}(\mathbf{y}, \mathbf{x}, \mathbf{r}) = \frac{1}{M} \sum_{m=1}^{M} \sum_{s=1}^{L} \sqrt{(\mathbf{y}_s - \mathbf{x}_s - \mathbf{r}_s)^2 + \epsilon^2}, \tag{7}$$

where for layer s, chunk residual is denoted as $\mathbf{r}_s$, upsampled lower-resolution chunk as $\mathbf{x}_s$ and ground truth higher-resolution chunk $\mathbf{y}_s$ and ϵ (set to small value like 10^{-3}) controls similarity to L_1 loss while staying differentiable.

2.5   Postprocessing

Synthetic CT and high-resolution CT obtained from pipeline components are in the $[0, 1]$ range and it is desired to have them in the Hounsfield scale, to perform threshold-based skull segmentation. This can be easily achieved by performing a histogram matching with one of the real CTs from the train set for CUT, and low-resolution CT for LapSRN. Furthermore, it was experimentally found that following Hounsfield-based thresholding with binary opening and binary closing is beneficial for artifacts removal. It is also important to note, that training both modules, CUT and LapSRN on the original Hounsfield scale can yield training instabilities due to the wide range of values and potential outliers, which can lead to gradient issues and slow convergence.

3   Experiments

3.1   Datasets

To address the generalization possibilities we create a huge dataset of 1,521 MR and 879 CT 3-D images from publicly available datasets [1, 3–6, 8, 9, 14–16, 23, 27–31, 36, 37] (See Supplementary materials for more information). Importantly, we extract a portion of the dataset from the SynthRAD 2023 Challenge [37] for validation purposes (25 images), as it is the only dataset that includes paired MR and CT samples, hence it can be used for evaluation with metrics like Dice coefficient. We also utilize the Han-Seg Dataset [28] for testing purposes (42 images), as it includes images of heads in both MR and CT formats. However, it is important to note that the images in this dataset are not registered. Therefore, we perform image registration before using this dataset for testing. The Han-Seg dataset demonstrates the generalization capabilities of the proposed method. To fully address the requirements of generalization, we construct the dataset of skull segmentation by including also other anatomical structures, such as the pelvis, aorta, or kidneys. This setup enables a more robust capturing of style rather than context, as commonly different institutions use various imaging devices, making single-institution models hard to generalize. To address this, we combined datasets with images of different anatomical structures to enhance cross-institution generalizability via different anatomical structures and different acquisition settings. We train CUT with the use of all datasets, and for LapSRN we only use a dataset of high-resolution CT skulls, namely CQ500 [4] (300 images for training and 50 images for testing).

3.2   Networks

We implement the proposed networks with the use of PyTorch library [26]. CUT and LapSRN subnetworks use 3-D convolutional layers. What's important, for the CUT model we use instance normalization [38] as it prevents instance-specific mean and covariance shifts, hence it is highly beneficial for modality translation tasks. CUT's generator is a 9-layer ResNet-based network, with residual connections between downsampling and upsampling blocks, it uses instance normalization and ReLU nonlinearity besides the decision layer for which we use hyperbolic tangent. The discriminator is a 3-layer convolutional network with Leaky ReLU and instance normalization (like PatchGAN discriminator [13]). CUT's feature extractor is a simple 2-layer multilayer perceptron with ReLU which operates on 64 patches of features extracted from the generator's flow. Hence, regarding Eq. 4 and Eq. 5, we operate on a fixed amount of patches equal to 64, and a fixed amount of G_{enc} layers equal to 9. LapSRN's image reconstructor is a 2-layer convolution/deconvolution network and the feature extractor is an 8-layer convolution/deconvolution network with Leaky ReLU nonlinearities. LapSRN's convolutional layers use $3{\times}3{\times}3$ kernels with 64 filters with He initialization [11], deconvolutions use the kernel of $4{\times}4{\times}4$ (upsampling by a factor of 2) and weights are initialized from a trilinear filter, as suggested in the original implementation [19].

Table 1. Comparison of methods with DSC and SDSC, evaluated on the subset of SynthRAD 2023 training dataset [37], extracted for a test set in this work.

Method	DSC ↑	SDSC ↑
SynthStrip [12] + subtraction	0.281	0.580
Bet [35] + subtraction	0.273	0.563
DeepBet [7] + subtraction	0.298	0.606
MedSAM [21]	-	-
Ours	**0.512**	**0.770**

3.3   Setup

We resize the data to $128 \times 128 \times 128$ for training both the CUT and LapSRN models. We train CUT using ADAM optimizer [17] with $\beta_1 = 0.5$ and $\beta_2 = 0.999$. We set the initial learning rate to $2 \cdot 10^{-4}$ and decrease it linearly every 50 epochs if the loss doesn't decrease. We set λ_{GAN}, λ_{syn} and λ_{idt} all equal to 1 and train CUT with the real batch size of 1 (batch size of 8 distributed across 8 GPUs). LapSRN is trained using SGD with momentum term set to 0.9 and weight decay of 10^{-4}. The learning rate is initialized to 10^{-5} and decreases linearly every 5 epochs if the loss doesn't decrease. The chunk size for splitting the inputs is set to 8 with 8 voxels overlapping ($128^3 \rightarrow 8 \times 64(+8)^3$), the batch size is 1 and we accumulate gradients every 16 batches. Models were trained until convergence, with the use of 8 NVIDIA A100 40GB GPUs.

4   Results

We evaluate the proposed method on the subset of samples extracted from the training dataset from SynthRAD 2023 Challenge [37] as it consists of paired MR and CT images. We present the Dice coefficient (DSC) and surface Dice coefficient (SDSC) results, calculated from segmentation masks obtained using Hounsfield scale thresholding in Table 1. Importantly, it should be noted that, as previously mentioned, skull stripping differs from skull segmentation. The results of skull stripping methods were followed by postprocessing, which involved manual thresholding for skull subtraction. Furthermore, we demonstrate that the medical imaging foundation model, MedSAM [21], fails to perform the complex task of skull segmentation from MR images. The failed results are presented in Fig. 4a for the bounding box approach and Fig. 4b for the experimental point prompt approach. We investigate the quality of generated synthetic CT via a small set of ablation studies which are presented in Fig. 5. They showcase the following settings: (*i*) the performance of LapSRN using a test sample from the CUT module, (*ii*) generalization capabilities to the downstream task of translation on the MR image of the child's skull, and (*iii*) the translation of the MR image of a defected (and reconstructed) skull. Finally, to demonstrate the generalization potential of the proposed solution from a quantitative perspective, we investigate additionally the MR subset of the HaN-Seg dataset [28] which was not present in the original training dataset. As the availability of paired MR and CT datasets is highly limited, besides SynthRAD 2023 dataset [37], we find HaN-Seg to be the most suitable option for this evaluation. Importantly, additional preprocessing in the form of image registration between CT and MR samples in this dataset was required to enable the quantitative evaluation. For comparison, we decided to train a state-of-the-art segmentation network, SwinUNETR [10], on the SynthRAD 2023 dataset [37], where the input was an MR image and the target was a skull mask derived from Hounsfield thresholding of its paired CT image. We also trained the model as a standard paired MR-to-CT translation and applied the same Hounsfield thresholding on the resulting synthetic CTs for consistency with the pipeline from Fig. 1. Table 2 shows that the proposed solution presents superior results in terms of the generalization capabilities in comparison to both SwinUNETR setups.

5 Discussion

Quantitative analysis of the CUT method shows that it can provide better results in terms of DSC and SDSC than the combination of skull stripping with subtraction and masking as shown in Table 1. Importantly, we find out that Med-SAM [21] struggles with the skull segmentation in MR images and fails to propagate the segmentation mask in both bounding-box and point-prompt (experimental) approaches. This can be potentially solved via task-specific fine-tuning (as the authors mention this MedSAM capability in their work). Good results of our method suggest that it is a promising direction for designing a fully unsupervised framework of skull segmentation from MR images, that can be used for downstream tasks (such as defect segmentation), and with the use of superresolution LapSRN module, the requirement of high-resolution CT images for tasks like craniectomy or surgery planning is also met. What's more, both CUT and LapSRN are relatively simple in terms of architectural design and also their learning objectives, hence they can be adapted to other translation tasks, or finetuned for other anatomical structures. With CUT we were able to obtain good generalization capabilities due to the very high diversity of the used dataset. Importantly, we also note several limitations of our methodology and leave it as a potential direction for further work. First of all, produced results sometimes include some blurriness; this can be attributed to using convolutional networks

Table 2. Evaluation of the generalization capabilities on the HaN-Seg dataset [28] in terms of DSC and SDSC.

Method	DSC ↑	SDSC ↑
SwinUNETR (MR-Mask)	0.084	0.135
SwinUNETR (MR-CT)	0.113	0.149
Ours	**0.310**	**0.396**

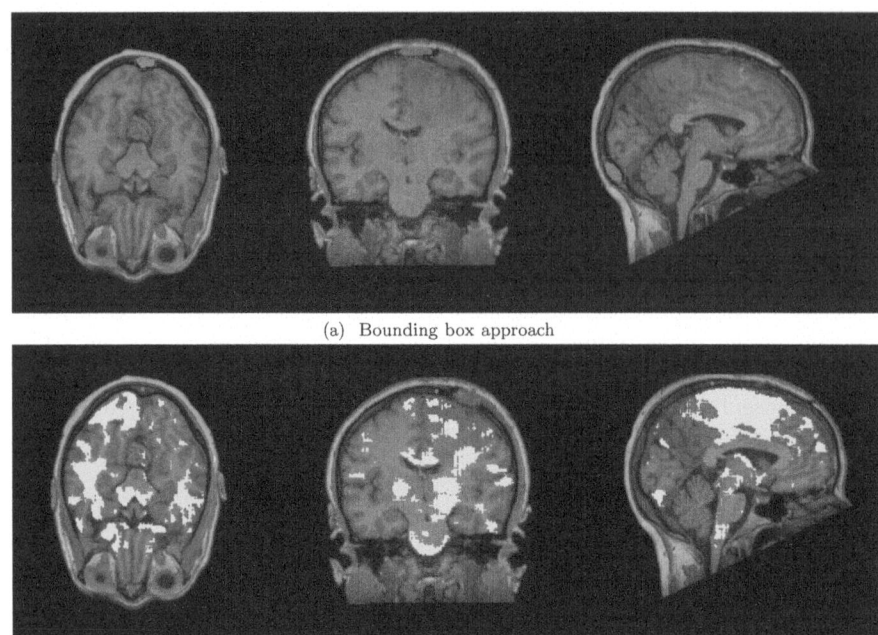

(a) Bounding box approach

(b) Point prompt approach

Fig. 4. Results of skull segmentation from MR images with the use of MedSAM: (a) bounding box and (b) point prompt approaches. Bounding box approach failure stems from a fact that segmented skull structure is not propagated through the whole image, and only small parts are captured. For point prompt, the model is unable to identify the skull and propagates segmentation into brain.

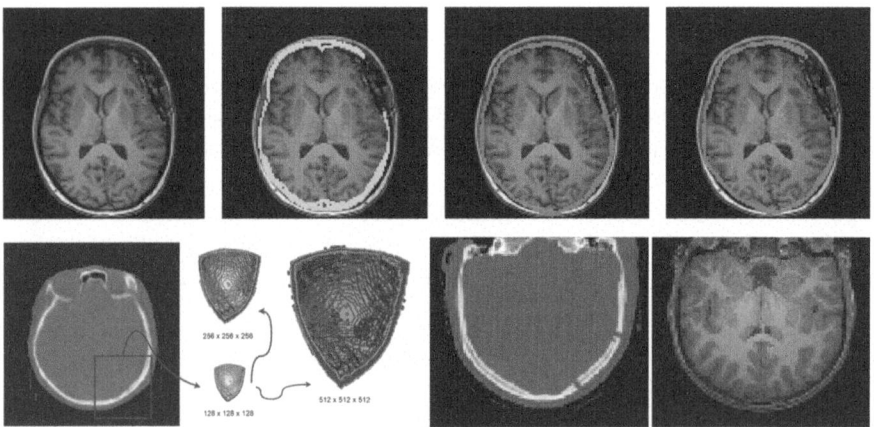

Fig. 5. Top: Results of translation and segmentation on defected skulls (from left to right: input MR, matched CT mask, synthetic CT mask, synthetic CT mask with removed implant area). Bottom left: Super-resolution of synthetic skull. Bottom right: MR-to-CT translation on child's skull.

with a relatively small number of filters and a shallow network depth. This issue stems from the challenges of designing large networks for high-resolution volumetric data. Even the latest GPUs struggle with VRAM capacities when trying to fit large tensors and extensive 3-D networks. Nevertheless, we are able to partially reduce this issue as the whole pipeline (presented in Fig. 1) employs postprocessing in the form of morphological operations which diminishes this effect in the task of segmentation. Secondly, while DSC results are better than other methods, they still require improvement to meet the demands of potential real-world medical applications. Finally, the primary motivation of this work was to achieve generalization capabilities for the segmentation task in an unsupervised and unpaired manner. Therefore, we do not compare our results with other methods used in MR-to-CT translation, as these are not primarily focused on segmentation. Nevertheless, we acknowledge that some existing translation methods may be more effective than CUT, and other super-resolution modules could outperform LapSRN. We consider this a potential area for further improvement, research, and investigation.

6   Conclusion

In conclusion, we presented a novel generative AI-based methodology for synthetic CT generation, specifically designed for skull segmentation from MR images. Our approach operates on 3-D data, is trained in an unsupervised manner, and demonstrates strong generalization capabilities while also super-resolving into higher resolutions. We analyze its performance utilizing quantitative metrics on two separate datasets to present its performance and generalization. We believe this opens up new research opportunities in this field, and we plan to further enhance the proposed solution.

Acknowledgements. The project was funded by The National Centre for Research and Development, Poland under Lider Grant no: LIDER13/0038/2022 (DeepImplant). We gratefully acknowledge Polish HPC infrastructure PLGrid support within computational grants no. PLG/2023/016239 and PLG/2024/017079.

References

1. Banfi, C., et al.: Reading-related functional activity in children with isolated spelling deficits and dyslexia. Lang. Cognit. Neurosci. **36**(5), 543–561 (2021)
2. Bosch de Basea Gomez, M., et al.: Risk of hematological malignancies from CT radiation exposure in children, adolescents and young adults. Nat. Med. **29**(12), 3111–3119 (2023)
3. Burleigh, L., Greening, S.G.: Fear in the mind's eye: the neural correlates of differential fear acquisition to imagined conditioned stimuli. Soc. Cognit. Affect. Neurosci. **18**(1), nsac063 (2023)
4. Chilamkurthy, S., et al.: Deep learning algorithms for detection of critical findings in head CT scans: a retrospective study. Lancet **392**(10162), 2388–2396 (2018)

5. Etzel, J.A., et al.: The dual mechanisms of cognitive control dataset, a theoretically-guided within-subject task FMRI battery. Scientific Data **9**(1), 114 (2022)
6. Fialkowski, K.P., Bush, K.A.: Identifying the neural correlates of resting state affect processing dynamics. Front. Neuroimaging **1**, 825105 (2022)
7. Fisch, L., et al.: DeepBET: fast brain extraction of t1-weighted MRI using convolutional neural networks. arXiv preprint arXiv:2308.07003 (2023)
8. Gaesser, B., et al.: A role for the medial temporal lobe subsystem in guiding prosociality: the effect of episodic processes on willingness to help others. Social Cognit. Affective Neurosci. **14**(4), 397–410 (2019)
9. Greene, D.J., et al.: Behavioral interventions for reducing head motion during MRI scans in children. Neuroimage **171**, 234–245 (2018)
10. Hatamizadeh, A., et al.: Swin UNETR: Swin transformers for semantic segmentation of brain tumors in MRI images. In: International MICCAI Brainlesion Workshop, pp. 272–284. Springer (2021). https://doi.org/10.1007/978-3-031-08999-2_22
11. He, K., Zhang, X., Ren, S., Sun, J.: Delving deep into rectifiers: Surpassing human-level performance on ImageNet classification. In: Proceedings of the IEEE International Conference on Computer Vision, pp. 1026–1034 (2015)
12. Hoopes, A., et al.: SynthStrip: skull-stripping for any brain image. Neuroimage **260**, 119474 (2022)
13. Isola, P., et al.: Image-to-image translation with conditional adversarial networks. In: Proceedings of the IEEE Conference on Computer Vision and Pattern Recognition, pp. 1125–1134 (2017)
14. Jo, H., et al.: A brain network that supports consensus-seeking and conflict-resolving of college couples' shopping interaction. Sci. Rep. **10**(1), 17601 (2020)
15. Kavur, A.E., et al.: Chaos challenge-combined (CT-MR) healthy abdominal organ segmentation. Med. Image Anal. **69**, 101950 (2021)
16. Keane, B.P., et al.: Brain network mechanisms of visual shape completion. Neuroimage **236**, 118069 (2021)
17. Kingma, D.P., Ba, J.: Adam: a method for stochastic optimization. arXiv preprint arXiv:1412.6980 (2014)
18. Kirillov, A., et al.: Segment anything. arXiv preprint arXiv:2304.02643 (2023)
19. Lai, W.S., et al.: Deep Laplacian pyramid networks for fast and accurate super-resolution. In: Proceedings of the IEEE Conference on Computer Vision and Pattern Recognition, pp. 624–632 (2017)
20. Liu, Y., et al.: CT synthesis from MRI using multi-cycle GAN for head-and-neck radiation therapy. Comput. Med. Imaging Graph. **91**, 101953 (2021)
21. Ma, J., et al.: Segment anything in medical images. Nat. Commun. **15**(1), 654 (2024)
22. Oord, A.v.d., Li, Y., Vinyals, O.: Representation learning with contrastive predictive coding. arXiv preprint arXiv:1807.03748 (2018)
23. Ottesen, T.D., et al.: Differences in chemo-signaling compound-evoked brain activity in male and female young adults: a pilot study in the role of sexual dimorphism in olfactory chemo-signaling. bioRxiv, pp. 2020–09 (2020)
24. Pan, S., et al.: Synthetic CT generation from MRI using 3D transformer-based denoising diffusion model. arXiv preprint arXiv:2305.19467 (2023)
25. Park, T., Efros, A.A., Zhang, R., Zhu, J.-Y.: Contrastive learning for unpaired image-to-image translation. In: Vedaldi, A., Bischof, H., Brox, T., Frahm, J.-M. (eds.) ECCV 2020. LNCS, vol. 12354, pp. 319–345. Springer, Cham (2020). https://doi.org/10.1007/978-3-030-58545-7_19
26. Paszke, A., et al.: Pytorch: an imperative style, high-performance deep learning library. In: Advances in Neural Information Processing Systems, vol. 32 (2019)

27. Peelle, J.E., et al.: Increased connectivity among sensory and motor regions during visual and audiovisual speech perception. J. Neurosci. **42**(3), 435–442 (2022)
28. Podobnik, G., et al.: Han-Seg: the head and neck organ-at-risk CT and MR segmentation dataset. Med. Phys. **50**(3), 1917–1927 (2023)
29. Racey, C., et al.: An open science MRI database of over 100 synaesthetic brains and accompanying deep phenotypic information. Scientific Data **10**(1), 766 (2023)
30. Radl, L., et al.: AVT: multicenter aortic vessel tree CTA dataset collection with ground truth segmentation masks. Data Brief **40**, 107801 (2022)
31. Rogers, C.S., et al.: Real-time feedback reduces participant motion during task-based FMRI. bioRxiv (2023)
32. Ronneberger, O., Fischer, P., Brox, T.: U-Net: convolutional networks for biomedical image segmentation. In: Navab, N., Hornegger, J., Wells, W.M., Frangi, A.F. (eds.) MICCAI 2015. LNCS, vol. 9351, pp. 234–241. Springer, Cham (2015). https://doi.org/10.1007/978-3-319-24574-4_28
33. Rulaningtyas, R., et al.: Ct scan image segmentation based on hounsfield unit values using OTSU thresholding method. J. Phys. Conf. Ser. **1816**, 012080 (2021). IOP Publishing
34. Sasaki, H., Willcocks, C.G., Breckon, T.P.: Unit-DDPM: unpaired image translation with denoising diffusion probabilistic models. arXiv preprint arXiv:2104.05358 (2021)
35. Smith, S.M.: Fast robust automated brain extraction. Hum. Brain Mapp. **17**(3), 143–155 (2002)
36. Thornton, M.A., et al.: People represent their own mental states more distinctly than those of others. Nat. Commun. **10**(1), 2117 (2019)
37. Thummerer, A., et al.: SynthRAD2023 grand challenge dataset: Generating synthetic CT for radiotherapy. Med. Phys. **50**, 4664–4674 (2023)
38. Ulyanov, D., Vedaldi, A., Lempitsky, V.: Instance normalization: the missing ingredient for fast stylization. arXiv preprint arXiv:1607.08022 (2016)
39. Wodzinski, M., Daniol, M., Hemmerling, D.: Improving the automatic cranial implant design in cranioplasty by linking different datasets. In: Li, J., Egger, J. (eds.) AutoImplant 2021. LNCS, vol. 13123, pp. 29–44. Springer, Cham (2021). https://doi.org/10.1007/978-3-030-92652-6_4
40. Wolterink, J.M., Dinkla, A.M., Savenije, M.H.F., Seevinck, P.R., van den Berg, C.A.T., Išgum, I.: Deep MR to CT synthesis using unpaired data. In: Tsaftaris, S.A., Gooya, A., Frangi, A.F., Prince, J.L. (eds.) SASHIMI 2017. LNCS, vol. 10557, pp. 14–23. Springer, Cham (2017). https://doi.org/10.1007/978-3-319-68127-6_2
41. Zhu, J.Y., et al.: Unpaired image-to-image translation using cycle-consistent adversarial networks. In: Proceedings of the IEEE International Conference on Computer Vision, pp. 2223–2232 (2017)

CleftLipGAN : Interactive GAN-Inpainting for Post-Operative Cleft Lip Reconstruction

Daniel Anojan Atputharuban[1]([✉]) [iD], Christoph Theopold[2] [iD],
and Aonghus Lawlor[1] [iD]

[1] The Insight Centre for Data Analytics, School of Computer Science,
University College Dublin, Dublin, Ireland
{daniel.anojan,aonghus.lawlor}@insight-centre.org
[2] Children's Health Ireland at Temple Street, Dublin, Ireland

Abstract. Synthetic generation of post-surgical outcomes holds significant value in the clinical domain, particularly for Cleft lip and Palate surgery. These synthetic images can be utilized for surgical planning, serve as reference points to evaluate surgical success and assist in educating patients and caretakers about potential outcomes. Image inpainting is effective for selectively generating Cleft-affected regions, making it a promising technique for this task. However, due to the lack of publicly available Cleft-specific datasets, Cleft inpainting models are typically trained on healthy data and applied to Cleft conditions to generate post-surgical lip appearances. Existing Cleft inpainting methods often struggle to capture the complexities of Cleft deformities, leading to implausible outcomes that fail to reflect the unique structural characteristics of Cleft-affected regions. To address this, we propose a Structural Guided Pluralistic Inpainting model, trained on healthy images, which allows for real-time, interactive adjustments to synthesize Cleft-specific images. We demonstrate the model's effectiveness by generating images that closely resemble Cleft conditions and benchmarking it against existing GAN-Inpainting methods. Additionally, we provide a user-friendly interface designed as a tool for post-surgical visualization of Cleft conditions. The source code is available at https://github.com/danielanojan/CleftLipGAN.git.

Keywords: Synthetic Lips · Pluralistic Inpainting · Cleft Lip and Palate

1 Introduction

Cleft lip and Palate is a prevalent congenital facial anomaly, affecting approximately 1 in 700 childbirths [40]. Cleft condition is primarily caused by genetic and environmental factors, which contribute to varying degrees of craniofacial malformations during fetal development. These malformations result in incomplete fusion of the upper lip and/or palate, leading to the formation of Cleft

M. Cho et al. (Eds.): ACCV 2024 Workshops, LNCS 15482, pp. 180–196, 2025.
https://doi.org/10.1007/978-981-96-2641-0_12

conditions at birth. [23]. Cleft conditions can significantly impact speech, feeding and dental health, and they also impose socioeconomic and psychological burdens on families and caregivers [4,13,44]. Treatment for Cleft repair typically involves a series of complex surgical procedures and rehabilitation, starting with initial repair surgery performed at 6 to 12 months of age. The primary goal of these surgeries are to restore essential facial functions and enhance both facial appearance and symmetry [54]. Despite advancements in surgical techniques, the inherent complexity of Cleft conditions and the variability of surgical outcomes pose significant challenges in both surgical planning and surgical outcome assessment [36]. Cleft surgical success is measured through functional outcomes such as speech tests, and aesthetic evaluations using facial markers. However, it is noteworthy that there are no universally accepted surgical planning protocols for Cleft lip repair, and the aesthetic assessments remain subjective, leading to inconsistencies and potential bias in clinical outcomes [12,46,47,49]. These challenges underscore the need for standardized methods for surgical planning and surgical evaluation, where machine learning techniques could enhance precision and consistency.

Synthetically generating variations of post-operative lip regions for patients with Cleft conditions plays a pivotal role in surgical planning and evaluation. By visualizing potential outcomes before surgery, surgeons can better tailor their approach, improving preoperative planning accuracy. These synthetic images serve as consistent and objective reference points for assessing surgical success. Additionally, they serve as educational tools, helping surgeons explain outcomes to parents and caregivers, enhancing their understanding of the surgical process. Thus, synthetic generation of post-operative lip images holds substantial value in improving both clinical outcomes and patient education.

Image inpainting models are particularly well-suited for addressing selective image region synthesis, as they focus on reconstructing missing areas of an image while preserving the surrounding context [51]. In particular, facial inpainting models have shown great promise in generating realistic and semantically coherent facial regions, which are used in real world applications such as facial restoration and digital forensics [10,45]. However, a major challenge arises from the lack of publicly available datasets specific to Cleft condition, which hinders the development of facial inpainting models tailored for Cleft condition [1]. To overcome this limitation, Cleft inpainting models are typically trained on images of healthy individuals and later applied to Cleft-affected regions to generate non-Cleft facial areas or simulate ideal post-surgical outcomes resembling those of children without Cleft conditions [1,6].

However, it has been observed that models trained on healthy facial datasets struggle to capture the complexity and variability of Cleft deformities, making it difficult to produce realistic post-surgical results. These models often generate overly smoothed upper lips, resembling average healthy lips [1], without capturing the unique features of Cleft conditions. The underlying issue lies in the fact that inpainting models pre-trained on healthy faces fail to capture the semantic and structural details specific to Cleft-affected faces.

On the other hand, research in Cleft conditions has demonstrated that Cleft deformities can be effectively modeled using facial landmarks [2,19]. Motivated by this observation, we have developed a structural guided pluralistic inpainting model with real-time editing capabilities. Our inpainting model, trained on images of healthy individuals, effectively reconstructs the facial features of patients with Cleft conditions and can be applied to Cleft-affected areas to generate realistic non-Cleft regions. Additionally, we demonstrate the model's ability to produce plausible results in real time by manipulation of facial keypoints. To enhance usability, we have developed an interactive user interface that facilitates real-time modeling of these adjustments, offering a valuable tool for both surgical planning and outcome assessment.

2   Related Work

In this study, we categorize the relevant prior work into two distinct areas: facial inpainting (Sect. 2.1) and the application of machine learning techniques for Cleft-related conditions (Sect. 2.2).

2.1   Facial Inpainting

Image inpainting refers to the process of synthetically generating missing parts of an image, ensuring that the result is both visually realistic and contextually meaningful. GAN-based inpainting models have demonstrated state-of-the-art performance by generating plausible content for large corrupted regions. These models typically follow an encoder-decoder architecture. Different methods have explored to improve the representation capability in inpainting process such as coarse-to-fine training [14,25,55], structural guidance [14,35,52,55,56], local and global discriminators [17], and semantic segmentation-based losses to preserve facial structure [26]. Specialized network modules, such as partial [29], Fourier [45], and gated [55] convolutions, are employed to effectively handle corrupted image regions. Additionally, techniques like pixel shuffle upsampling [7] and attention-based upsampling [58] help mitigate feature degradation during image propagation. With the success of transformers in machine translation tasks, vision transformer-based GAN-Inpainting models have been proposed to enhance long range dependencies present in images, surpassing the performance of traditional convolution-based GAN-Inpainting frameworks [11,21,25,59,59]. However, the quadratic complexity of self-attention poses a bottleneck for implementing end-to-end vision transformers in inpainting. To overcome this limitation, variations of transformer-based inpainting models have been developed with linear computational complexity, integrating the strengths of both convolution and attention mechanisms [7,9,58].

Structure-guided inpainting methods incorporate prior guidance to generate semantically consistent inpainting results. Traditional GAN-Inpainting models employ various forms of structural guidance, such as facial keypoints [52], canny edges [14,35,58], segmentation maps [5], and gradient maps [11]. While edge, segmentation, and gradient information generally offer more robust guidance compared to landmark-based methods, the accuracy of the structural predictions is critical for achieving high-quality inpainting results. Inaccurate or redundant structural information can significantly degrade the performance of inpainting models, leading to artifacts or inconsistencies in the generated regions [52]. In parallel, pluralistic inpainting methods aim to generate multiple plausible solutions for a missing region, enhancing the diversity of inpainting outputs [3,63]. Guided pluralistic inpainting models further refine these outputs by incorporating user input [28,62] or prior information [22,32].

Building on these approaches, in this work we incorporate edge priors to guide the inpainting process, ensuring semantically consistent results while enabling the generation of pluralistic outputs that can represent Cleft conditions. During the training phase, we use edge contours derived from a face parsing module for structural guidance. In the inference phase, we leverage facial keypoints, connecting them to form edge contours which guide the inpainting process. Facial keypoints are employed in inference phase for their ease of prediction and flexibility in modifying landmark positions, which facilitates the generation of diverse and realistic outcomes.

In GAN-based inpainting models, discriminators are crucial for enhancing the realism of generated textures by guiding the generator to produce visually coherent and perceptually pleasing results. Traditionally, GAN inpainting models employ DCGAN based [39] image level discriminators. However, image level discriminators fails to generate inpainting regions that maintain local consistency with surrounding areas. To address this limitation, global and local discriminators have been introduced to ensure overall image consistency and local coherence [17].

Building on this concept, patch-wise discriminators were proposed as a more generalized approach, focusing on small patches rather than the entire image. This design enhances the model's ability to distinguish between real and synthesized regions locally, contributing to more precise inpainting results [30,54,61]. Further advancements in patch-wise discriminators include the incorporation of spectral normalization, which constrains the spectral norm of the discriminator's weight matrix, leading to smoother gradient updates and more stable training processes [35,55,61]. Yang et al. [52] improved this approach by integrating attention layers into the discriminator to adaptively manage features, thus enhancing image consistency. Additionally, WGAN [53,57] and LSGAN [41] based discriminators have been explored for faster and more stable training. Despite these innovations, the design of discriminators in GAN-inpainting models remains comparatively under explored, especially when contrasted with the extensive modifications applied to generators.

Inpainting models are often trained with free-form masks, where irregularly shaped masks are randomly placed on images during training [8,29,35,52]. In contrast, our method employs fixed-form masks [14,48,56], specifically conditioning the inpainting of Cleft-prone upper lip and philtrum regions based on lower lip and surrounding regions which remains unaffected in Cleft conditions.

This targeted approach ensures a more focused reconstruction of the affected areas. Our proposed inpainting mask covers between 10% and 30% of the total image area.

2.2   Machine Learning Applications for Cleft Condition

Machine learning applications for Cleft Lip and Palate analysis can be broadly divided into three subdomains: Cleft surgical marker localization [27,42,43], Cleft severity prediction [33,37] and synthetic generation of Cleft-specific facial regions [1,6,15]. Facial landmark models are instrumental in predicting Cleft-specific landmarks for tasks such as prenatal diagnosis, Cleft severity assessment, surgical planning, and Cleft surgical success assessment. Petcas et al. [37] used pretrained Facial Beauty Prediction (FBP) models to predict Cleft severity, benchmarking their results against human ratings. However, their study shows that FBP models trained on healthy individuals fail to capture the key visual indicators specific for Cleft severity.

Due to the lack of publicly available Cleft-specific facial datasets, models for generating Cleft-specific synthetic faces are trained using publicly accessible facial datasets of healthy individuals. These synthetic faces are designed to resemble post-operative Cleft lips, which appear similar to normal lips, using GAN-based inpainting or inversion techniques. Hayajneh et al. [15] have deployed GAN inversion technique on Stylegan2 [18] latent space to generate natural looking normal lips for Cleft patients. They further extend the work to generate a synthetic Cleft dataset using GAN inversion techniques. While this work has explored the development of a synthetic Cleft dataset to address the lack of publicly available data, the authors report an inherent limitation in the model's ability to accurately categorize Cleft faces within StyleGAN's latent space. This limitation makes it challenging to modify the severity of the Cleft condition without altering surrounding facial features [15]. A GAN-Inpainting approach has been proposed by Chen et al. to generate post-operative Cleft faces [6]. This model is trained on the CelebA dataset [31] using free-form masks and processes the entire facial image at a resolution of 256×256 to generate post-operative Cleft faces. Atputharuban et al. [1] trained a GAN-inpainting model to generate upper lip and philtrum regions that resemble post-operative Cleft conditions and developed a surgical success score using the GAN-inpainted images as reference points to assess Cleft surgery outcomes. However, It can be observed that the inpainting models generate flat, asymmetric lips that do not accurately represent Cleft conditions. These models struggle to capture the unique anatomical features of Cleft-affected lips. We build upon this work by generating realistic Cleft lips with real-time editable capabilities to model varying Cleft severities.

3   Methodology

The proposed approach consists of three key components. First, the dataset acquisition pipeline, which involves obtaining Cleft-prone orofacial regions of the face, the inpainting mask, and the corresponding edge contour (Sect. 3.1). Next, CleftLipGAN inpainting model is described in detail (Sect. 3.2). Finally, the custom landmark detector module, for localizing keypoints in the orofacial region, is presented (Sect. 3.3).

3.1 Dataset Acquisition

Experiments in this study are focused on Cleft-prone orofacial regions, as this is the area primarily affected by Cleft conditions. Clinicians use these specific cropped regions to assess surgical outcomes, making it highly relevant for our study. However, due to the unavailability of Cleft-specific datasets capturing orofacial regions, we created a dataset from the high resolution FFHQ facial dataset [18] to support our experiments.

FFHQ dataset [18] comprises 70,000 in-the-wild images of healthy individuals, representing a wide variation in age and racial demographics with the resolution of 1024×1024. We cropped the orofacial regions and generated custom fixed-form masks following the approach outlined in [1]. We utilized the facial keypoint detector and the face parsing model proposed by FaRL [64] for generating masks cropping the region of interest (ROI) and generating edge contour maps. Resulting images are resized to the resolution of 256×256. The dataset was then manually curated to remove images with occlusions, extreme poses, and blurred facial regions. The final dataset consists of 51,000 images, of which 46,000 are used for training inpainting models and remaining 5000 images are used for testing.

We use a post-operative Cleft dataset to evaluate the performance of inpainting models on Cleft condition. This fully anonymized dataset, consisting of 164 post-operative Cleft repair surgery images, was collected at Children's Health Ireland at Temple Street between 2009 and 2012 under a research agreement between Children's Health Ireland at Temple Street and University College Dublin. We refer to this post-operative Cleft dataset as Cleft164 dataset. Cleft surgeries were performed on children aged between 6 months to 1 year, with post-operative images captured during follow-up consultations when the children were 5 years old. Cleft164 dataset was used as a test set to assess the performance of CleftLipGAN model on abnormal post-operative Cleft images. Inpainting masks for the Cleft164 dataset was manually generated, encompassing the upper lip and philtrum region.

3.2 CleftLipGAN Model for Post-Operative Cleft Lip Reconstruction

We have proposed CleftLipGAN, an interactive inpainting model designed for generating post-operative Cleft images. Formally, the inpainting pipeline is formulated as follows: the input for the inpainting module, I_{input}, is obtained by concatenating the masked image, $I_{masked} = I \odot M$, the mask image M and the corresponding edge contour I_{edge}. The input image, I_{input}, is then processed by the proposed inpainting model, CleftLipGAN, to produce a semantically appealing inpainted image, I_{out}. The overall formulation of Cleft lip reconstruction is denoted as $I_{out} = \text{CleftLipGAN}(I_{input})$. The following subsections describe sub modules of CleftLipGAN: namely generator, discriminator and the structural guidance pipeline. CleftLipGAN inpainting module pipeline is illustrated in Fig. 1.

Generator. We adapt the generator module proposed by Xiankang et al. [58] for our inpainting model. The module is built in encoder-decoder fashion with U-shaped skip connections to pass shallow features from the encoder to decoder. The generator architecture is composed of Spatial Attention Based Gated Convolution(SAGC) module and Channel Attention based Gated Convolution(CAGC) module. The SAGC module is employed in both the encoder and decoder blocks of the network, aiding in the extraction of structural information. In the SACG module, spatial attention is used instead of self-attention to achieve linear complexity when handling high-resolution features in both the encoder and decoder. The bottleneck layers of the network are built using the CAGC block, which employs Squeeze-and-Excitation attention for semantic feature extraction. SAGC and CAGC blocks, along with the gated convolution layer, produce semantically consistent images without blurring or watermark artifacts. Gated convolution layers are employed for their ability to dynamically extract features using learned gating mechanisms, which help guide the weights of each pixel based on prior information. We choose the number of the bottleneck layers to be 4 for this task. The model inputs for the generator are masked RGB image, grayscale inpainting mask and grayscale edge contour.

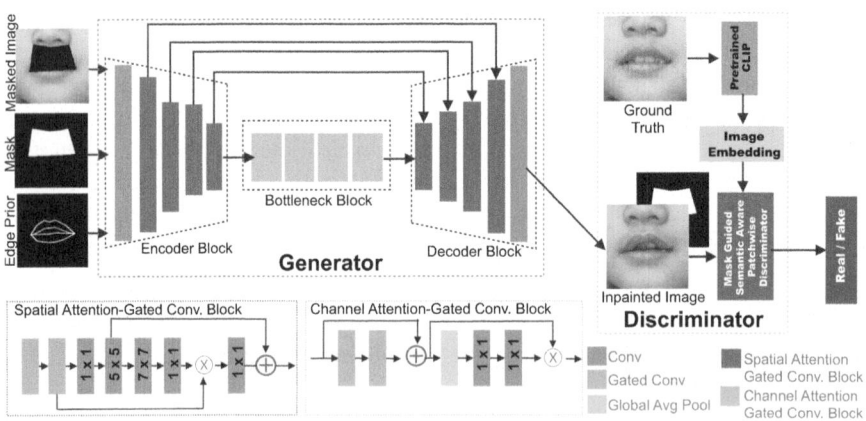

Fig. 1. Proposed CleftLipGAN inpainting model for reconstructing post-operative Cleft images: The model takes a masked image, inpainting mask, and edge prior as inputs to synthetically generate Cleft-prone regions.

Discriminator. We adapt the discriminator proposed by [24] for image super-resolution task. This discriminator module incorporates semantic guidance by fusing features extracted by a pretrained feature extractor, allowing the discriminator to learn fine-grained distributions. The feature extraction branch built using a pretrained CLIP 'RN50' module [38], is selected for its robust representation capabilities. As illustrated in Fig. 2, the ground truth image is passed through the CLIP model to obtain semantic details, which are then fused with the patch-wise discriminator via the semantic-aware fusion block to guide the assessment of realism. Unlike vanilla discriminators that focus on coarse-grained image distributions, this enhanced module leverages self-attention and cross-attention mechanisms to discriminate fine-grained details, such as textures, aiding the inpainting process. The inpainted image and mask are input into the discriminator to condition the evaluation specifically on the inpainted regions.

Structural Guidance. We train the inpainting network using edge contours corresponding to the lip region. We used the FaRL face parsing model [64] to obtain segmentation maps, from which we extract lip contours by outlining the segmented regions. During inference, we use facial keypoints generated by a custom keypoint detection model to construct lip contours by connecting the keypoints, following the approach proposed in [55]. This approach facilitates more flexible modifications, and by utilizing contours, richer structural information can be provided for inpainting.

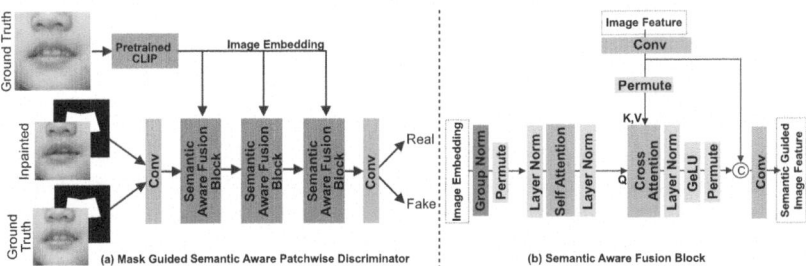

Fig. 2. Mask-guided semantic-aware patch-wise discriminator for the CleftLipGAN inpainting model: Image embedding features are extracted using a pretrained CLIP model. These features are fused with image features from the discriminator network through a Semantic-Aware Fusion block. This approach enhances the semantic extraction of patch-wise features, helping the discriminator better assess the realism of inpainted regions.

Loss Functions. To produce high quality inpainting results we have incorporated a combination of loss functions in line with GAN-Inpainting literature [58]. We train the inpainting network with L_1 loss to ensure pixel-wise consistency and ensure realness in generated pixel values. Perceptual loss L_{perc} is used to enforce high level structural and semantic features, while adversarial loss is employed L_{adv} to improve overall quality of the output. The Loss function is denoted as

$$L_{\text{total}}(I_{\text{gen}}, I_{\text{gt}}) = \lambda_1 L_1 + \lambda_2 L_{\text{perc}} + \lambda_3 L_{\text{adv}}$$

where $\lambda_1 = 5$, $\lambda_2 = 0.4$ and $\lambda_3 = 0.05$ are chosen based on experiments.

3.3   Landmark Detection for Orofacial Region

We also developed a custom lips keypoint detector module specifically designed to predict keypoints in cropped orofacial region. In our approach, the predicted keypoints act as guidance during inference, enabling users to adjust and refine them to create edge contours, facilitating the generation of pluralistic results.

Our facial keypoint detector model is trained to predict 20 keypoints specifically from the cropped facial region. For this purpose, we have used a subset of 6000 images randomly obtained from CelebA-HQ [31] dataset, in which 5000 images were used for training and 1000 images were used for validation. We have preprocessed the images as outlined in Sect. 3.1 and resize the images to 256×256 resolution. The ground truth keypoints for the lip region were obtained using the FaRL facial landmark model [64]. The keypoint detection model is built with a MobileNet [16] backbone and trained using L1 Loss. We measure the performance of keypoint detector model against the test dataset with Normalized Mean Squared Error(NSME), with the lip width used as the normalization factor. NMSE on test set is reported to be 0.23.

4   Experiments and Discussion

We have performed the experiments on a single NVIDIA GForce RTX 4090 GPU with the batch size of 4. We train the CleftLipGAN inpainting model with Adam optimizer [20] with $\beta_1 = 0.99$ and $\beta_2 = 0.5$. Model is trained with a learning rate of 5×10^{-4} and trained for 300000 iterations. It utilizes a warm-up phase consisting of 2 epochs, where only the generator is trained. Following this, generator and discriminator are concurrently trained to fine tune the adversarial loss.

We benchmark the performance of the CleftLipGAN inpainting model against state-of-the-art inpainting models using both quantitative and qualitative evaluations. For quantitative evaluation, we employ PSNR, SSIM [50], and Brisque [34] to assess the performance of the inpainting models. PSNR measures pixel-wise reconstruction accuracy, indicating how closely the inpainted result matches the original image in terms of visual fidelity. SSIM [50] evaluates structural similarity, ensuring that the inpainted regions maintain continuity with the surrounding pixels and remain contextually aligned. Brisque [34] evaluates the perceptual quality of an image by identifying distortions and visual artifacts that affect overall image quality.

Table 1. Quantitative comparison on the FFHQ test set shows that the proposed CleftLipGAN model outperforms in all three metrics employed.

Model	PSNR ↑	SSIM [50] ↑	BRISQUE [34] ↓
Ours	**31.46**	**0.9382**	**21.89**
AGG-Net [58]	30.33	0.9246	23.37
HINT [7]	26.25	0.9048	25.49
E2F-GAN [14]	27.89	0.9087	26.67
HourglassAttention [8]	29.12	0.9178	25.89
DeepfillV2 [55]	25.39	0.8979	26.17

For comprehensive evaluation, we compare our model against three structural guided inpainting models, E2F-Net [14], DeepfillV2 [55] and AGGNet [58]. We retrained the models using the hyperparameters specified by the authors, but incorporated our edge contour-based guidance for the inpainting process. We also benchmark against HINT [7] inpainting model, a vision transformer based model which has demonstrated superior performance compared to contemporary state-of-the-art inpainting models. Additionally, we evaluate our model's performance against the Cleft inpainting model proposed by Atputharuban et al. [1] for synthesizing Cleft-prone region, which is built on the HourglassAttention [8] architecture. We benchmark the performance of CleftLipGAN against this model to assess its effectiveness in Cleft inpainting.

Quantitative studies are performed on the test set proposed in Sect. 3.1 consisting 5000 images and the results are presented in Table 1. It can be observed that our model outperforms the other state-of-the-art inpainting models on all evaluation metrics. This demonstrates that incorporating semantic-aware discriminators enhances inpainting results when combined with carefully selected generator architecture.

This observation can be further validated with qualitative evaluation. We conduct qualitative evaluation on Cleft164 dataset specified in Sect. 3.1. Fig. 5 illustrates the post operative Cleft lip reconstruction results. It can be observed that non structure based inpainting models produce flat upper lips that lack symmetry and are not semantically accurate. In contrast, by adjusting facial keypoints in inference stage, structure based inpainting models which were trained on healthy individuals can generate lips which resemble that of Cleft condition. When compared to other structure guided inpainting models, CleftLipGAN model surpasses them by generating structurally coherent and visually appealing lips for Cleft conditions, with fine textural details preserved. Although E2F-GAN [14] can produce accurate structural results, it consistently fails to capture fine details and struggles to produce smooth textures. Additionally, the DeepFillV2 [55] model generates blurry results with visible artifacts.

We also evaluate the possibility of generating images resembling Cleft conditions, with the results presented in Fig. 3. It can be observed that CleftLip-GAN produces images closely resembling Cleft conditions with fewer artifacts

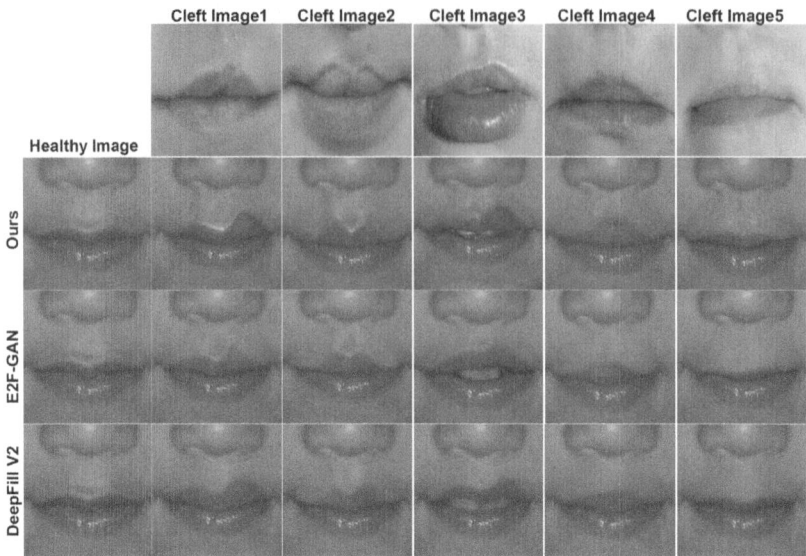

Fig. 3. Synthetic generation of Cleft lips from healthy lips. We generate lips that resemble the Cleft condition from healthy lip images by fine tuning facial landmark locations. All three structure guided inpainting models successfully produce Cleft-like lips; however, our model captures finer details and achieves a more accurate structural representation of Cleft lips, enhancing the realism and anatomical correctness of the generated results.

compared to other structure-guided inpainting models. We believe that the Cleft-LipGAN model can be employed to generate a synthetic dataset resembling Cleft conditions, particularly in this domain where no publicly available Cleft datasets exist.

Additionally, we have developed an interactive user interface, as illustrated in Figure. 4. The user interface integrates lip landmark prediction module

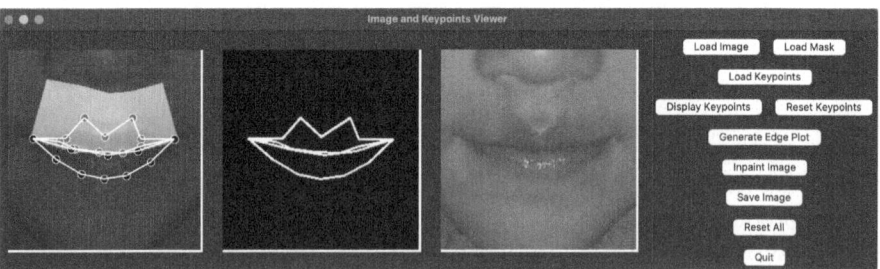

Fig. 4. User interface for selective inpainting for Cleft region. Lip keypoints can be manually adjusted to obtain a edge contour map, which in turn guides CleftLipGAN model to produce pluralistic inpainting results.

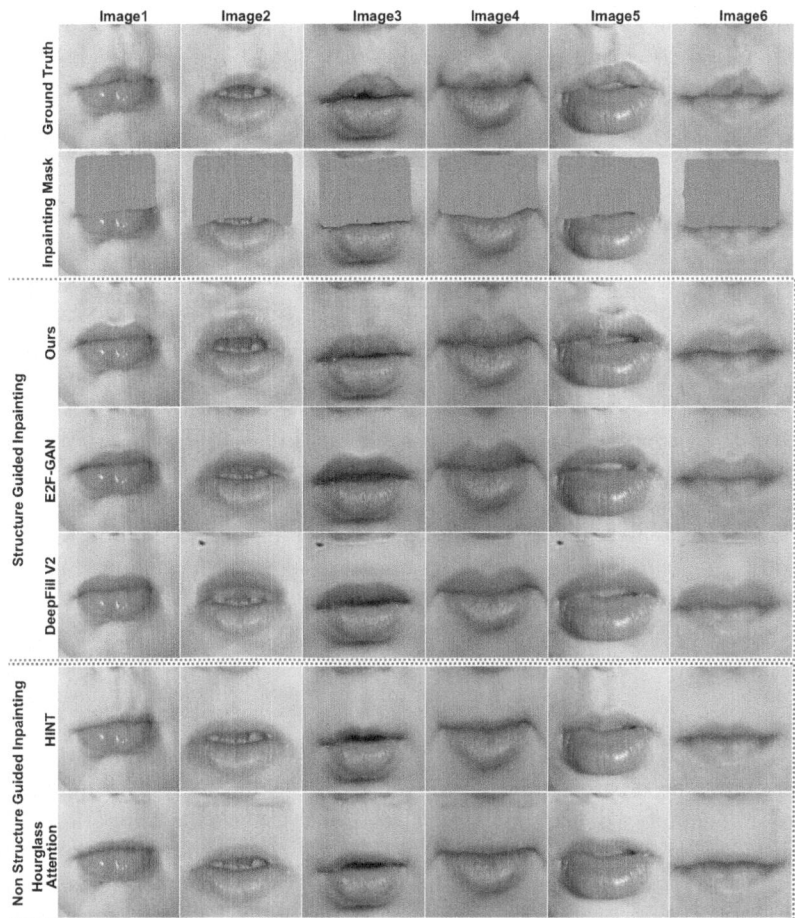

Fig. 5. Synthetic Non-Cleft Lip generation for patients with Cleft condition: Notably, we can observe that non structure based inpainting models, HourglassAttention [8] and HINT [7] fail to capture the specific semantics of Cleft conditions. In contrast, structure guided inpainting models can generate semantically appealing results specific for Cleft condition with user guidance. Additionally, DeepFillV2 [55] model produces visible artifacts making it unsuitable for Cleft reconstruction. In comparison, E2F-GAN [14] and our model generates semantically appealing lips with minimal artifacts. But E2F-GAN [14] fails to capture fine details in the inpainted region. On comparison to both structure-guided and non-structure-guided inpainting models, our model generates lips that are both semantically and structurally accurate for Cleft conditions.

(Sect. 3.3) for identifying keypoints in the orofacial region and the inpainting model (Sect. 3.2) for reconstruction of Cleft prone areas. The interface receives the image of the orofacial region and the corresponding mask, where the inpainting will be applied, as inputs. Baseline lip keypoints are predicted using the keypoint detector module, which can be interactively adjusted by the user to generate pluralistic results. Once the adjustments are finalized, an edge contour map is generated by connecting the keypoints. The masked image, inpainting mask, and edge contour map are then fed into the inpainting model to generate the inpainted image, which is displayed in the interface for further fine-tuning.

5   Conclusion and Future Work

In this study, we demonstrate the ability of image inpainting models trained on images of healthy individuals to capture the semantics of Cleft conditions and generate anatomically accurate lips for Cleft patients using structural guidance. To achieve this, we propose the CleftLipGAN model, which features a novel mask-guided, semantic-aware, patch-wise discriminator. Our results indicate that the proposed model outperforms existing state-of-the-art inpainting methods in producing semantically coherent Cleft lips for both normal and post-operative condition. These findings are supported by comprehensive quantitative and qualitative analyses. Our pipeline, which includes facial landmark localization and reconstruction, is applicable to privacy-sensitive Cleft images, focusing on the cropped orofacial region to generate post-operative outcomes. Additionally, the interactive user interface enables the generation of synthetic Cleft faces, addressing the challenge of limited Cleft-specific datasets by providing synthetic Cleft data, which can be used, for example, in training facial landmark models tailored to Cleft conditions.

Future work will focus on clinically validating the results and expanding the evaluation to a broader distribution of Cleft faces. Additionally, we aim to further explore the semantics of Cleft facial features and automate the generation of Cleft lips based on specific Cleft conditions. With thorough validation, this approach could serve as a tool for generating post-operative Cleft facial reconstructions, aiding in patient and caretaker education about potential outcomes, as well as providing an objective measure to assess surgical success. Furthermore, we plan to explore alternative conditional inpainting methods to produce high-resolution, plausible lip reconstructions for Cleft conditions.

Acknowledgement. This publication has emanated from research conducted with the financial support of Science Foundation Ireland under Grant number [12/RC/2289_P2]. For the purpose of Open Access, the author has applied a CC BY public copyright licence to any Author Accepted Manuscript version arising from this submission.

References

1. Atputharuban, D., Theopold, C., Lawlor, A.: Enhancing surgical visualization: feasibility study on GAN-based image generation for post operative cleft palate images. In: Proceedings of the 13th International Conference on Pattern Recognition Applications and Methods, pp. 939–945 (2024). https://doi.org/10.5220/0012576900003654

2. Brief, D.J., Behle, D.J.H., Stellzig-Eisenhauer, D.A., Hassfeld, D.S.: Precision of landmark positioning on digitized models from patients with cleft lip and palate. Cleft Palate Craniofac. J. **43**(2), 168–173 (2006). https://doi.org/10.1597/04-106.1

3. Cai, W., Wei, Z.: PiiGAN: generative adversarial networks for pluralistic image inpainting. IEEE Access **8**, 48451–48463 (2020). https://doi.org/10.1109/ACCESS.2020.2979348

4. Chaudhari, P.K., et al.: Factors affecting high caries risk in children with and without cleft lip and/or palate: a cross-sectional study. Cleft Palate Craniofac. J. **58**(9), 1150–1159 (2021). https://doi.org/10.1177/1055665620980206

5. Chen, Q., Qiang, Z., Zhao, Y., Lin, H., He, L., Dai, F.: Rdfinet: reference-guided directional diverse face inpainting network. In: Complex & Intelligent Systems, pp. 1–12 (2024)

6. Chen, S., Atapour-Abarghouei, A., Ho, E.S., Shum, H.P.: INCLG: inpainting for non-cleft lip generation with a multi-task image processing network. Softw. Impacts **17**, 100517 (2023). https://doi.org/10.1016/j.simpa.2023.100517

7. Chen, S., Atapour-Abarghouei, A., Shum, H.P.H.: HINT: high-quality inpainting transformer with mask-aware encoding and enhanced attention. arXiv preprint arXiv:2402.14185 (2024)

8. Deng, Y., Hui, S., Meng, R., Zhou, S., Wang, J.: Hourglass attention network for image inpainting. In: ECCV, pp. 483–501 (2022). https://doi.org/10.1007/978-3-031-19797-0_28

9. Deng, Y., Hui, S., Zhou, S., Meng, D., Wang, J.: T-former: an efficient transformer for image inpainting. In: Proceedings of the 30th ACM International Conference on Multimedia. MM '22. ACM (2022). https://doi.org/10.1145/3503161.3548446

10. Ding, F., Zhu, G., Li, Y., Zhang, X., Atrey, P.K., Lyu, S.: Anti-forensics for face swapping videos via adversarial training. IEEE Trans. Multimedia **24**, 3429–3441 (2022). https://doi.org/10.1109/TMM.2021.3098422

11. Dong, Q., Cao, C., Fu, Y.: Incremental transformer structure enhanced image inpainting with masking positional encoding. In: Proceedings of the IEEE/CVF Conference on Computer Vision and Pattern Recognition (CVPR), pp. 11358–11368 (2022)

12. Duggal, I., Talwar, A., Duggal, R., Chaudhari, P., Samrit, V.: Comparative evaluation of nasolabial appearance of unilateral cleft lip and palate patients by professional, patient and layperson using 2 aesthetic scoring systems: a cross sectional study. Orthod. Craniofac. Res. **26** (2023). https://doi.org/10.1111/ocr.12663

13. Grollemund, B., et al.: The impact of having a baby with cleft lip and palate on parents and on parent-baby relationship: the first French prospective multicentre study. BMC Pediatr. **20**, 1–11 (2020)

14. Daryani, A.E., et al.: E2F-GAN: eyes-to-face inpainting via edge-aware coarse-to-fine GANs. IEEE Access **10**, 32406–32417 (2022)

15. Hayajneh, A., Serpedin, E., Shaqfeh, M., Glass, G., Stotland, M.A.: CleftGAN: adapting a style-based generative adversarial network to create images depicting cleft lip deformity. arXiv preprint arXiv:2310.07969 (2023)

16. Howard, A.G., et al.: MobileNets: efficient convolutional neural networks for mobile vision applications. arXiv preprint arXiv:1704.04861 (2017)

17. Iizuka, S., Simo-Serra, E., Ishikawa, H.: Globally and locally consistent image completion. ACM Trans. Graph. Proc. SIGGRAPH **36**(4), 107 (2017)

18. Karras, T., Laine, S., Aila, T.: A style-based generator architecture for generative adversarial networks. arXiv preprint arXiv:1812.04948 (2018)

19. Kau, C.H., Medina, L., English, J.D., Xia, J., Gateno, J., Teichgraber, J.: A comparison between landmark and surface shape measurements in a sample of cleft lip and palate patients after secondary alveolar bone grafting. Orthod. Art Pract. Dentofac. Enhancement **12**(3), 188 (2011)
20. Kingma, D.P., Ba, J.: Adam: a method for stochastic optimization. arXiv preprint arXiv:1412.6980 (2017)
21. Ko, K., Kim, C.S.: Continuously masked transformer for image inpainting. In: Proceedings of the IEEE/CVF International Conference on Computer Vision (ICCV), pp. 13169–13178 (2023)
22. Lahiri, A., Jain, A.K., Agrawal, S., Mitra, P., Biswas, P.K.: Prior guided GAN based semantic inpainting. In: Proceedings of the IEEE/CVF Conference on Computer Vision and Pattern Recognition (CVPR) (2020)
23. Leslie, E.J., Marazita, M.L.: Genetics of cleft lip and cleft palate. Am. J. Med. Genet. C Semin. Med. Genet. **163**(4), 246–258 (2013). https://doi.org/10.1002/ajmg.c.31381
24. Li, B., et al.: SeD: semantic-aware discriminator for image super-resolution. arXiv preprint arXiv:2402.19387 (2024)
25. Li, W., Lin, Z., Zhou, K., Qi, L., Wang, Y., Jia, J.: MAT: mask-aware transformer for large hole image inpainting. In: Proceedings of the IEEE/CVF Conference on Computer Vision and Pattern Recognition (2022)
26. Li, Y., Liu, S., Yang, J., Yang, M.H.: Generative face completion. arXiv preprint arXiv:1704.05838 (2017)
27. Li, Y., Cheng, J., Mei, H., Ma, H., Chen, Z., Li, Y.: CLPNet: cleft lip and palate surgery support with deep learning. In: 2019 41st Annual International Conference of the IEEE Engineering in Medicine and Biology Society (EMBC), pp. 3666–3672 (2019). https://doi.org/10.1109/embc.2019.8857799
28. Liu, C., Xu, S., Peng, J., Zhang, K., Liu, D.: Toward interactive image inpainting via robust sketch refinement. IEEE Trans. Multimedia **26**, 9973–9987 (2024). https://doi.org/10.1109/TMM.2024.3402620
29. Liu, G., Reda, F.A., Shih, K.J., Wang, T.C., Tao, A., Catanzaro, B.: Image inpainting for irregular holes using partial convolutions. arXiv preprint arXiv:1804.07723 (2018)
30. Liu, H., Jiang, B., Song, Y., Huang, W., Yang, C.: Rethinking image inpainting via a mutual encoder-decoder with feature equalizations. arXiv preprint arXiv:2007.06929 (2020)
31. Liu, Z., Luo, P., Wang, X., Tang, X.: Deep learning face attributes in the wild. arXiv preprint arXiv:1411.7766 (2014)
32. Lu, W., et al.: Do inpainting yourself: generative facial inpainting guided by exemplars. arXiv preprint arXiv:2202.06358 (2022)
33. McCullough, M., et al.: Convolutional neural network models for automatic preoperative severity assessment in unilateral cleft lip. Plast. Reconstr. Surg. **148**(1), 162–169 (2021). https://doi.org/10.1097/prs.0000000000008063
34. Mittal, A., Moorthy, A.K., Bovik, A.C.: Blind/referenceless image spatial quality evaluator. In: 2011 Conference Record of the Forty Fifth Asilomar Conference on Signals, Systems and Computers (ASILOMAR), pp. 723–727. IEEE (2011)
35. Nazeri, K., Ng, E., Joseph, T., Qureshi, F., Ebrahimi, M.: EdgeConnect: structure guided image inpainting using edge prediction. In: Proceedings of the IEEE/CVF International Conference on Computer Vision Workshops (2019)
36. Paradowska-Stolarz, A., Mikulewicz, M., Duś-Ilnicka, I.: Current concepts and challenges in the treatment of cleft lip and palate patients–a comprehensive review. J. Personalized Med. **12**(12), 2089 (2022). https://doi.org/10.3390/jpm12122089

37. Patcas, R., et al.: Facial attractiveness of cleft patients: a direct comparison between artificial-intelligence-based scoring and conventional rater groups. Eur. J. Orthod. (2019). https://doi.org/10.1093/ejo/cjz007

38. Radford, A., et al.: Learning transferable visual models from natural language supervision. arXiv preprint arXiv:2103.00020 (2021)

39. Radford, A., Metz, L., Chintala, S.: Unsupervised representation learning with deep convolutional generative adversarial networks. arXiv preprint arXiv:1511.06434 (2016)

40. Rahimov, F., Jugessur, A., Murray, J.C.: Genetics of nonsyndromic orofacial clefts. Cleft Palate Craniofac. J. **49**(1), 73–91 (2012)

41. Ren, K., Meng, L., Fan, C., Wang, P.: Least squares DCGAN based semantic image inpainting. In: 2018 5th IEEE International Conference on Cloud Computing and Intelligence Systems (CCIS), pp. 890–894. IEEE (2018)

42. Rosero, K., Salman, A., Sisman, B., Hallac, R., Busso, C.: Enhanced facial landmarks detection for patients with repaired cleft lip and palate, pp. 1–10 (2024). https://doi.org/10.1109/FG59268.2024.10582022

43. Sayadi, L.R., Hamdan, U.S., Zhangli, Q., Hu, J., Vyas, R.M.: Harnessing the power of artificial intelligence to teach cleft lip surgery. Plast. Reconstr. Surg. Glob. Open (2022). https://doi.org/10.1097/gox.0000000000004451

44. Stiernman, M., Österlind, K., Rumsey, N., Becker, M., Persson, M.: Parental and health care professional views on psychosocial and educational outcomes in patients with cleft lip and/or cleft palate. Eur. J. Plast. Surg. **42**(4), 325–336 (2019). https://doi.org/10.1007/s00238-019-01530-0

45. Suvorov, R., et al.: Resolution-robust large mask inpainting with fourier convolutions. In: 2022 IEEE/CVF Winter Conference on Applications of Computer Vision (WACV), pp. 3172–3182 (2022). https://doi.org/10.1109/WACV51458.2022.00323

46. Trotman, C.A.: Faces in 4 dimensions: why do we care, and why the fourth dimension? Am. J. Orthod. Dentofac. Orthop. **140**(6), 895–899 (2011)

47. Trotman, C.A., et al.: Functional outcomes of cleft lip surgery. part i: study design and surgeon ratings of lip disability and need for lip revision. Cleft Palate-Craniofac. J. **44**(6), 598–606 (2007)

48. Ud Din, N., Javed, K., Bae, S., Yi, J.: A novel GAN-based network for unmasking of masked face. IEEE Access **8**, 44276–44287 (2020). https://doi.org/10.1109/ACCESS.2020.2977386

49. Wadde, K., Chowdhar, A., Venkatakrishnan, L., Ghodake, M., Sachdev, S.S., Chhapane, A.: Protocols in the management of cleft lip and palate: a systematic review. J. Stomatol. Oral Maxillofac. Surg. **124**(2), 101338 (2023)

50. Wang, Z., Bovik, A., Sheikh, H., Simoncelli, E.: Image quality assessment: from error visibility to structural similarity. IEEE Trans. Image Process. **13**(4), 600–612 (2004). https://doi.org/10.1109/TIP.2003.819861

51. Xiang, H., Zou, Q., Nawaz, M.A., Huang, X., Zhang, F., Yu, H.: Deep learning for image inpainting: a survey. Pattern Recogn. **134**, 109046 (2023)

52. Yang, Y., Guo, X.: Generative landmark guided face inpainting. In: Chinese Conference on Pattern Recognition and Computer Vision (PRCV), pp. 14–26. Springer (2020)

53. Yi, Z., Tang, Q., Azizi, S., Jang, D., Xu, Z.: Contextual residual aggregation for ultra high-resolution image inpainting. arXiv preprint arXiv:2005.09704 (2020)

54. Yilmaz, H.N., Ozbilen, E.O., Üstün, T.: The prevalence of cleft lip and palate patients: a single-center experience for 17 years. Turk. J. Orthod. **32**(3), 139–144 (2019)

55. Yu, J., Lin, Z., Yang, J., Shen, X., Lu, X., Huang, T.: Free-form image inpainting with gated convolution. arXiv preprint arXiv:1806.03589 (2019)
56. Yu, J., Lin, Z., Yang, J., Shen, X., Lu, X., Huang, T.S.: Generative image inpainting with contextual attention. arXiv preprint arXiv:1801.07892 (2018)
57. Yu, J., Lin, Z., Yang, J., Shen, X., Lu, X., Huang, T.S.: Generative image inpainting with contextual attention. In: 2018 IEEE/CVF Conference on Computer Vision and Pattern Recognition, pp. 5505–5514 (2018). https://doi.org/10.1109/CVPR.2018.00577
58. Yu, X., Dai, L., Chen, Z., Sheng, B.: AGG: attention-based gated convolutional GAN with prior guidance for image inpainting. Neural Comput. Appl. **36**(20), 12589–12604 (2024)
59. Yu, Y., et al.: Diverse image inpainting with bidirectional and autoregressive transformers. In: Proceedings of the 29th ACM International Conference on Multimedia, pp. 69–78. MM '21, Association for Computing Machinery, New York, NY, USA (2021). https://doi.org/10.1145/3474085.3475436
60. Zeng, Y., Fu, J., Chao, H., Guo, B.: Learning pyramid-context encoder network for high-quality image inpainting. arXiv preprint arXiv:1904.07475 (2019)
61. Zeng, Y., Lin, Z., Yang, J., Zhang, J., Shechtman, E., Lu, H.: High-resolution image inpainting with iterative confidence feedback and guided upsampling. arXiv preprint arXiv:2005.11742 (2020)
62. Zhang, X., Ma, W., Varinlioglu, G., Rauh, N., He, L., Aliaga, D.: Guided pluralistic building contour completion. Vis. Comput. **38**(9), 3205–3216 (2022)
63. Zheng, C., Cham, T.J., Cai, J.: Pluralistic free-from image completion. Int. J. Comput. Vis., 1–20 (2021)
64. Zheng, Y., et al.: General facial representation learning in a visual-linguistic manner. arXiv preprint arXiv:2112.03109 (2021)

A Comparative Study on Diffusion Sampling Methods Across Diverse Medical Imaging Modalities

Muhammad Ali Farooq[1]([⊠]) [iD], Ayman Abaid[2] [iD], Ihsan Ullah[2,3] [iD],
and Peter Corcoran[1] [iD]

[1] School of Engineering, University of Galway, Galway H91TK33, Ireland
{muhammadali.farooq,peter.corcoran}@universityofgalway.ie
[2] School of Computer Science, University of Galway, Galway H91TK33, Ireland
{a.abaid1,ihsan.ullah}@universityofgalway.ie
[3] Insight SFI Research Centre for Data Analytics, University of Galway,
Galway, Ireland

Abstract. The evaluation of diffusion-based image sampling methods is pivotal in improving the quality and reliability of synthetic data generation, particularly in medical imaging applications. Medical imaging requires high precision and fidelity, as even subtle artifacts or inconsistencies can significantly impact clinical decision-making. This study examines the effectiveness of four different image sampling techniques across various medical imaging modalities, focusing on dermoscopic skin lesion data, computed tomography angiography for Type B Aortic Dissection, and chest X-ray imaging. By systematically assessing these methods, we aim to enhance the fidelity of synthetic datasets, ensuring they more closely resemble real-world clinical data thereby supporting more accurate diagnostics, treatment planning, and prognostic predictions. In this work, we evaluate the performance of four different sampling techniques by incorporating Euler, Euler A, Denoising Diffusion Implicit Mode (DDIM), and Pseudolinear Multistep (PLMS) approaches for medical image synthesis. The study utilizes quantitative metrics including Structural Similarity Index Measure (SSIM), and Learned Perceptual Image Patch Similarity (LPIPS) to assess the realism and structural integrity of the generated images. Additionally, we employed t-SNE visualization to illustrate the latent feature representations of rendered synthetic medical images, providing an intuitive understanding of the underlying structure. We also analyzed and compared the computational complexity associated with each image sampling technique, offering insights into the efficiency of different approaches. The generated medical images are available at Diffusion-Sampling-for-Medical-Image-Synthesis (GitHub Link: https://github.com/MAli-Farooq/Diffusion-Sampling-for-Medical-Image-Synthesis-).

Keywords: Synthesis · Stable Diffusion · CTA · Dermoscopic · X-ray · Text to Image

M. Cho et al. (Eds.): ACCV 2024 Workshops, LNCS 15482, pp. 197–210, 2025.
https://doi.org/10.1007/978-981-96-2641-0_13

1   Introduction

Medical imaging plays a critical role in modern healthcare, providing clinicians with invaluable insights into the structure, function, and pathology of the human body. However, the medical data privacy concern and further acquisition of high-quality medical imaging data often relies on expensive and time-consuming processes, limiting the availability of large datasets for training machine learning models. Additionally, the diversity and complexity of medical imaging data present unique challenges for traditional machine learning approaches. To address these challenges, recent advancements in artificial intelligence (AI) have shown promise in generating synthetic medical imaging data using generative models. In recent years, there has been a growing interest in the application of image diffusion techniques across different modalities, including X-ray, magnetic resonance imaging (MRI), computed tomography (CT), ultrasound, and positron emission tomography (PET), among others [1,4,7,8].

This work focuses on the synthesis of three specific medical imaging modalities: computed tomography angiography (CTA), X-ray, and dermoscopic images. The first application concentrates on synthesizing dermoscopic images to enhance the diagnosis of malignant skin cancer. Dermoscopic imaging is particularly vital in dermatology, as it provides critical insights into the microscopic structures and pigmentation patterns of skin lesions. This research aims to generate synthetic dermoscopic images that closely resemble real-world examples, thereby facilitating the development and validation of machine learning algorithms specifically designed for skin cancer diagnosis. The second application involves the rendering of synthetic cardiac computed tomography angiography (CTA) images to aid in the diagnosis and prognosis of Type B Aortic Dissection (TBAD). TBAD is a life-threatening condition characterized by a tear in the inner layer of the aorta, necessitating prompt diagnosis and intervention. The generation of synthetic CTA images aims to enhance diagnostic capabilities for this critical condition, ultimately improving patient outcomes. The third application focuses on generating synthetic chest X-ray images for pediatric patients to assist in the diagnosis of pneumonia. Given the challenges associated with obtaining sufficient training data for machine learning models in pediatric populations, the synthesis of these images is crucial for developing effective diagnostic tools.

This study aims to provide a comprehensive exploration of adapting image diffusion modeling in medical imaging [16], encompassing its underlying principles, methodologies, applications, and future directions. While substantial work has been done in generating medical imaging data using diffusion models, this research seeks to evaluate how sampling methodologies impact each modality.

2   Medical Image Synthesis Using Diffusion Modeling

This section explores the stable diffusion modeling techniques and integration of image sampling methods, for synthesizing high-quality medical images, enabling the generation of realistic and clinically relevant datasets.

2.1   Text-to-Image Stable Diffusion Model

Text-to-image synthesis in medical imaging refers to the process of generating realistic medical images from textual descriptions or medical reports, which has significant potential for enhancing diagnostic accuracy and facilitating the training of AI-driven systems. Various generative models, including Variational Autoencoders (VAEs) [11], Generative Adversarial Networks (GANs) [15], and diffusion models [3], have been employed to achieve this task. Among these, diffusion-based models are particularly notable for their enhanced training stability and superior alignment with textual descriptions, while generating high-fidelity images with realistic textures and intricate anatomical details. These characteristics make diffusion models particularly well-suited for medical imaging, where the accurate portrayal of subtle anatomical structures is crucial for clinical utility. Diffusion's ability to maintain image quality with limited data availability further strengthens its applicability in scenarios where access to large annotated datasets is restricted, a common challenge in medical research.

Diffusion models, such as Stable Diffusion [12] and DALL-E [9,10], have demonstrated remarkable capabilities in generating images based on textual prompts, having been trained on vast datasets containing billions of images. However, training a diffusion model from scratch poses significant challenges in the medical domain due to the limited availability of large, annotated datasets, primarily due to ethical and logistical constraints. This is where few-shot learning techniques become invaluable. Few-shot learning enables models to be trained with a minimal number of examples, making it particularly advantageous in scenarios where acquiring extensive annotated datasets is impractical. In this research, we propose an integrated approach that combines few-shot learning with Stable Diffusion to generate synthetic medical imaging data that accurately reflects real-world conditions. This augmentation of training data is essential for downstream tasks such as image segmentation, classification, and disease diagnosis.

By leveraging this synergistic approach, we aim to enhance the performance of AI-driven medical imaging models and improve diagnostic workflows, ultimately advancing the field of medical AI.

2.2   Sampling Techniques

Image diffusion modeling, as discussed in [2], is grounded in stochastic processes, where the goal is to propagate and diffuse information across image pixels or voxels in a structured and controlled manner. A key component of this methodology is the image samplers, which are responsible for generating new image samples that maintain a distribution closely aligned with the original seed data. The process begins by generating a random image within a latent space as demonstrated in Fig. 1. A noise predictor, typically a U-Net architecture, is employed to estimate the noise present in the image. At each step of the process, the predicted noise is subtracted from the image, and this denoising procedure is iteratively repeated over multiple timesteps. Through this repeated refinement,

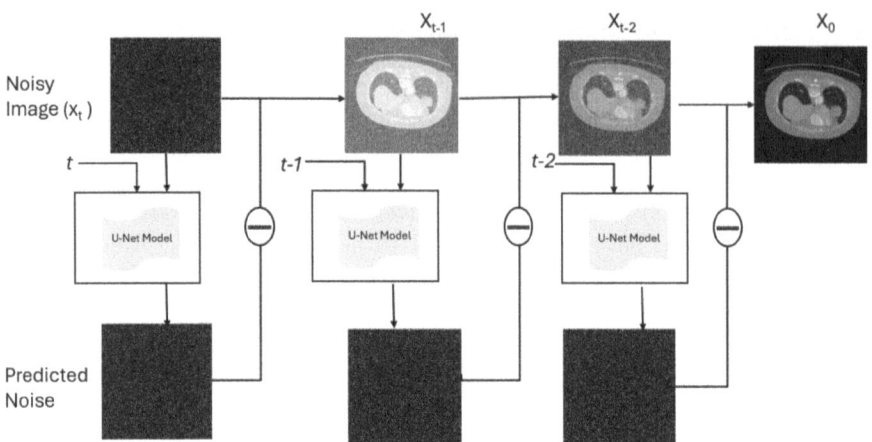

Fig. 1. Schematic representation of the reverse diffusion process using a U-Net in a Diffusion Model

the model gradually transforms the noisy image into a clean, coherent output. Formally, the process starts with the latent variable $\mathbf{x}_T$ sampled from a normal distribution $\mathcal{N}(0, \mathbf{I})$. At each timestep $T = t, t - 1, \ldots, 1$, the next state $\mathbf{x}_{t-1}$ is computed as follows:

$$\mathbf{x}_{t-1} = \frac{1}{\sqrt{\alpha_t}} \mathbf{x}_t - \sqrt{\frac{1 - \alpha_t}{1 - \bar{\alpha}_t}} \epsilon_\theta(\mathbf{x}_t, t) + \sigma_t \mathbf{z}$$

where the noise term $\mathbf{z}$ is sampled from a standard normal distribution $\mathcal{N}(0, \mathbf{I})$ for $t > 1$, and is set to 0 when $t = 1$. Here, α_t and $\bar{\alpha}_t$ are predefined schedule parameters, typically derived from noise schedules such as linear or cosine schedules, while σ_t is the variance controlling the noise scale at timestep t. The function $\epsilon_\theta(\mathbf{x}_t, t)$ is a learned denoising model, usually implemented as a neural network. The denoising procedure iterates until $t = 1$, at which point the final denoised output $\mathbf{x}_0$ is obtained and returned. The below subsections provide details about different image sampling methods employed for rendering distinct medical modality data.

2.2.1   Euler A

Euler A, where 'A' denotes Ancestral, represents an ancestral variant of the Euler sampling method, commonly employed in diffusion models. The term "Ancestral" indicates that this approach incorporates randomness from earlier steps within the sampling process, reintroducing noise that had previously been reduced or removed. This step-wise reintroduction of noise adds an element of unpredictability, as the process revisits earlier states while progressing. Notably, Euler A maintains a constant presence of noise throughout the process, which prevents the sampler from fully converging to a single, deterministic solution. Instead, this method introduces a degree of flexibility, enabling small variations in the results with each step, even in later stages of the sampling procedure. This stochastic nature makes Euler A a useful tool for exploring the solution space in a more comprehensive manner, ultimately fostering innovation through the generation of multiple, slightly varied outputs.

2.2.2 Euler

Euler is a numerical sampling method used in diffusion-based generative models, particularly in Stable Diffusion and other generative models like Denoising Diffusion Probabilistic Models (DDPMs). These methods are originally derived from Euler's method, which is based on numerical method for differential equations. For critical applications like medical imaging, where anatomical accuracy is crucial, the choice between Euler and Euler A depends on the balance between speed and image fidelity. Euler is preferable for generating more controlled outputs, while Euler A might be useful when slight variations in images are beneficial for tasks like data augmentation.

2.2.3 Denoising Diffusion Implicit Mode

The Denoising Diffusion Implicit Mode (DDIM) [14] method introduces a more efficient sampling process by using deterministic, implicit denoising steps. It leverages a non-Markovian forward process, meaning the noise added at each time-step depends not just on the previous step, but also on other steps in the process. This results in the same final data distribution with far fewer steps, making the process faster and more computationally efficient.

2.2.4 Pseudolinear Multistep

Pseudolinear Multistep (PLMS) derives from numerical methods known as Linear Multistep Methods (LMS) [6]. It is a multistep method, such that it uses information from multiple previous time steps to predict the current image state. Instead of relying only on the current time-step's noise and gradient, PLMS takes advantage of the history of the reverse diffusion process, making it more efficient and accurate. By using a pseudolinear strategy, PLMS allows for faster convergence compared to single-step methods like Euler. PLMS reduces the number of timesteps needed to transform noise into a clear image, making it more computationally efficient without compromising quality.

3 Dataset

In this study, we acquired three different public medical imaging datasets as input data for fine-tuning diffusion models and defined short text prompts for each data class used during the training phase. The dermoscopic images were sourced from the International Skin Imaging Collaboration (ISIC) Archive, consisting of 2,800 images evenly distributed between benign and malignant lesions (1,400 images in each category). For the ImageTBAD dataset, which contains 100 annotated 3D computed tomography angiography (CTA) images labeled

for true lumen (TL), false lumen (FL), and false lumen thrombosis (FLT), the
3D volumes were converted into 2D by treating each axial slice as an individual image. The dataset was divided into training and testing sets, with three
primary classes: Class 1 for TL, Class 2 for FLT, and Class 3 containing both
TL and FL. Additionally, a chest X-ray dataset consisting of anterior-posterior
images from pediatric patients aged one to five years was employed, with images
collected from Guangzhou Women and Children's Medical Center. After initial
quality control, the images were diagnostically validated by two board-certified
physicians, and a third expert independently reviewed the evaluation set to minimize labeling errors. The detailed train-test split for each of these datasets is
presented in Table 1. The train data samples were used for fine-tuning diffusion
model whereas the test data samples were used for the quantitative evaluation
and data visualization phase.

Table 1. Overview of the dataset with the number of training and test samples for
each class of three different medical imaging modalities.

Type	Classes	Train	Test
Dermoscopic	**Benign**	1440	360
	Malignant	1197	300
CTA	**True Lumen**	501	235
	True Lumen + False Lumen	16616	3395
	False Lumen Thrombus	121	52
Chest X-Ray	**Normal**	1342	242
	Pneumonia	3876	398

4   Methodology for Synthesizing Dermoscopic, CTA TBAD, and Chest X-Ray Imaging Data

This section will discuss the adapted methodology to synthesize high quality
medical imaging data as a reliable source to enlarge the existing datasets and
indulge more additional variety and diversity. In this context the recent success in
text to image translation using pretrained stable diffusion models by employing
language encoders such as CLIP model has gained enough popularity. By leveraging diffusion modelling in text-to-image translation can assist in tasks such as
generating synthetic medical images for training machine learning algorithms,
augmenting medical image datasets, or aiding in medical education and communication. Additionally, these models have the potential to facilitate interdis-
ciplinary collaboration by enabling seamless communication between clinicians
and imaging experts through textual descriptions of medical findings or diagnostic impressions translated into visual representations. Figure 2 shows the adapted

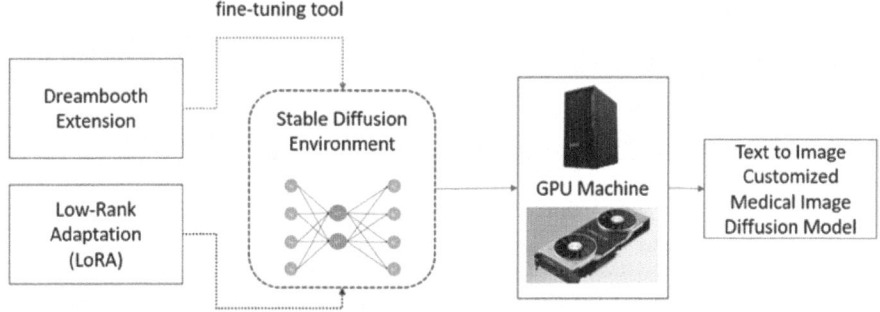

Fig. 2. Block diagram of the proposed methodology. DreamBooth [13] with LoRA (Low-Rank Adaptation) [5] utilized for fine-tuning pre-trained diffusion models in medical imaging, enabling the generation of highly specialized and domain-specific synthetic data, including dermoscopic, CTA and X-ray images, while maintaining high fidelity and anatomical accuracy.

methodology for tuning large-scale diffusion models using transfer learning, tailored for customized medical image rendering tasks. As it can be observed from Fig. 2 that we have employed Dreambooth tool [13] for the purpose of transfer learning and fine-tuning the pretrained stable diffusion model. Further Low Rank Adaptation (LoRA) [5] is used as a lightweight training methodology to fine-tune Large Language and Stable Diffusion Models without necessitating complete model retraining. The conventional approach of fully fine-tuning larger models, characterized by billions of parameters, entails inherent resource intensiveness and temporal demands. LoRA operates by introducing a reduced set of new weights to the model during training, thereby bypassing the requirement to retrain the entirety of the model's parameter space. Consequently, this strategy significantly reduces the number of trainable parameters, leading to expedited training durations and more manageable file sizes, typically spanning a few hundred megabytes. The next stage includes passing on the trained models in the inference pipelines. The image inference pipeline in stable diffusion models comprises a series of sequential steps aimed at generating high-quality images from textual prompts/ descriptions. Figure 3 shows four sequential steps for image rendering using optimized Text to Image Diffusion models. The same pipeline has been adapted in our experimental work for generating synthetic dermoscopic, CTA, and X-ray medical imaging data. The further details on data sourcing and training the models via few shot learning methodology are available in our accepted papers [1,4].

5 Experimental Results

The complete experimental work was carried on workstation machine equipped with A6000 graphic card with 48 GB of graphical video memory. Further we have used Pytorch framework for code implementation.

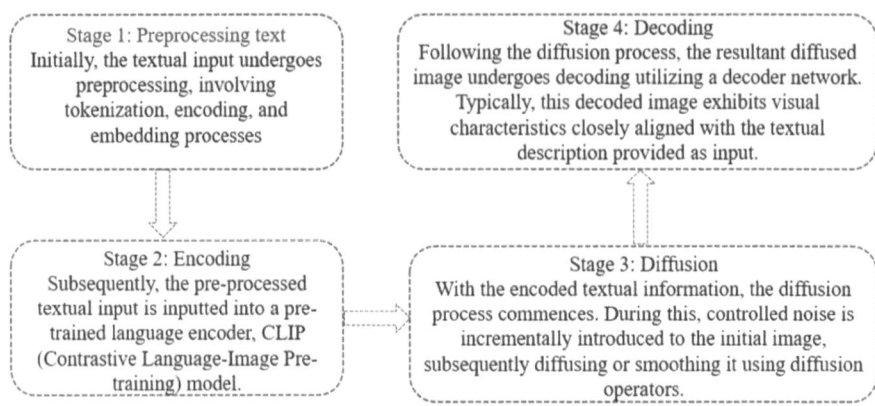

Fig. 3. Four stage pipeline for fine-tuning large scale pretrained stable diffusion model via transfer learning for customized medical imaging application.

5.1 Inference Results Using Text Guided Prompts with Various Diffusion Sampling Methods

The first phase of experimental results focuses on image rendering results using four different sampling methods as discussed in Subsect. 2.2.1, 2.2.2, 2.2.3, and 2.2.4. All the images were generated in 512×512 image resolution and stored in lossless PNG format. Figure 4 shows the rendered benign and malignant dermoscopic data samples. Whereas Fig. 5 and Fig. 6 shows the synthetic CTA TBAD and chest X-ray imaging data generated via tuned diffusion pipeline elaborated in Fig. 2. The complete inference results for each of the medical imaging modality is available at out GitHub repository Diffusion-Sampling-for-Medical-Image-Synthesis.

5.2 Performance Comparison of Sampling Methods for Medical Image Generation

In this section, we conduct a comprehensive evaluation of the impact of various sampling methods on different medical imaging modalities. Our assessment is based on two critical factors: (1) the quality of the generated data, and (2) computational complexity.

5.2.1 Image Quality Evaluation

The visual samples from each imaging modality, generated using four distinct sampling techniques, are presented in Figs. 4, 5, and 6. Notably, as seen in Fig. 5, most sampling methods yielded robust and realistic results across modalities. However, PLMS sampler exhibited considerable variability, particularly in the case of CTA data, where random sampling noise led to inconsistent outputs. This inconsistency suggests that PLMS struggled to generate realistic CTA images,

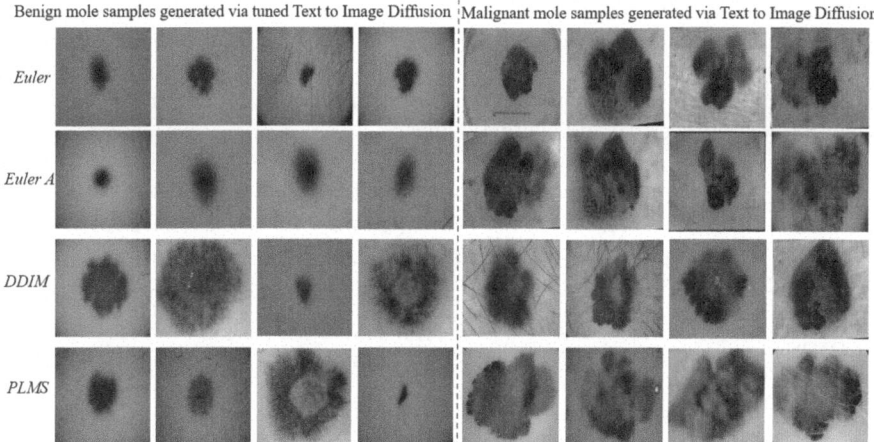

Fig. 4. Synthetic bening and malignant mole samples generated via Derm-T2IM model using four different sampling methods which includes Euler, Euler A, DDIM and PLMS.

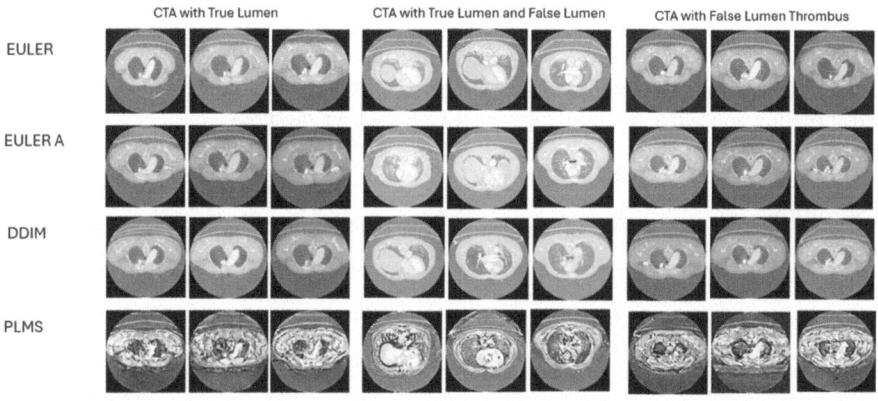

Fig. 5. Synthetic CTA images of Type B aortic dissection showing the true lumen in the first column, both the true lumen and false lumen in the second column, and the false lumen thrombus in the third column. Images were generated using four different sampling methods: Euler, Euler A, DDIM, and PLMS.

especially from a non-expert observer's perspective. The sampler's limitation can likely be attributed to the intricate anatomical details required in CTA data, which PLMS failed to accurately capture. This observation is further reinforced by the t-SNE visualizations in Fig. 7, where synthetic CTA images generated by PLMS cluster significantly farther from their real counterparts, compared to the closer clustering of images generated by other samplers, which more closely align with the real data. A similar trend is noted with dermoscopic images, where malignant samples generated by PLMS lacked diversity, producing similar

Table 2. Quantitative comparison of sampling methods (DDIM, Euler, Euler A, PLMS) across three datasets, using LPIPS and SSIM to measure image similarity to real data for each modality. Green colored numbers represent higher perceptual similarity (lower LPIPS), while blue colored numbers indicate better structural similarity (higher SSIM) between real and synthetic images.

		Dermascopic (n=665)		ImageTBAD (n=52)			Chest X-Ray (n=242)	
		Benign	Malignant	TL	TL+FL	FLT	Normal	Pneumonia
LPIPS	DDIM	0.575	0.586	0.446	0.411	0.418	0.481	0.493
	Euler	0.571	0.606	0.447	0.445	0.413	0.480	0.492
	Euler A	0.558	0.597	0.442	0.447	0.417	0.479	0.497
	PLMS	0.583	0.586	0.564	0.530	0.540	0.486	0.492
SSIM	DDIM	0.621	0.503	0.245	0.259	0.244	0.305	0.416
	Euler	0.631	0.463	0.234	0.261	0.259	0.310	0.422
	Euler A	0.642	0.498	0.265	0.261	0.268	0.305	0.400
	PLMS	0.581	0.556	0.144	0.171	0.153	0.291	0.404

Table 3. Inference time (in seconds per image) across various sampling step intervals for four distinct sampling methods. The green values denote the lowest inference times achieved for each medical imaging modality, emphasizing the superior efficiency of the corresponding sampling method compared to the other methods.

Sampling Steps	Sampler	Dermoscopy		CTA			X-Ray	
		Benign	Malignant	TL	TL+FL	FLT	Normal	Pneumonia
20	Euler	1.6	1.6	1.5	1.6	1.5	1.4	1.4
	Euler A	1.5	1.6	1.6	1.6	1.5	1.3	1.3
	DDIM	1.5	1.5	1.5	1.5	1.5	1.9	1.9
	PLMS	1.5	1.6	1.5	1.6	1.5	1.4	1.4
22	Euler	1.7	1.7	1.6	1.7	1.6	1.6	1.5
	Euler A	1.8	1.7	1.7	1.6	1.7	1.5	1.5
	DDIM	1.7	1.8	1.7	1.6	1.7	2.2	2.2
	PLMS	1.8	1.8	1.7	1.8	1.7	1.6	1.6
24	Euler	1.9	1.8	1.8	1.8	1.6	1.6	1.6
	Euler A	1.9	1.8	1.8	1.6	1.8	1.6	1.6
	DDIM	1.9	1.9	1.8	1.8	1.8	2.1	2.4
	PLMS	1.9	1.9	1.8	1.9	1.9	1.7	1.7

structures repeatedly. Interestingly, despite these limitations, PLMS performed adequately when applied to X-ray data, generating outputs that were visually plausible to untrained observers. The t-SNE analysis further shows that syn-

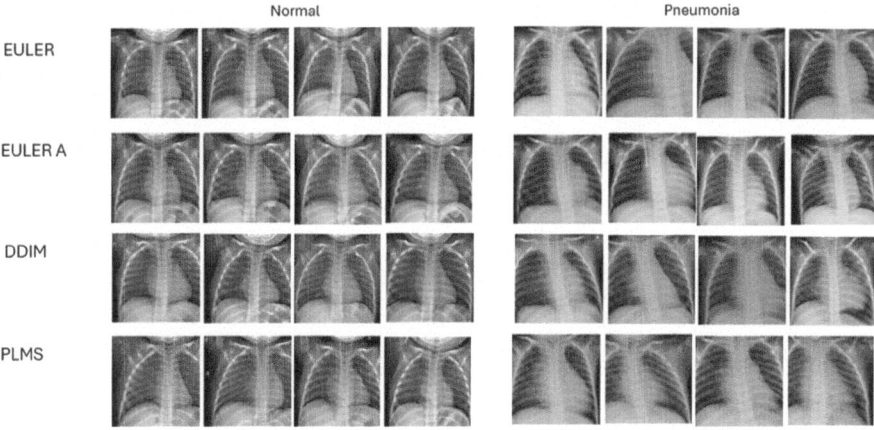

Fig. 6. Synthetic chest X-ray images of pediatric patients with and without pneumonia highlight lung conditions. The first column shows healthy lungs, while the second depicts pneumonia-related opacities. Images were generated using four sampling methods—Euler, Euler A, DDIM, and PLMS.

thetic dermoscopic and CTA images from most samplers clustered more closely with real images, with synthetic dermoscopic images from all samplers aligning more closely with their real counterparts compared to X-ray data. This can be attributed to the richer feature set provided by the three-channel RGB nature of dermoscopic images, which enhances the model's ability to generate more realistic outputs. In contrast, the grayscale and anatomically specialized features of CTA and X-ray images present greater challenges for realistic synthesis.

For the quantitative evaluation of generated images, we employed Learned Perceptual Image Patch Similarity (LPIPS) [17] and Structural Similarity Index Measure (SSIM) to assess the quality of images across different classes within each imaging modality. LPIPS evaluates perceptual similarity by focusing on how closely the generated images align with human visual quality perception, with lower LPIPS values indicating greater perceived similarity. In contrast, SSIM measures similarity based on structural information, luminance, and contrast, offering a more comprehensive assessment of perceived image quality than traditional pixel-wise comparisons. Higher SSIM values reflect better structural similarity, with values approaching 1 indicating high similarity and values closer to 0 indicating greater dissimilarity. As shown in Table 2, the lowest LPIPS score and highest SSIM score for benign dermoscopic images was achieved using the Euler A method. These results demonstrate that benign samples generated using the Euler A method exhibit the highest similarity to real data, both in terms of perceptual and structural quality. For the CTA modality, the PLMS method yielded the highest LPIPS score especially when rendering the TL class data, thus indicating the lowest perceptual similarity when compared to real-world CTA data. Similarly the lowest SSIM score was also achieved on all classes of CTA data using the PLMS method which means that rendered data samples

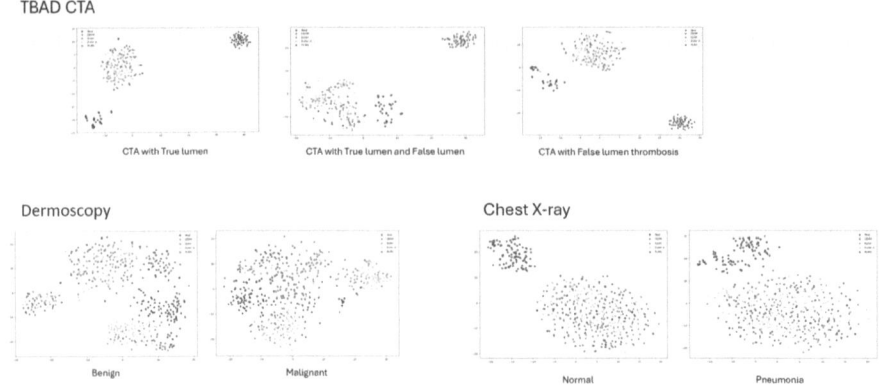

Fig. 7. t-SNE visualization comparing synthetic and real images across each imaging modality, showcasing results for all four sampling methods.

have not enough similarity with real world CTA data samples when generated via PLMS sampling method.

5.2.2   Computational Efficiency and Resource Usage

We assess the computational complexity of diffusion-based sampling methods by evaluating their time and space complexities and analyzing the trade-offs among various samplers. Our focus is on key parameters such as the number of sampling steps, Classifier-Free Guidance (CFG) scale, and inference time. The CFG scale modulates the influence of conditional information (e.g., a text prompt) on the model; higher CFG values place greater emphasis on the prompt, while lower values introduce more diversity and randomness. For our experiments, we set the CFG scale to 7 and investigated the effects of increasing sampling steps. In this evaluation, we utilized time required in seconds to generate single image as critical metrics to gauge the efficiency of each method. Table 3 shows the inference time per image by selecting the sampling steps ranging from 20 to 24 to render robust imaging output for each medical imaging modality. It can be observed from Table 3 that DDIM requires the highest inference time, with larger sampling steps especially in case of X-ray imaging data whereas Euler and Euler A relatively require less inference time for all the imaging modalities thus making this computationally less expensive.

6   Conclusion and Future Work

This study comprehensively evaluates the performance of four distinct diffusion sampling methods for three different medical imaging modalities. By employing a combination of structural and perceptual similarity metrics along with t-SNE visualization, and inference time per image, we have provided a balanced and thorough assessment of each sampling method's effectiveness in rendering diversified medical data. The evaluation highlights differences in image quality, perceptual fidelity, and computational efficiency, ensuring a robust and fair comparison between the sampling methods. From our experimental findings, we concluded that different image sampling methods perform variably across different medical imaging modalities. The Euler A method yields the highest perceptual quality (lower LPIPS score) and highest structural similarity index measure (SSIM) score for benign dermoscopic images. In CTA imaging, the PLMS sampler introduces the most noise across all subclasses, as evidenced by its high LPIPS scores (indicating low perceptual similarity) and low SSIM scores (reflecting poor structural similarity). Notably, both Euler and Euler A require less inference time across all modalities, making them efficient for generating high-quality synthetic data. Thus we concluded that Euler sampling method performs good by generating robust quality data for most of the selected medical imaging modalities and further Euler and Euler A also requires least inference time for all the medical imaging modalities. This highlights the importance of selecting the right sampling method for balancing data quality and computational efficiency.

For future work, we aim to explore more advanced sampling techniques and integrate additional evaluation metrics, particularly those focusing on clinical relevance and diagnostic accuracy. Expanding the study to cover more diverse medical imaging modalities and datasets will further strengthen the generalizability of the findings. Additionally, optimizing sampling methods for improved computational performance without compromising image quality remains a priority, with the goal of enhancing real-time medical image generation in clinical settings.

Acknowledgements. The first author would like to thank the research funding from the College of Science and Engineering. In addition, the research conducted in this publication was jointly supported by ADAPT - Centre for Digital Content Technology, Enterprise Ireland, Irish Research Council under grant number IRCLA/2023/1992 and with the financial support of Science Foundation Ireland under Grant Agreement No SFI/12/RC/2289_P2.

References

1. Abaid, A., Farooq, M.A., Hynes, N., Corcoran, P., Ullah, I.: Synthesizing CTA image data for type-b aortic dissection using stable diffusion models. arXiv preprint arXiv:2402.06969 (2024)
2. Croitoru, F.A., Hondru, V., Ionescu, R.T., Shah, M.: Diffusion models in vision: a survey. IEEE Trans. Pattern Anal. Mach. Intell. **45**(9), 10850–10869 (2023)
3. Dhariwal, P., Nichol, A.: Diffusion models beat GANs on image synthesis. Adv. Neural. Inf. Process. Syst. **34**, 8780–8794 (2021)
4. Farooq, M.A., Yao, W., Schukat, M., Little, M.A., Corcoran, P.: DERM-T2IM: harnessing synthetic skin lesion data via stable diffusion models for enhanced skin disease classification using VIT and CNN. arXiv preprint arXiv:2401.05159 (2024)

5. Hu, E.J., et al.: LORA: low-rank adaptation of large language models. arXiv preprint arXiv:2106.09685 (2021)
6. Liu, L., Ren, Y., Lin, Z., Zhao, Z.: Pseudo numerical methods for diffusion models on manifolds. arXiv preprint arXiv:2202.09778 (2022)
7. Müller-Franzes, G., et al.: A multimodal comparison of latent denoising diffusion probabilistic models and generative adversarial networks for medical image synthesis. Sci. Rep. **13**(1), 12098 (2023)
8. Pan, S., et al.: Synthetic CT generation from MRI using 3D transformer-based denoising diffusion model. Med. Phys. **51**(4), 2538–2548 (2024)
9. Ramesh, A., Dhariwal, P., Nichol, A., Chu, C., Chen, M.: Hierarchical text-conditional image generation with clip latents. **1**(2), 3 (2022). arXiv preprint arXiv:2204.06125
10. Ramesh, A., et al.: Zero-shot text-to-image generation. In: International Conference on Machine Learning, pp. 8821–8831. PMLR (2021)
11. Razavi, A., Van den Oord, A., Vinyals, O.: Generating diverse high-fidelity images with VQ-VAE-2. In: Advances in Neural Information Processing Systems, vol. 32 (2019)
12. Rombach, R., Blattmann, A., Lorenz, D., Esser, P., Ommer, B.: High-resolution image synthesis with latent diffusion models. In: Proceedings of the IEEE/CVF Conference on Computer Vision and Pattern Recognition, pp. 10684–10695 (2022)
13. Ruiz, N., Li, Y., Jampani, V., Pritch, Y., Rubinstein, M., Aberman, K.: Dream-Booth: fine tuning text-to-image diffusion models for subject-driven generation. In: Proceedings of the IEEE/CVF Conference on Computer Vision and Pattern Recognition, pp. 22500–22510 (2023)
14. Song, J., Meng, C., Ermon, S.: Denoising diffusion implicit models. arXiv preprint arXiv:2010.02502 (2020)
15. Wang, T.C., Liu, M.Y., Zhu, J.Y., Tao, A., Kautz, J., Catanzaro, B.: High-resolution image synthesis and semantic manipulation with conditional GANs. In: Proceedings of the IEEE Conference on Computer Vision and Pattern Recognition, pp. 8798–8807 (2018)
16. Yang, L., et al.: Diffusion models: a comprehensive survey of methods and applications. ACM Comput. Surv. **56**(4), 1–39 (2023)
17. Zhang, R., Isola, P., Efros, A.A., Shechtman, E., Wang, O.: The unreasonable effectiveness of deep features as a perceptual metric. In: Proceedings of the IEEE Conference on Computer Vision and Pattern Recognition, pp. 586–595 (2018)

Medical Imaging Complexity and Its Effects on GAN Performance

William Cagas, Chan Ko, Blake Hsiao, Shryuk Grandhi, Rishi Bhattacharya,
Kevin Zhu[✉], and Michael Lam

Algoverse AI Research, Palo Alto, USA
kevin@algoverse.us

Abstract. The proliferation of machine learning models in diverse clinical applications has led to a growing need for high-fidelity, medical image training data. Such data is often scarce due to cost constraints and privacy concerns. Alleviating this burden, medical image synthesis via generative adversarial networks (GANs) emerged as a powerful method for synthetically generating photo-realistic images based on existing sets of real medical images. However, the exact image set size required to efficiently train such a GAN is unclear. In this work, we experimentally establish benchmarks that measure the relationship between a sample dataset size and the fidelity of the generated images, given the dataset's distribution of image complexities. We analyze statistical metrics based on delentropy, an image complexity measure rooted in Shannon's entropy in information theory. For our pipeline, we conduct experiments with two state-of-the-art GANs, StyleGAN 3 and SPADE-GAN, trained on multiple medical imaging datasets with variable sample sizes. Across both GANs, general performance improved with increasing training set size but suffered with increasing complexity.

Keywords: GAN · Entropy · Synthetic Data Generation

1 Introduction

Machine learning in healthcare is a rapidly growing field with countless applications [28] including disease diagnosis [24], clinical treatment [26], drug development [19], and mental health [7]. The machine learning models driving these advances require the collection of high-quality, annotated medical training data, which persists as an arduous task due to privacy concerns surrounding sensitive patient data [23] and the time-intensive nature of labeling [5]. To address these issues, synthetic data—artificially generated information mimicking real-world data—has surfaced as a promising solution [8].

Currently, generative adversarial networks (GANs) remain one of the leading approaches to synthetic data generation [17]. Since its inception in 2014 [6],

W. Cagas—Lead Author.

K. Zhu and M. Lam—Senior Author.

GANs have gained increasing attention in the medical research community due to their ability to synthesize medical images [29]. However, achieving results with high fidelity remains a difficult task factoring the lack of medical data and prevalence of smaller datasets in the medical domain. With limited data, a GAN's efficacy is directly affected with consequences including mode collapse, where the generator produces a limited variety of outputs [20], and overfitting, where the GAN replicates training data rather than generalizing from it [32]. Various papers such as Wang *et al.* 's [31] transfer learning and Robb *et al.* 's [25] Few-Shot GAN (FSGAN) have tried addressing these issues as architecture-centric approaches, achieving increased training efficiency only as a result of the changes in a GAN's structure. However, such approaches are ineffective when making alterations to a GAN's internal structure are not feasible and when time constraints are present. As such, a data-centric approach by providing the GAN with the optimal amount of data to produce high-quality results is more appropriate. Nevertheless, the exact sample set size required to train state-of-the-art GANs is obscure.

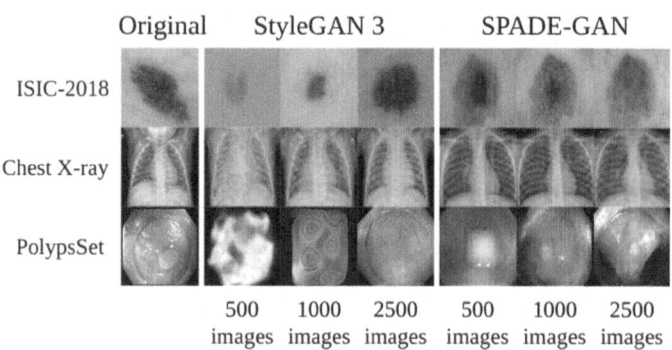

Fig. 1. Comparison between original images and synthetic images from StyleGAN 3 and SPADE-GAN based on variable image set sizes.

In this study, we introduce a data-centric optimization method to create efficient GAN training for medical image synthesis. Our approach investigates how the image complexity distribution of a medical image dataset can be utilized as a measure of training difficulty for a GAN. By doing so, we can ascertain a correlation between the image complexities of the training images and the optimal training set sizes by establishing benchmarks that evaluate the relationship between a sample training set size and the fidelity of the generated images. We hypothesize that given a dataset of a specific image complexity distribution, healthcare professionals can reference the closest image fidelity curve to identify the optimal amount of experimental trials to produce superlative results. Ultimately, our approach can avoid both undertraining and wasteful overtraining by constructing a data-efficient, GAN training pipeline.

2   Background

Generative Adversarial Networks (GANs). Introduced by Goodfellow *et al.* [6], GANs are a class of generative models that consist of two neural networks: a generator G, which aims to transform its latent variable distribution $p(z)$ to closely resemble the training data distribution $p(x)$, and a discriminator D, which differentiates between the ground truth and data generated by G. Training is an adversarial process where G attempts to deceive D into classifying its outputs as real. This two-player minimax game is represented by the following loss function:

$$\min_{G} \max_{D} V(D, G) = \mathbb{E}_{\mathbf{x} \sim p(\mathbf{x})} \left[\log D(\mathbf{x}) \right] \mathbb{E}_{\mathbf{z} \sim p(\mathbf{z})} \left[\log(1 - D(G(\mathbf{z}))) \right]. \tag{1}$$

Many papers have tried to address data scarcity and computational costs in GAN training architecturally. One proposed approach was transfer learning [31], which consists of fine-tuning a pre-trained generator and discriminator to the desired domain. However, if the pre-trained models do not align well with the target domain, this could result in even higher data and computational demands [33]. Another approach, Few-Shot GAN (FSGAN) [25], achieved impressive adaptation even with extremely few training examples, albeit at the cost of prolonged training times. This results in the reduced quality and diversity of the synthetic data when time constraints are present.

Image Complexity. Objectively, image complexity can be defined as the variety of features and details within an image. It has been shown that information entropy is a traditional, heuristic-based method of calculating the complexities of images in small-scale datasets [16].

Traditional entropy is a foundational abstraction in information theory introduced by Shannon [27]. Used as a measure of uncertainty or "surprise" in data, it is the variation in the distribution of pixel intensities of an image in grayscale format. The equation is defined as

$$H = -\sum_{i=0}^{n-1} p_i \log_b p_i, \tag{2}$$

where n denotes the number of gray levels (256 for 8-bit images), b stands for the logarithmic base (returning bits when $b = 2$), and p_i is the probability of a pixel having gray level i. However, although Shannon entropy considers compositional image information, it fails to account for spatial information, specifically the relationship between neighbouring pixels [4].

Another entropy-based metric, the Gray Level Co-Occurrence Matrix (GLCM) entropy, unlike Shannon entropy, is a measure of how often pairs of pixel values occur in a grayscale image distribution [9]. Taking into account this local spatial information, the GLCM is useful for various textural analysis tasks such as feature extraction for medical image segmentation [14]. The GLCM entropy can be represented as

$$H_g = -\sum_{i=0}^{n-1} \sum_{j=0}^{n-1} p_{(i,j)} \log_b p_{(i,j)}, \tag{3}$$

where $p_{(i,j)}$ is the probability of two pixels having gray levels i and j at a certain angle θ and distance d away from each other. Despite GLCM better-capturing complexities within an image, it does not consider spatial patterns and global pixel relationships beyond its adjacent pairing.

3   Methodology

3.1   Image Complexity Metric

Our approach utilizes Larkin's delentropy, a metric identical to both the Shannon entropy and the GLCM entropy, but incorporating a new density function known as the *deledensity* [15]. By analyzing the relationship between the local and global features of an image, delentropy accounts for both an image's gradient vector field and pixel co-occurrence, encapsulating its spatial information as a whole. The deledensity, as a joint probability function, is formulated as

$$p_{(i,j)} = \frac{1}{4WH} \sum_{w=0}^{W-1} \sum_{h=0}^{H-1} \delta_{i,d_x(w,h)} \delta_{j,d_y(w,h)}, \tag{4}$$

where d_x and d_y denote the derivative kernels in the x and y direction, δ is the Kronecker delta to describe the binning operation required to generate a histogram, and H and W is the image's dimensions (height and width) [13]. By obtaining this, we can then calculate delentropy as

$$DE = -\frac{1}{2} \sum_{i=0}^{I-1} \sum_{j=0}^{J-1} p_{(i,j)} \log_b p_{(i,j)}, \tag{5}$$

such that I and J represent the number of bins (discrete cells) in the 2D distribution, and the $\frac{1}{2}$ is derived from Papoulis' generalized sampling expansion [21].

To interpret this measure, yielding a high delentropy suggests an image has a high range of variation in pixel intensities and more sophisticated details. A low delentropy can be interpreted as a result of having a uniform distribution of pixel intensities, indicating simple structure and a less-detailed image.

Prior to any calculations, each image was preprocessed into an 8-bit, grayscale image. This ensured delentropy was calculated in a consistent, single-channel format throughout each dataset.

3.2   GAN Selection

Core to the experimental approach was the selection of two state-of-the-art GANs, SPADE-GAN [22] and StyleGAN 3 [11] on which to run the experimental pipeline. These networks have been widely adopted by the medical image synthesis community and empirically observed to produce superior-quality medical images when compared to predecessor GANs [29]. StyleGAN 3's large community support and wide availability of its code repository along with its numerous configurations for different training settings were taken into account as well.

3.3   GAN Pipeline

For experiments, given our data-centric approach, StyleGAN 3 and SPADE-GAN were run with the official, publicly available implementations with default hyperparameters and no augmentations to each network's architecture.

Preprocessing. We first set all images to a consistent 512×512 resolution. As such, training parameters were based on the size of the preprocessed images, as documented in the official implementations. SPADE-GAN additionally relies on segmentation masks to produce synthetic data. We used pre-existing annotations for ISIC-2018 and the Polyps Set. Because the Chest X-ray dataset did not have such annotations, masks were generated using TorchXRayVision [3]. All experiments were performed on one NVIDIA A100 and three NVIDIA A40 GPUs.

Training and Generation. The experimental pipeline was designed to identify the role of image dataset size in the image generation fidelity of selected GANs, for which to be compared to the image complexity distribution of each dataset. To that end, for each GAN training run, all parameters were held constant with the exception of the image set size, which was subsequently set to 500, 1000, and 2500 images, randomly sampled from the same dataset for each experimental run, respectively. The pipeline was designed to incorporate 500 images as a baseline for GANs to train with limited data. We facilitated experiments with 2500 images for a more comprehensive training run to better capture the underlying image distribution and use an intermediary set of 1000 images serving as a middle ground. For StyleGAN 3, all experimental runs were trained for 100 epochs; for SPADE-GAN, training iterated 50 epochs. The trained adversarial network was then used to generate synthetic images, the fidelity of which was then evaluated for each training set size.

Evaluation. The Fréchet Inception Distance (FID) [10] is a common metric used to evaluate the fidelity of the synthetically generated images for GANs [1]. Defined as the distance between the distributions of the ground truth and the generated images respectively, in our paper, we use the FID to assess the performance of the GAN (i.e. image fidelity) for each experimental run across both GANs. **A lower FID score signifies that a GAN is more proficient at generating synthetic data close to its target distribution**. From these data, we obtained fidelity curves for each dataset that describe how FID scores trend with increasing training set size.

4   Experimental Results

Datasets. We employed three medical image datasets: International Skin Imaging Collaboration 2018 Challenge (ISIC-2018) [2], Chest X-Ray Images (Chest X-ray) [12], and Colonoscopy Polyp Detection and Classification (Polyps Set)

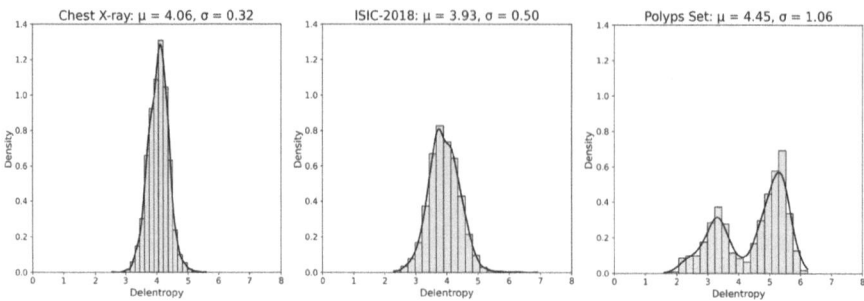

Fig. 2. Delentropy distributions across each medical image dataset. A higher mean delentropy μ indicates a dataset with more complex images.

[30]. These datasets were chosen for their diversity in both perceptual complexity, ranging from relatively skin lesions to complex colon polyps, and imaging modality (dermoscopy vs. x-ray vs. colonoscopy).

We carried out delentropy calculations as described in Sect. 3 by using a publicly available implementation from Marchesoni [18]. To effectively capture the overall complexity of each image dataset, we captured each dataset's delentropy distribution as displayed in Fig. 2.

Across the experimental runs, FID scores consistently decreased with increasing dataset size. On StyleGAN 3, synthesized images that had been generated by a GAN trained on 2500 images exhibited an average FID score reduction of 48% when compared to those generated by a StyleGAN 3 that had been trained on a mere 500 images (Fig. 3). SPADE-GAN experienced an analogous 31% FID score reduction on average, though it is worth noting that FID reduction plateaued after only 1000 training images.

Comparing both Fig. 2 and Fig. 3, one can see a general relationship between the delentropy distribution and the training performance of both GANs. As the spread of image complexities increases from a slender, peaked distribution to a broader, bimodal one, we see a corresponding increase in FID scores for each dataset sample size. The Chest X-ray dataset with the most homogeneous image complexities shown by a tall and narrow distribution, yields the lowest FID score after being trained for 2500 images, indicating that both GANs had easier training runs with this dataset. On the contrast, the Polyps Set—the dataset with the widest distribution and multiple complexity peaks—correlates with the highest FID scores for each dataset sample size, which suggests that the GAN was faced with a more challenging and unstable training run. Ultimately, this pattern shows a general inverse relationship—GAN performance decreases with an increasing spread of image complexities within a dataset.

5 Discussion

SPADE-GAN outperformed StyleGAN 3 across *all datasets and training sizes*, with FID scores averaging 33% lower, likely due to its architecture that

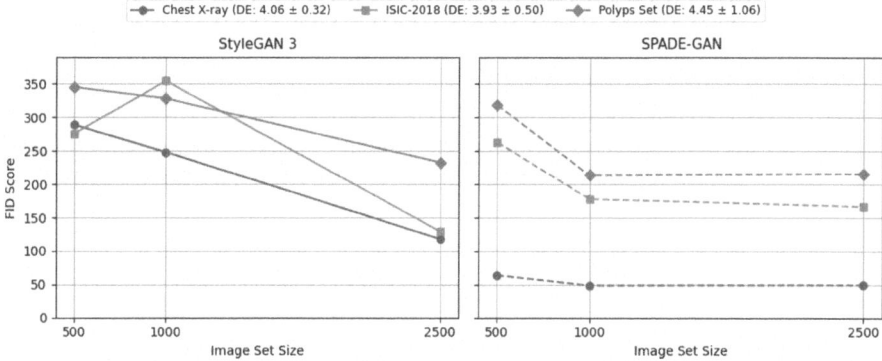

Fig. 3. Fréchet Inception Distance (FID) curves comparing StyleGAN 3 and SPADE-GAN across each medical image dataset with varying sample sizes. Lower FID scores correspond to higher fidelity synthetic images.

incorporates segmentation masks for structural information, whereas StyleGAN 3 trained on raw image data alone, making it more difficult to generalize to high-delentropy datasets. Moreover, ISIC-2018 being an outlier can be attributed to its fluctuations in image complexity, reflected by the standard deviation in delentropy (Fig. 2). Despite having a lower mean delentropy, its spread likely resulted in difficulties in GAN training and learning the images' distribution, contrasting with the Chest X-ray dataset.

While the experimental results generally reflected an intuitive understanding of how image complexity and training data influence GAN training, the FID curves provide insightful details, offering a deeper perspective on these effects. SPADE-GAN exhibits both better quality results than StyleGAN 3 in the form of lower FID scores and more consistent training as evidenced by the smooth, non-overlapping FID curves (Fig. 3). As aforementioned, performance plateaued after 1000 training images, suggesting that additional training data past that point may not help increase GAN performance as measured by FID score. This is also apparent in the generated images themselves, which exhibit little perceptual difference between those generated after 1000 training images and those generated after 2500 (Fig. 1). Contrast this with the StyleGAN 3 curves, which do not reach any noticeable plateau between 500 and 2500 training images. In fact, the increasingly negative slope values of the StyleGAN 3 graphs imply that StyleGAN 3 begins to better capture the images' features at a point past 1000 images, the exact whereabouts of which would need to be determined by a separate study.

The FID curves generated by this set of experiments set up a useful benchmark to which other potential training image data sets can be compared. For training sets that are of similar delentropy distributions and used to train Style-GAN 3 or SPADE-GAN, it is not unreasonable to predict that their training curves will be similar to those represented in Fig. 3, though many more training set sizes and image sets are required before a truly comprehensive representation can be reached.

Broader Impacts. Our research on GANs for medical image synthesis may have positive and negative societal implications. On the positive side, it can enhance healthcare outcomes by improving the training of machine learning models with realistic synthetic data, therefore protecting patient privacy. Contrarily, potential negative impacts include the risk of maliciously generating fraudulent synthetic data and the possibility of reinforcing biases due to a lack of diversity of representing patient populations. These considerations demonstrate the importance of addressing both the benefits and potential risks associated with the use of GANs in the medical domain.

6   Conclusion

In this work, we highlight the impact of image complexity on GAN performance in medical image synthesis. We empirically demonstrate a general inverse relationship: higher image complexity leads to poorer image fidelity results and lesser performance in GANs. Furthermore, we demonstrate FID curves showing healthcare professionals the possibility for the use of our benchmarks to gauge an estimate of data training requirements to achieve desirable results based on the image complexity distribution of a medical image dataset.

7   Limitations

Due to limited resources, experiments were only run on 500, 1000, and 2500 training images, leading to coarse-grained results. An extended study with a larger range and finer-grained increments would better elucidate exactly how FID scores respond to changes in training image dataset size. The use of FID scores as a sole evaluation metric also has its limitations, not necessarily correlating with human perceptual interpretations, something that is extremely important in the medical field where human doctors are still largely the source of truth. Skandarani *et al.* [29] shows that a lower FID score may not be a good measure of how well synthetic images can perform on a downstream task as well. Although this study specifically focused on GANs as the primary generative model, similar experiments extrapolated across other generative models such as stable diffusion may prove more relevant in the context of recent advancements in generative AI. More research with larger resources involving multiple evaluations of a similar experimental setup is required.

Acknowledgements. We are grateful to Michael Lam and Kevin Zhu for their excellent mentorship, constructive feedback, and unwavering support throughout our research.

References

1. Borji, A.: Pros and cons of GAN evaluation measures: new developments. Comput. Vis. Image Underst. **215**, 103329 (2022). https://doi.org/10.1016/j.cviu.2021.103329
2. Codella, N., et al.: Skin lesion analysis toward melanoma detection 2018: a challenge hosted by the international skin imaging collaboration (ISIC), licensed under the Creative Commons Attribution-NonCommercial (CC-BY-NC) license. arXiv preprint arxiv:1902.03368 (2018)
3. Cohen, J.P., et al.: TorchXRayVision: a library of chest X-ray datasets and models. In: Medical Imaging with Deep Learning (2022). https://github.com/mlmed/torchxrayvision
4. Gao, P., Li, Z., Zhang, H.: Thermodynamics-based evaluation of various improved Shannon entropies for configurational information of gray-level images. Entropy **20**(1) (2018). https://doi.org/10.3390/e20010019
5. Gilbert, A., Marciniak, M., Rodero, C., Lamata, P., Samset, E., Mcleod, K.: Generating synthetic labeled data from existing anatomical models: an example with echocardiography segmentation. IEEE Trans. Med. Imag. **40**(10), 2783–2794 (2021). https://doi.org/10.1109/TMI.2021.3051806
6. Goodfellow, I., et al.: Generative adversarial NETs. Adv. Neural Inf. Process. Syst. **27** (2014)
7. Graham, S., et al.: Artificial intelligence for mental health and mental illnesses: an overview. Curr. Psychiatry Rep. **21**, 1–18 (2019)
8. Guo, X., Chen, Y.: Generative ai for synthetic data generation: methods, challenges and the future. arXiv preprint arXiv:2403.04190 (2024)
9. Haralick, R.M., Shanmugam, K., Dinstein, I.: Textural features for image classification. IEEE Trans. Syst. Man Cybern. **SMC-3**(6), 610–621 (1973). https://doi.org/10.1109/TSMC.1973.4309314
10. Heusel, M., Ramsauer, H., Unterthiner, T., Nessler, B., Hochreiter, S.: GANs trained by a two time-scale update rule converge to a local Nash equilibrium. In: Advances in Neural Information Processing Systems (NeurIPS), vol. 30 (2017). https://arxiv.org/abs/1706.08500
11. Karras, T., Laine, S., Aittala, M., Hellsten, J., Lehtinen, J., Aila, T.: Analyzing and improving the image quality of StyleGAN, licensed under the Nvidia Source Code License . arXiv preprint arXiv:1912.04958 (2020)
12. Kermany, D., Zhang, K., Goldbaum, M.: Labeled optical coherence tomography (Oct) and chest X-ray images for classification, licensed under the Creative Commons Attribution 4.0 International (CC BY 4.0) license (2018). https://doi.org/10.17632/rscbjbr9sj.2
13. Khan, T.M., Naqvi, S.S., Meijering, E.: Leveraging image complexity in macro-level neural network design for medical image segmentation. Sci. Rep. **12** (2022). https://doi.org/10.1038/s41598-022-26482-7
14. Khan, Z.F., Alotaibi, S.R.: Computerised segmentation of medical images using neural networks and GLCM. In: 2019 International Conference on Advances in the Emerging Computing Technologies (AECT), pp. 1–5 (2020). https://doi.org/10.1109/AECT47998.2020.9194196
15. Larkin, K.G.: Reflections on Shannon information: in search of a natural information-entropy for images. arXiv preprint arXiv:1609.01117 (2016)
16. Liu, S., Zhao, L., Chen, D., Song, Z.: Contrastive learning for image complexity representation. arXiv preprint arXiv:2408.03230 (2024)

17. Lu, Y., et al.: Machine learning for synthetic data generation: a review. arXiv preprint arXiv:2302.04062 (2024)
18. Marchesoni, F.: How do I know if an image after image enhancement is better than before? (2023). https://ai.stackexchange.com/questions/39483/how-do-i-know-if-image-after-image-enhancement-is-better-than-before-image-pre
19. Nordon, G., Koren, G., Shalev, V., Horvitz, E., Radinsky, K.: Separating wheat from chaff: joining biomedical knowledge and patient data for repurposing medications. In: Proceedings of the AAAI Conference on Artificial Intelligence, vol. 33, pp. 9565–9572 (2019)
20. Pan, Z., Yu, W., Yi, X., Khan, A., Yuan, F., Zheng, Y.: Recent progress on generative adversarial networks (GANs): a survey. IEEE Access **7**, 36322–36333 (2019)
21. Papoulis, A.: Generalized sampling expansion. IEEE Trans. Cir. Syst. **24**(11), 652–654 (1977). https://doi.org/10.1109/TCS.1977.1084284
22. Park, T., Liu, M.Y., Wang, T.C., Zhu, J.Y.: Semantic image synthesis with spatially-adaptive normalization, licensed under the Creative Commons Attribution-NonCommercial-ShareAlike 4.0 International (CC BY-NC-SA 4.0) license. arXiv preprint arXiv:1903.07291 (2019)
23. Pezoulas, V.C., et al.: Synthetic data generation methods in healthcare: a review on open-source tools and methods. Comput. Struct. Biotechnol. J. **23**, 2892–2910 (2024). https://doi.org/10.1016/j.csbj.2024.07.005
24. Rahman, S., Ibtisum, S., Bazgir, E., Barai, T.: The significance of machine learning in clinical disease diagnosis: a review. arXiv preprint arXiv:2310.16978 (2023)
25. Robb, E., Chu, W.S., Kumar, A., Huang, J.B.: Few-shot adaptation of generative adversarial networks. arXiv preprint arXiv:2010.11943 (2020)
26. Shang, J., Ma, T., Xiao, C., Sun, J.: Pre-training of graph augmented transformers for medication recommendation. arXiv preprint arXiv:1906.00346 (2019)
27. Shannon, C.E.: A mathematical theory of communication. Bell Syst. Techn. J. **27**(3), 379–423 (1948). https://doi.org/10.1002/j.1538-7305.1948.tb01338.x
28. Shi, Z.R., Wang, C., Fang, F.: Artificial intelligence for social good: a survey. arXiv preprint arXiv:2001.01818 (2020)
29. Skandarani, Y., Jodoin, P.M., Lalande, A.: GANs for medical image synthesis: an empirical study. J. Imag. **9**(3), 69 (2023). https://doi.org/10.3390/jimaging9030069
30. Wang, G.: Replication data for: colonoscopy polyp detection and classification: dataset creation and comparative evaluations, licensed under the CC0 1.0 Universal (CC0 1.0) Public Domain Dedication (2021). https://doi.org/10.7910/DVN/FCBUOR
31. Wang, Y., Wu, C., Herranz, L., Van de Weijer, J., Gonzalez-Garcia, A., Raducanu, B.: Transferring gans: generating images from limited data. In: Proceedings of the European Conference on Computer Vision (ECCV), pp. 218–234 (2018)
32. Webster, R., Rabin, J., Simon, L., Jurie, F.: Detecting overfitting of deep generative networks via latent recovery. In: Proceedings of the IEEE/CVF Conference on Computer Vision and Pattern Recognition, pp. 11273–11282 (2019)
33. Zhuang, F., et al.: A comprehensive survey on transfer learning. Proc. IEEE **109**(1), 43–76 (2020)

LAMM

RSSep: Sequence-to-Sequence Model for Simultaneous Referring Remote Sensing Segmentation and Detection

Ngoc-Vuong Ho[✉], Thinh Phan, Meredith Adkins, Chase Rainwater, Jackson Cothren, and Ngan Le

University of Arkansas, Fayetteville, AR, USA
vuongh@uark.edu

Abstract. Semantic segmentation in remote sensing images plays a crucial role in a wide range of geographic information applications. Despite the abundance of data, this field faces limitations due to the restricted set of categories and the inability of existing methods to accurately describe and localize individual or multiple objects within scenes. Addressing this challenge, the emerging fields of referring remote sensing image segmentation (RRSIS) and referring remote sensing object detection (RRSOD) have recently garnered attention. Both tasks, RRSIS and RRSOD, combine computer vision and natural language processing to localize objects based on a text query, with the outputs being segmentation masks and bounding boxes. Additionally, boundary information in remote sensing images, such as land-cover delineations, is crucial for segmentation tasks. To tackle this novel challenge, we introduce **RSSep**, a Sequence-to-Sequence model designed for simultaneous RRSIS and RRSOD. Unlike conventional approaches that use encoder-decoder blocks for pixel-level classification, our network leverages a sequence-to-sequence model to estimate polygonal boundaries, represented as sequences of vertices. Furthermore, we enhanced the network by improving the text encoder using both query and object noun features, employing the same architecture to extract these features. Our network is benchmarked on the recently introduced RRSIS-D dataset, notable for its extensive collection of image-caption-mask triplets across diverse scales and variations. Experimental results demonstrate the superiority of our method over existing techniques in both the RRSIS and RRSOD fields, underscoring its efficacy in semantic segmentation and object detection tasks in remote sensing imagery.

Supplementary Information The online version contains supplementary material available at https://doi.org/10.1007/978-981-96-2641-0_15.

1 Introduction

Owing to the advancement and pervasiveness of satellites and aerial vehicles, remote sensing data collection has surged, leading to an increasing need for analyzing and understanding scene images. Remote sensing image segmentation (RSIS) has become one of the key tasks, applied in multiple fields such as urban planning [34], land resource management [21], environmental monitoring [22], disaster monitoring [39], agricultural planning [46], street view extraction [14,40], land change detection [32,38,48], land cover classification [42], climate change studies [36], and deforested region monitoring [2], among others. Although current approaches in RSIS are proficient at identifying objects in a scene, they struggle to specify areas based on descriptions or to perceive the spatial and orientational features of objects. To address these limitations, Referring Remote Sensing Image Segmentation (RRSIS) has emerged. The goal of RRSIS is to take both an aerial image and a natural language query describing appearance, position, and direction, and return a pixel mask of the relevant objects or areas in the scene.

Unlike Referring Remote Sensing Image Segmentation (RRSIS), referring image segmentation (RIS) has been a well-known task for some time, with the majority of RIS approaches [18,20,23,51] following an encoder-decoder framework where the encoder is responsible for extracting visual and language features using two separate networks and then aligning them through recurrent interaction [26], cross-modal attention [45], graph reasoning [19], or cross-attention from a Transformer [5], and subsequently, the decoder unravels the combined features and performs pixel-level classification, outputting the segmentation mask for the desired objects. Despite their success on multiple datasets, general RIS methods are sub-optimal when fully applied to specific data types like remote sensing images due to the aerial viewpoint reducing the noticeable discrepancies in color and appearance between objects and backgrounds, while the scales and sizes of objects vary greatly depending on the distance between the camera and the ground, resulting in weak contrast between object boundaries and backgrounds in low spatial resolution images, which often causes the predicted masks to appear smeared. To address these challenges, we adopt a sequence-to-sequence (Seq2Seq) framework to indirectly infer the segmentation mask where the input remains the same as in the encoder-decoder framework but the decoder performs a regression task and outputs the polygonal boundary instead, making this method better at recognizing object geometry and leading to more precise masks where the segmentation mask is converted into a sequence of polygon vertices (with unrestricted length) and a bounding box is also output by the Seq2Seq module, transforming each vertex into a coordinate embedding token, and in cases where multiple objects are queried, different sequences can be merged into a longer sequence and distinguished by separator tokens while the model learns to predict the next coordinate token based on the visual features, text features, and previous tokens, with the polygon sequence being extended iteratively until the end-of-sequence token is predicted, which allows our approach to be less susceptible to inconspicuous boundaries, scale variations, and omni-directional objects.

Our main contributions are summarized as follows:

- We present an effective RSSeq network for simultaneous RRSIS and RRSOD. Our RSSeq is built upon the Seq2Seq framework, modeling the object boundary as a sequence of vertices. Our network is designed not only to focus on the object boundary but also to handle an arbitrary number of objects.
- To effectively train the model to handle both boundary and segmentation mask tasks, our RSSeq model is trained using a combination of weighted L_1 and L_2 regression losses, cross-entropy (CE) loss, and *Dice* loss. This comprehensive loss function aims to optimize the prediction of polygon vertices, vertex types, and segmentation masks.
- We benchmarked the proposed RSSeq on the newly introduced RRSIS-D dataset, demonstrating superior performance over all existing state-of-the-art methods in the RRSIS task.

2   Related Work

2.1   Mask and Polygon-Based Image Segmentation

Mask-based image segmentation has still been on the growth and the primary technique for object segmentation. Fully Convolutional Neural (FCN) [31] established the baseline for semantic segmentation field by replace all fully-connected layers with convolution layers in classification network. For the purposed of accumulating multi-scale contextual information, DeepLab series [7] upgraded FCN with dilated convolutions. With the same intention, PSPNet [53] introduced the pyramid pooling operations. Latest work such as Mask2Former [9] utilized the end-to-end Transformer [3] encoder-decoder network and multi-scale high resolution features, deducing the each object mask from corresponding embedding query. Treating segmentation mask as set of polygon vertices is also considered because this task simulates how human annotates the mask. The boundary is refined or sequentially predicted until we reach the initial point. The early work [4] made use of the Recurrent Neural Network (RNN) and was extended by [1] with the application of graph neural network. Ling et al. [25] initiated with a circle and tried to deform it into the boundary. Done et al. [13] extended this task to spline curve prediction and did the multitasking training on edge detection and object segmentation. PolyTransform [24] predicted the mask first and forwarded it as polygon type to deforming network for final polygon prediction.

2.2   Remote Sensing Image Segmentation (RSIS)

RSIS aims to segment and classify the objects such as building, vehicle, road or field on the earth surface from the aerial viewpoint. In the early period of deep learning application on this topic, FCN was the standard approach on many datasets [8]. The methods improved along with the development of deep learning segmentation model. ResUNet-a [12] combined U-Net with other CNN to eliminate the problem of gradient disappearance and explosion. S-RA-FCN [33] enhanced the global contextual information by adding the spatial and channel relational reasoning modules. HMANet [35] proposing three attention modules to better obtain correlation features in space, channel and category. The efficiency of self-attention in transformer-based network set a new model trend in this field. Due to the low contrast between the foreground saliency and background noise, RSSFormer [49] was designed with the Adaptive Transformer Fusion Module and Detail-aware Attention Layer. [43] introduced the a densely connected feature aggregation module for precise segmentation. While transformer-based methods are good at capturing long-range dependencies, intricate and tiny objects are still an obstacle for them.

2.3   Referring Image Segmentation (RIS)

RIS has been one of most active topics in the field of visual-language understanding and interaction. The main objective has been the fusion mechanism of visual and language features. Straightforward feature concatenation [41] was first implemented as the fusion operation. Chen et al. [6] followed this technique but applied recurrent refinement to polish the feature maps at different scales. Later works employed several types of attention mechanism [51] to model the visual-textual co-embeddings. CMPC [19] used graph-based reasoning to localize the image region that were highly related to the linguistic features of entity words and attribute words. BRINet [18] computed the relevance among each word and each image area in a bi-directional relationship modeling through vision and language-guided attention modules. As a result of the success of vision-language model such as CLIP [37], recent models [44] tried to transfer this rich knowledge to their fusion model. LAVT [50] replaced the complicated cross-modal decoder by early language-aware encoding module. PolyFormer [27] proposed the regression-based Tranformer decoder which directly output 2D coordinates from concatenated image feature and textual feature.

3   Methodology

3.1   Overall Methodology

Network Architecture: At the core of our approach is the idea of feature fusion between natural language processing and computer vision. Figure 1.b illustrates the network architecture: the inputs are the image and a text query, and the outputs are the polygon covering to the mask needed for segmentation and the bounding box for object detection as per the text description. Motivated by recent advancements in multimodal architectures such as CLIP [37], we use two separate encoder branches to extract visual and textual features from both the image and text prompt. The image encoder is based on Swin transformer [30], whereas the text encoder is based on BERT [10]. Both the visual and textual

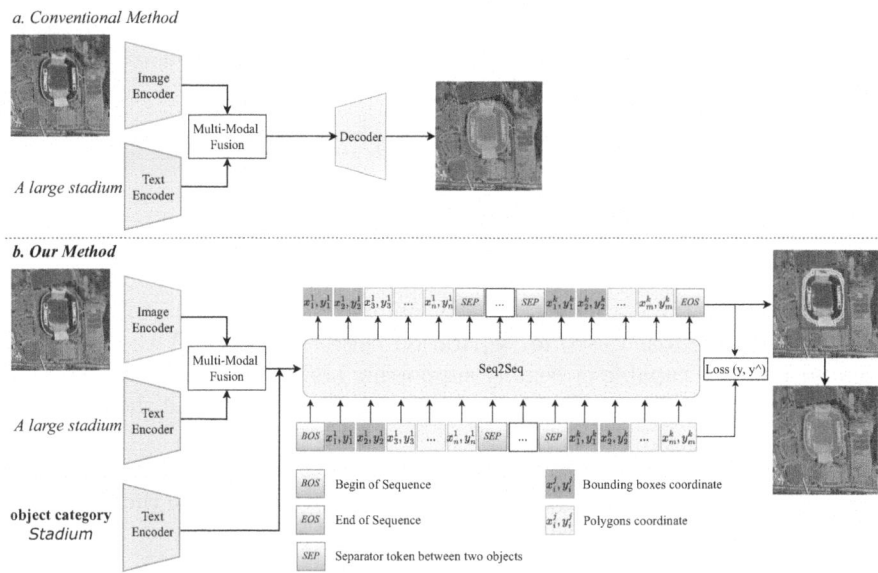

Fig. 1. Comparison between existing RRSIS approaches (top) and our proposed RSSeq (bottom). While conventional RRSIS methods directly generate a segmentation mask from a Decoder network, our RSSeq first produces a sequence of vertices and subsequently converts them into a polygonal segmentation mask.

features are then concatenated by a fusion module before passing through a Seq2Seq network to obtain a sequence of vertexes of a polygon. To effectively handle floating-point coordinates, vertexes are passed through a regression network. Finally, the polygon will be converted to a segmentation mask.

Encoder: We use Swin Transformer $f_v(.|\theta_v)$, defined by weights θ_v, to extract visual feature F_v from a given image $I \in \mathbb{R}^{H \times W \times 3}$, i.e. $F_v = f_v(I|\theta_v)$. We select the feature at stage-4, thus the feature $F_v \in \mathbb{R}^{\frac{H}{32} \times \frac{W}{32} \times C_v}$. For the textual description extraction, we use BERT [10] $f_t(.|\theta_t)$, defined by weights θ_t, to extract textual feature from a given text prompt T, i.e. $F_t = f_t(T|\theta_t)$ and $F_t \in \mathbb{R}^{N \times C_n}$, where N is the number of words.

Fusion module: The visual feature F_v and textual feature F_t into new feature F_{vt}. To achieve this, the visual feature $F_v \in \mathbb{R}^{\frac{H}{32} \times \frac{W}{32} \times C_v}$ is first flatten into 2D space $\hat{F}_v \in \mathbb{R}^{(\frac{H}{32} * \frac{W}{32}) \times C_v}$. Subsequently, both $\hat{F}_v$ and F_t are passed through two separate fully-connected (FC) layers to project them into the same embedding space. These projected features are then concatenated into F_{vl}, i.e., $F_{vl} = \left[FC(\hat{F}_v), FC(F_t) \right]$. Finally, F_{vl} is fed into a Transformer encoder consisting of multi-head self attention layer to object a visual-textual feature F.

Seq2Seq: The Seq2Seq takes the 2 inputs: (i) visual-textual feature F from the fusion module and (ii) input token, which represents the polygon vertexes. The input token of the module are define as format:

$$[\text{<BOS>}, (x_1^1, y_1^1), (x_2^1, y_2^1), (x_3^1, y_3^1), ..., (x_n^1, y_n^1), \text{<SEP>}$$

$$(x_1^2, y_1^2), ..., (x_m^1, y_m^1)\text{<SEP>}, (x_k^1, y_k^1), ..., \text{<EOS>}], \tag{1}$$

where (x_1^i, y_1^i) and (x_2^i, y_2^i) denote the bounding box coordinates, and $(x_3^i, y_3^i), ..., (x_n^i, y_n^i)$ represent the coordinates of the bounding polygons. The tokens <BOS> and <EOS> indicate the beginning and end of the tokens, while <SEP> denotes the token used for separation between objects. With this definition, our model is capable of flexibly supporting multiple objects.

The prediction of vertex (x_t, y_t) at time step t depends on all the preceding tokens $(x_1, y_1), (x_2, y_2), .., (x_{t-1}, y_{t-1})$.

The model accommodates an arbitrary number of input tokens to handle an arbitrary number of objects. During training, we process based on the last token, while during inference, the procedure is based on the EOS token.

Decoder: To predict continuous coordinate values for vertex coordinates, we apply bilinear interpolation [15] into the decoder. Let $f(x, y)$ denote the coordinate embedding of (x, y). To capture the relations between the visual-textual feature F and the coordinate embedding $f(x, y)$, we apply N transformer decoder layers with multi-head cross-attention mechanism. Consequently, we obtain F^N as the output feature of the last decoder layer.

Prediction Heads: There are two output heads, i.e., token-based head and coordinate-based head. Both are built on the last feature F^N. The token-based head consists of a linear layer, which predicts the token type. Token types can be either coordinate token $((x_1^1, y_1^1), (x_2^1, y_2^1), (x_3^1, y_3^1), ..., (x_n^1, y_n^1))$, which is labeled as 0 or separate token <SEP>, which is labeled as 1, or ending token <EOS>, which is labeled as 1. The coordinate-based head is defined as a 3-layer feed-forward network (FFN). It aims to predict the 2D coordinates of the bounding box corner points $((x_1^i, y_1^i), (x_2^i, y_2^i))$ and polygon vertices $((x_3^i, y_3^i), ..., (x_n^i, y_n^i))$ for the reference object i.

Mask to Polygons Converter: The current available dataset provides masks only. Therefore, to train the model, we need to convert these masks into polygons. Given a mask $M \in \mathbb{R}^{H \times W}$, we obtain the contour, a set of points $(x, y) \in \mathbb{R}^2$, which are then converted into polygons. Due to the large number of points on the contour, we sample a subset of points, typically ranging from 100 to 200 points for each object. We select the top-left point as the starting point of the sequence $\{(x_i, y_i)\}_{i=1}^P$, $(x_i, y_i) \in \mathbb{R}^2$ in the clock-wise order, $P \in [100, 200]$. Finally, we construct the input tokens for the polygon coordinates as <BOS>$(x_1^1, y_1^1), (x_2^1, y_2^1)...(x_n^1, y_n^1)$ <SEP>... <SEP>$(x_2^k, y_2^k), (x_2^k, y_2^k)...(x_m^k, y_m^k)$ <EOS>, where k is the number of object, n and m are the number of token for each object. By using <SEP>, the number of objects and the number of tokens for each object are flexible.

Polygon to Mask Converter: The output of the network contains two components: a token-based head and a coordinate-based head. To convert from the polygon to a mask, we utilize both the token-type and coordinate outputs. By combining the outputs of both heads, we can obtain:

<BOS>$(x_1^1, y_1^1), (x_2^1, y_2^1)...(x_n^1, y_n^1)$ <SEP>...   <SEP>$(x_2^k, y_2^k), (x_2^k, y_2^k)...(x_m^k, y_m^k)$
<EOS>

Then, we can separate the coordinate output into multiple bounding boxes and boundary vertices based on <SEP>. Finally, the boundary vertices are converted to a binary mask to be compared with the ground truth.

3.2 Loss Function

Given an image I, a text prompt (referring description) T, and preceding tokens x_i, y_i, the model is trained to predict the next token x_t, y_t, its corresponding token type l, and the corresponding segmentation S. The prediction of the next token is guided by a combination of weighted $L1$ and $L2$ regression losses. The token type is is determined by the cross-entropy (CE) loss. The segmentation mask is based on $Dice$ loss.

$$\mathcal{L} = L1\left((x_t, y_t), (\hat{x}_t, \hat{y}_t)\right) + L2((x_t, y_t), (\hat{x}_t, \hat{y}_t)) \\ + CE(l, \hat{l}) + Dice(S, \hat{S}) \tag{2}$$

The weighted $L1$ and $L2$ loss employs different weights to balance the importance of box coordinates and polygon coordinates, and is defined as follows.

$$L1\left((x_t, y_t), (\hat{x}_t, \hat{y}_t)\right) = 0.1 \times L1\left((x_t, y_t), (\hat{x}_t, \hat{y}_t)\right)_{t=1,2} \\ + 0.9 \times L1\left((x_t, y_t), (\hat{x}_t, \hat{y}_t)\right)_{t>2} \tag{3}$$

Similar weights are employed in the weighted $L2$ loss.

4 Experiments

4.1 Datasets and Metrics

RRSIS-D Dataset: In this work, we utilize the RRSIS-D dataset, which was made public at CVPR 2024, for our experiments. This dataset contains 17,402 image-caption-mask triplets, split into train/validation/test sets with 12,181/1,740/3,481 samples, respectively. The remote sensing images have various spatial resolutions ranging from 0.3 to 30.0 m/pixel, with each image having a size of 800 × 800 pixels. The dataset comprises 20 categories: `ariplane`, `airport`, `basketball court`, `bridge`, `baseball field`, `chimney`, `dam`, `expressway service area`, `expressway toll station`, `golf field`, `ground track field`, `harbor`, `overpass`, `ship`, `stadium`, `storage tank`, `tennis court`, `train station`, `vehicle`, `wind mill`.

Metrics: In the experiments, we use overall Intersection-over-Union (oIoU), which is the overall ratio of intersection to union areas between predicted and ground truth masks, while mean Intersection-over-Union (mIoU) calculates the average accuracy for all the predicted and ground truth masks in pairs. Additionally, we use Precision@X (P@X) as an evaluation metric to evaluate precision based on IoU thresholds, reflecting the method's accuracy in object targeting.

4.2   Implementation Details

We trained the model on both the base and large versions of the Swin Transformer [30], with the language backbone based on BERT from Hugging Face's library [47]. The model was trained on an NVIDIA RTX 4090 TI for 100 epochs over 1 day. We initialized the learning rate to 0.00003 using Adam optimization. For better performance, we first trained RSSeq with RRSOD to obtain the initial checkpoint, making it aware of the global features. Then, we reloaded the checkpoint for the RRSOD task to train the multi-task RRSOD and RRSIS. In our experiments, we found that initializing the model with RRSOD resulted in better performance compared to training the multi-task setup from the start.

4.3   Comparison Results

Table 1 presents a quantitative comparison between our RSSeq model and existing state-of-the-art methods in referring segmentation. It is noteworthy that RMSIN [29] represents the latest advancement in RRSIS and introduces the RRSIS-D dataset. We showcase the performance of our RSSeq model with two backbone architectures: Swin-B and Swin-L. Both RSSeq-B and RSSeq-L outperform existing state-of-the-art methods, with RSSeq-L achieving superior performance across all metrics except for P@0.9. Further investigation into this metric will be included in future analyses.

Table 1. Quantitative comparison with state-of-the-art methods on **validation set of RRSIS-D dataset** [29]. Our proposed The best result is bold.

Methods	Venues	Visual Encoder	Language Encoder	Performance						
				P@0.5↑	P@0.6↑	P@0.7↑	P@0.8↑	P@0.9↑	oIoU↑	mIoU↑
RRN [23]	CVPR 2018	ResNet-101 [16]	LSTM [17]	51.09	42.47	33.04	20.80	6.14	66.53	46.06
CSMA [51]	CVPR 2019	ResNet-101	-	55.68	48.04	38.27	26.55	9.02	69.68	48.85
LSCM [20]	ECCV 2020	ResNet-101	LSTM	57.12	48.04	37.87	26.37	7.93	69.28	50.36
CMPC [19]	CVPR 2020	ResNet-101	LSTM	57.93	48.85	38.50	25.28	9.31	70.15	50.41
BRINet [18]	CVPR 2020	ResNet-101	LSTM	58.79	49.54	39.65	28.21	9.19	70.73	51.14
CMPC+ [28]	TPAMI	ResNet-101	LSTM	59.19	49.36	38.67	25.91	8.16	70.14	51.41
LGCE [52]	-	Swin-B [30]	BERT [11]	68.10	60.52	52.24	42.24	23.85	76.68	60.16
LAVT [50]	CVPR 2022	Swin-B	BERT	69.54	63.51	53.16	43.97	24.25	77.59	61.46
RMSIN [29]	CVPR 2024	Swin-B	BERT	74.66	68.22	57.41	45.29	**24.43**	78.27	65.10
RSSeq - B (Ours)	-	Swin-B	BERT	79.13	71.03	61.56	46.87	17.85	81.08	67.33
RSSeq - L (Ours)	-	Swin-L	BERT	**80.25**	**73.29**	**62.43**	**48.91**	17.87	**82.10**	**69.23**

Figure 2 visually illustrates a qualitative comparison between our RSSeq model and the runner-up, RMSIN [29]. While RMSIN only provides the segmentation mask, our RSSeq offers both the boundary and segmentation mask, along with the corresponding bounding boxes. By leveraging polygons to focus on boundaries, our model adeptly localizes objects, particularly along their edges.

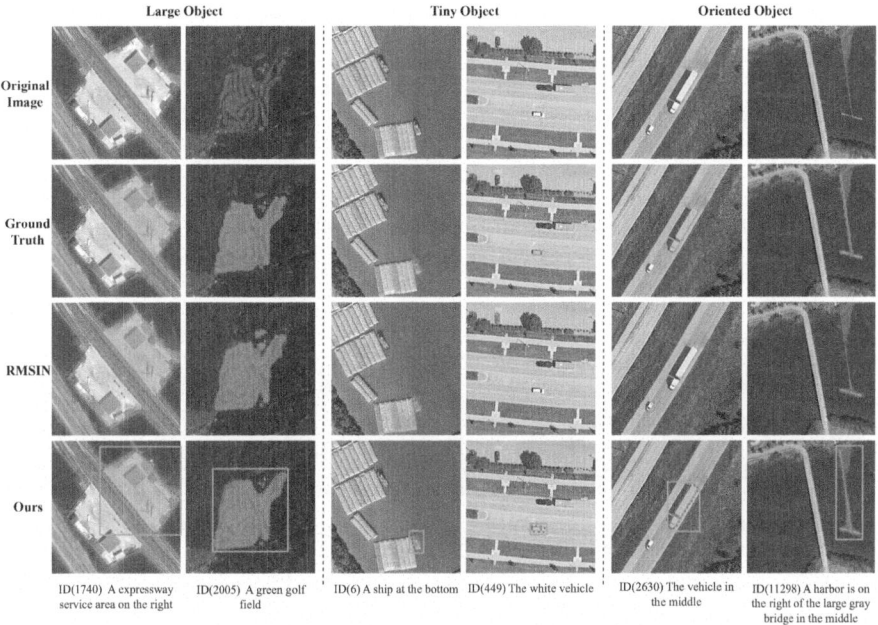

Fig. 2. Visualization comparison on the RRSIS-D Dataset [29]. From the top: 1st: Original images with various types of objects of interest; 2nd: Ground truth; 3rd: Segmentation mask by RMSIN [29]; 4th: Our proposed RSSeq, which simultaneously generates both segmentation and detection.

4.4 Ablation Studies

We further investigate the effectiveness of our proposed RSSeq by pre-training the model on the RMSIN-D dataset to obtain initial weights. Table 2 shows a comparison between two scenarios: one with and one without initializing weights on the RMSIN-D dataset across two different backbones.

Table 2. Ablation study of our proposed RSSeq on the **validation set of the RRSIS-D dataset** [29], comparing the performance with and without the pre-training procedure for initializing weights.

Language-Encoder	Visual-Encoder	Pre-train	Performance						
			P@0.5	P@0.6	P@0.7	P@0.8	P@0.9	oIoU	mIoU
BERT	Swin-B	✗	75.31	67.09	57.60	43.16	17.50	79.26	64.26
		✓	**79.13**	**71.03**	**61.56**	**46.87**	**17.85**	**81.08**	**67.33**
	Swin-L	✗	79.63	71.21	59.17	44.71	21.49	78.20	68.18
		✓	**80.25**	**73.29**	**62.43**	**48.91**	**17.87**	**82.10**	**69.23**

To better illustrate how our RSSeq predicts polygon vertices at inference time, we visualize some vertices around objects, as shown in Fig. 3, at different time steps.

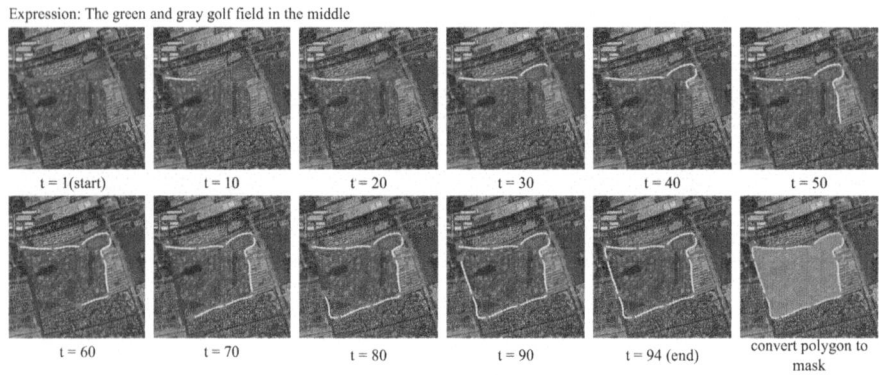

Fig. 3. Visualization of the steps during inference. The model begins with the BOS token. At step t = 1, it predicts the top-left vertex of the object. The model continues predicting until it reaches the EOS token. Finally, once the full polygon covers the object, the last step converts the polygon to a mask.

5   Conclusion

In this study, we introduce RSSeq, an end-to-end framework designed for referring remote sensing image segmentation (RRSIS) and object detection (RRSOD). RSSeq employs a multimodal approach within a sequence-to-sequence framework for multitask learning. By leveraging this architecture, RSSeq effectively segments object boundaries as sequences of vertices while supporting image segmentation for multiple objects. Additionally, RSSeq outperforms all existing state-of-the-art methods in RRSIS. We anticipate that this method can be extended to other remote sensing tasks, such as multi-label image segmentation and crop-type classification.

In future work, we plan to explore sequence-to-sequence methods for multi-label remote sensing image segmentation.

References

1. Acuna, D., Ling, H., Kar, A., Fidler, S.: Efficient interactive annotation of segmentation datasets with polygon-RNN++. In: Proceedings of the IEEE Conference on Computer Vision and Pattern Recognition, pp. 859–868 (2018)
2. Andrade, R., et al.: Evaluation of semantic segmentation methods for deforestation detection in the amazon. ISPRS Archives; 43, B3 **43**(B3), 1497–1505 (2020)
3. Carion, N., Massa, F., Synnaeve, G., Usunier, N., Kirillov, A., Zagoruyko, S.: End-to-end object detection with transformers. In: European conference on computer vision. pp. 213–229. Springer (2020)
4. Castrejon, L., Kundu, K., Urtasun, R., Fidler, S.: Annotating object instances with a polygon-RNN. In: Proceedings of the IEEE Conference on Computer Vision and Pattern Recognition, pp. 5230–5238 (2017)
5. Chen, C.F.R., Fan, Q., Panda, R.: CrossVIT: cross-attention multi-scale vision transformer for image classification. In: Proceedings of the IEEE/CVF International Conference on Computer Vision, pp. 357–366 (2021)
6. Chen, D.J., Jia, S., Lo, Y.C., Chen, H.T., Liu, T.L.: See-through-text grouping for referring image segmentation. In: Proceedings of the IEEE/CVF International Conference on Computer Vision, pp. 7454–7463 (2019)
7. Chen, L.C., Papandreou, G., Kokkinos, I., Murphy, K., Yuille, A.L.: DeepLab: aemantic image segmentation with deep convolutional nets, atrous convolution, and fully connected CRFS. IEEE Trans. Pattern Anal. Mach. Intell. **40**(4), 834–848 (2017)
8. Chen, L.C., Zhu, Y., Papandreou, G., Schroff, F., Adam, H.: Encoder-decoder with atrous separable convolution for semantic image segmentation. In: Proceedings of the European Conference on Computer Vision (ECCV), pp. 801–818 (2018)
9. Cheng, B., Misra, I., Schwing, A.G., Kirillov, A., Girdhar, R.: Masked-attention mask transformer for universal image segmentation. In: Proceedings of the IEEE/CVF Conference on Computer Vision and Pattern Recognition, pp. 1290–1299 (2022)
10. Devlin, J., Chang, M.W., Lee, K., Toutanova, K.: Bert: Pre-training of deep bidirectional transformers for language understanding. arXiv preprint arXiv:1810.04805 (2018)
11. Devlin, J., Chang, M.W., Lee, K., Toutanova, K.: BERT: pre-training of deep bidirectional transformers for language understanding. In: North American Chapter of the Association for Computational Linguistics (2019)
12. Diakogiannis, F.I., Waldner, F., Caccetta, P., Wu, C.: ResuNet-A: a deep learning framework for semantic segmentation of remotely sensed data. ISPRS J. Photogramm. Remote. Sens. **162**, 94–114 (2020)
13. Dong, Z., Zhang, R., Shao, X.: Automatic annotation and segmentation of object instances with deep active curve network. IEEE Access **7**, 147501–147512 (2019)
14. Griffiths, D., Boehm, J.: Improving public data for building segmentation from convolutional neural networks (CNNs) for fused airborne lidar and image data using active contours. ISPRS J. Photogramm. Remote. Sens. **154**, 70–83 (2019)

15. He, K., Gkioxari, G., Dollár, P., Girshick, R.: Mask R-CNN. In: Proceedings of the IEEE International Conference on Computer Vision, pp. 2961–2969 (2017)
16. He, K., Zhang, X., Ren, S., Sun, J.: Deep residual learning for image recognition. In: Proceedings of the IEEE Conference on Computer Vision and Pattern Recognition, pp. 770–778 (2016)
17. Hochreiter, S., Schmidhuber, J.: Long short-term memory. Neural Comput. 9(8), 1735–1780 (1997)
18. Hu, Z., Feng, G., Sun, J., Zhang, L., Lu, H.: Bi-directional relationship inferring network for referring image segmentation. In: Proceedings of the IEEE/CVF Conference on Computer Vision and Pattern Recognition (CVPR) (2020)
19. Huang, S., et al.: Referring image segmentation via cross-modal progressive comprehension. In: Proceedings of the IEEE/CVF Conference on Computer Vision and Pattern Recognition (CVPR) (2020)
20. Hui, T., et al.: Linguistic structure guided context modeling for referring image segmentation. In: Vedaldi, A., Bischof, H., Brox, T., Frahm, J.-M. (eds.) ECCV 2020. LNCS, vol. 12355, pp. 59–75. Springer, Cham (2020). https://doi.org/10.1007/978-3-030-58607-2_4
21. Kumar, S., et al.: Remote sensing for agriculture and resource management. In: Natural Resources Conservation and Advances for Sustainability, pp. 91–135. Elsevier (2022)
22. Li, J., Pei, Y., Zhao, S., Xiao, R., Sang, X., Zhang, C.: A review of remote sensing for environmental monitoring in china. Remote Sens. 12(7), 1130 (2020)
23. Li, R., et al.: Referring image segmentation via recurrent refinement networks. In: Proceedings of the IEEE Conference on Computer Vision and Pattern Recognition (CVPR) (2018)
24. Liang, J., Homayounfar, N., Ma, W.C., Xiong, Y., Hu, R., Urtasun, R.: Polytransform: Deep polygon transformer for instance segmentation. In: Proceedings of the IEEE/CVF Conference on Computer Vision and Pattern Recognition, pp. 9131–9140 (2020)
25. Ling, H., Gao, J., Kar, A., Chen, W., Fidler, S.: Fast interactive object annotation with Curve-GCN. In: Proceedings of the IEEE/CVF Conference on Computer Vision and Pattern Recognition, pp. 5257–5266 (2019)
26. Liu, C., Lin, Z., Shen, X., Yang, J., Lu, X., Yuille, A.: Recurrent multimodal interaction for referring image segmentation. In: Proceedings of the IEEE International Conference on Computer Vision, pp. 1271–1280 (2017)
27. Liu, J., Ding, H., Cai, Z., Zhang, Y., Satzoda, R.K., Mahadevan, V., Manmatha, R.: PolyFormer: referring image segmentation as sequential polygon generation. In: Proceedings of the IEEE/CVF Conference on Computer Vision and Pattern Recognition, pp. 18653–18663 (2023)
28. Liu, S., Hui, T., Huang, S., Wei, Y., Li, B., Li, G.: Cross-modal progressive comprehension for referring segmentation. IEEE Trans. Pattern Anal. Mach. Intell. 44, 4761–4775 (2021)
29. Liu, S., et al.: Rotated multi-scale interaction network for referring remote sensing image segmentation. arXiv preprint arXiv:2312.12470 (2023)
30. Liu, Z., et al.: Swin transformer: hierarchical vision transformer using shifted windows. In: Proceedings of the IEEE/CVF International Conference on Computer Vision (ICCV), pp. 10012–10022 (2021)
31. Long, J., Shelhamer, E., Darrell, T.: Fully convolutional networks for semantic segmentation. In: Proceedings of the IEEE Conference on Computer Vision and Pattern Recognition, pp. 3431–3440 (2015)

32. Marcos, D., Volpi, M., Kellenberger, B., Tuia, D.: Land cover mapping at very high resolution with rotation equivariant CNNs: towards small yet accurate models. ISPRS J. Photogramm. Remote. Sens. **145**, 96–107 (2018)

33. Mou, L., Hua, Y., Zhu, X.X.: Relation matters: relational context-aware fully convolutional network for semantic segmentation of high-resolution aerial images. IEEE Trans. Geosci. Remote Sens. **58**(11), 7557–7569 (2020)

34. Netzband, M., Stefanov, W.L., Redman, C.: Applied Remote Sensing for Urban Planning, Governance and Sustainability. Springer Science & Business Media (2007). https://doi.org/10.1007/978-3-540-68009-3

35. Niu, R., Sun, X., Tian, Y., Diao, W., Chen, K., Fu, K.: Hybrid multiple attention network for semantic segmentation in aerial images. IEEE Trans. Geosci. Remote Sens. **60**, 1–18 (2021)

36. O'neill, S.J., Boykoff, M., Niemeyer, S., Day, S.A.: On the use of imagery for climate change engagement. Global Environ. Change **23**(2), 413–421 (2013)

37. Radford, A., et al.: Learning transferable visual models from natural language supervision. In: International Conference on Machine Learning, pp. 8748–8763. PMLR (2021)

38. Samie, A., et al.: Examining the impacts of future land use/land cover changes on climate in Punjab province, Pakistan: implications for environmental sustainability and economic growth. Environ. Sci. Pollut. Res. **27**, 25415–25433 (2020)

39. Schumann, G.J., Brakenridge, G.R., Kettner, A.J., Kashif, R., Niebuhr, E.: Assisting flood disaster response with earth observation data and products: a critical assessment. Remote Sens. **10**(8), 1230 (2018)

40. Shamsolmoali, P., Zareapoor, M., Zhou, H., Wang, R., Yang, J.: Road segmentation for remote sensing images using adversarial spatial pyramid networks. IEEE Trans. Geosci. Remote Sens. **59**(6), 4673–4688 (2020)

41. Shi, H., Li, H., Meng, F., Wu, Q.: Key-word-aware network for referring expression image segmentation. In: Proceedings of the European Conference on Computer Vision (ECCV), pp. 38–54 (2018)

42. Wang, J., Zheng, Z., Ma, A., Lu, X., Zhong, Y.: Loveda: A remote sensing land-cover dataset for domain adaptive semantic segmentation. In: Vanschoren, J., Yeung, S. (eds.) Proceedings of the Neural Information Processing Systems Track on Datasets and Benchmarks, vol. 1. Curran Associates, Inc. (2021). https://datasets-benchmarks-proceedings.neurips.cc/paper_files/paper/2021/file/4e732ced3463d06de0ca9a15b6153677-Paper-round2.pdf

43. Wang, L., Li, R., Duan, C., Zhang, C., Meng, X., Fang, S.: A novel transformer based semantic segmentation scheme for fine-resolution remote sensing images. IEEE Geosci. Remote Sens. Lett. **19**, 1–5 (2022)

44. Wang, Z., et al.: CRIS: clip-driven referring image segmentation. In: Proceedings of the IEEE/CVF Conference on Computer Vision and Pattern Recognition, pp. 11686–11695 (2022)

45. Wei, X., Zhang, T., Li, Y., Zhang, Y., Wu, F.: Multi-modality cross attention network for image and sentence matching. In: Proceedings of the IEEE/CVF Conference on Computer Vision and Pattern Recognition (CVPR) (2020)

46. Weiss, M., Jacob, F., Duveiller, G.: Remote sensing for agricultural applications: a meta-review. Remote Sens. Environ. **236**, 111402 (2020)

47. Wolf, T., et al.: Transformers: state-of-the-art natural language processing. In: Conference on Empirical Methods in Natural Language Processing (2019)

48. Xia, J., Yokoya, N., Adriano, B., Broni-Bediako, C.: OpenEarthMap: a benchmark dataset for global high-resolution land cover mapping. In: Proceedings of the

IEEE/CVF Winter Conference on Applications of Computer Vision, pp. 6254–6264 (2023)

49. Xu, R., Wang, C., Zhang, J., Xu, S., Meng, W., Zhang, X.: RSSFormer: foreground saliency enhancement for remote sensing land-cover segmentation. IEEE Trans. Image Process. **32**, 1052–1064 (2023)

50. Yang, Z., Wang, J., Tang, Y., Chen, K., Zhao, H., Torr, P.H.: LAVT: language-aware vision transformer for referring image segmentation. In: Proceedings of the IEEE/CVF Conference on Computer Vision and Pattern Recognition, pp. 18155–18165 (2022)

51. Ye, L., Rochan, M., Liu, Z., Wang, Y.: Cross-modal self-attention network for referring image segmentation. In: Proceedings of the IEEE Conference on Computer Vision and Pattern Recognition, pp. 10502–10511 (2019)

52. Yuan, Z., Mou, L., Hua, Y., Zhu, X.X.: Rrsis: Referring remote sensing image segmentation. arXiv preprint arXiv:2306.08625 (2023)

53. Zhao, H., Shi, J., Qi, X., Wang, X., Jia, J.: Pyramid scene parsing network. In: Proceedings of the IEEE Conference on Computer Vision and Pattern Recognition, pp. 2881–2890 (2017)

Smart Camera Parking System with Auto Parking Spot Detection

Tuan T. Nguyen(✉) and Mina Sartipi

University of Tennessee at Chattanooga, Chattanooga 37402, USA
xwz778@mocs.utc.edu, mina-sartipi@utc.edu

Abstract. The proliferation of urban centers has exacerbated traffic congestion, underscoring the critical need for intelligent parking solutions. While computer vision approaches have gained traction, their reliance on manual spot labeling poses practical challenges. This study introduces PakLoc, a novel framework for automated parking spot detection, complemented by PakSke, a module that refines bounding box orientation and dimensions. Empirical evaluation on the PKLot dataset demonstrates a remarkable 94.25% reduction in manual labor requirements. Furthermore, we present PakSta, an innovative method leveraging PakLoc's object detector to concurrently assess occupancy across all parking spaces within a given frame. PakSta achieves an impressive AP75 of 93.6% on the PKLot dataset, surpassing the performance of the benchmark Yolo SPS (93.3% AP75) which relies on manually labeled data, and significantly outperforming other methods such as POD (61.8% AP75). These advancements offer a promising avenue for efficient, label-free smart parking systems, potentially revolutionizing urban parking management.

Keywords: Smart Parking System · SPS · Parking Spots Localization

1 Introduction

Urban population growth presents unprecedented challenges for city planners and residents alike. The United Nations projects that by 2050, a staggering 68% of the global population will reside in urban areas [3]. This demographic shift exacerbates existing issues, particularly in transportation and parking management. While recent research has focused on traffic simulation to mitigate congestion [24,30,32], the parking problem remains a significant concern. An INRIX survey reveals that American drivers spend an average of 17 h annually searching for parking, with this figure soaring to 107 h in densely populated cities like New York [2]. As autonomous vehicles become more prevalent, the need for accurate, real-time parking information becomes increasingly critical.

Smart parking systems (SPS) offer a promising solution, linking drivers and parking operators to optimize space utilization and reduce emissions. However, current sensor-based systems, while precise, are prohibitively expensive for large-scale deployment. The San Francisco SF Park program, for instance, spent $27 million to equip 19,250 parking spaces with sensors, at a cost of $1,400 per unit [1]. Computer vision (CV) approaches present a cost-effective alternative [7,14]. CV systems can monitor multiple spaces with a single camera, reducing installation and maintenance costs while offering additional benefits such as detecting improper parking and suspicious activities. Despite their promise, existing CV algorithms face three main challenges:

1. Time-consuming manual labeling of parking spots
2. Poor scalability due to the need for re-annotation in new environments
3. Slow real-time performance for large parking lots due to multiple forward passes in classification methods [17,27]

To address these issues, we propose three novel contributions:

- **PakLoc**: An automated parking space localization algorithm that reduces manual labeling effort by 94.25% (AR75) when deploying an SPS in new environments.
- **PakSke**: A module that automatically adjusts bounding box rotation and size, ensuring alignment with actual parking spot angles (Fig. 1).
- **PakSta**: A framework that simultaneously detects and monitors the status of all parking spots, utilizing the PakLoc detector to eliminate the need for additional model training.

Fig. 1. Parking spot detection challenges: varying angles and orientations (PKLot Dataset [10])

Our comprehensive experiments on the PKLot dataset demonstrate the efficacy of these approaches, offering a competitive and scalable solution for smart parking systems. The remainder of this paper is organized as follows: Sect. 2 discusses related work, Sect. 3 details our proposed methodology, Sect. 4 presents our experimental setup and results, and Sect. 5 offers concluding remarks and future directions.

2   Related Work

2.1   Automatic Parking Spots Localization

Parking spot localization, crucial during system installation or camera perspective changes, has evolved through three main approaches:

Traditional Image Processing: Early methods employed classical computer vision techniques like perspective transformation and edge detection [8, 31]. While effective in controlled environments, these struggled with real-world variability.

Chessboard Approaches: These methods use homography transformations to generate bird's-eye views of parking lots [16, 28]. However, they can produce false positives for passing vehicles.

Deep Learning Approaches: Recent advancements include CNN-based methods using Mask R-CNN [4, 9] and transformer-based models [19, 29]. Despite their sophistication, most generate perpendicular bounding boxes, misaligning with angled parking spots.

Our proposed PakSke module addresses this limitation by automatically aligning bounding boxes with actual parking spot angles. We employ deformable DETR [33] as our detection backbone for multi-scale object detection.

2.2   Parking Spots Status Identification

Parking spot status identification employs two main approaches: classification and detection.

Classification Methods treat each spot independently, using either feature extraction or deep learning techniques. Feature extraction methods, like those by Almeida et al. [5] and Suwignyo et al. [25], pre-process images to extract salient features before classification. Deep learning approaches, exemplified by Nyambal and Klein [17], integrate feature extraction and classification within a representation learning framework. While effective, these methods often rely on manually defined spots or perspective transformations, limiting scalability.

Detection Approaches unify localization and classification tasks, offering increased flexibility and faster inference. Two-stage detectors, like Faster R-CNN [23], first propose regions of interest before classification. One-stage detectors, such as RetinaNet adaptations [18], perform both tasks simultaneously. Recent advancements include attention mechanisms and custom network backbones [11] to improve accuracy.

Our proposed methodology innovates within the two-stage detector framework, applying the initial detector to ROI-filtered frames and using a mapping schema for efficient status determination across all parking slots.

3   Methodology

As describe in Sect. 1, our proposed method is divided into two modules: (1) PakLoc for automatic parking spots localization task and (2) PakSta for parking spots status identification task. The detail architecture is visualized in Fig. 2. The automobile detector plays a vital role in our proposed concept. The component in question holds significant importance in both the PakLoc and PakSta modules, since its performance directly influences the overall consequences of the architecture. Consequently, we have partitioned this section into three distinct components, namely the Vehicle Detector, PakLoc, and PakSta.

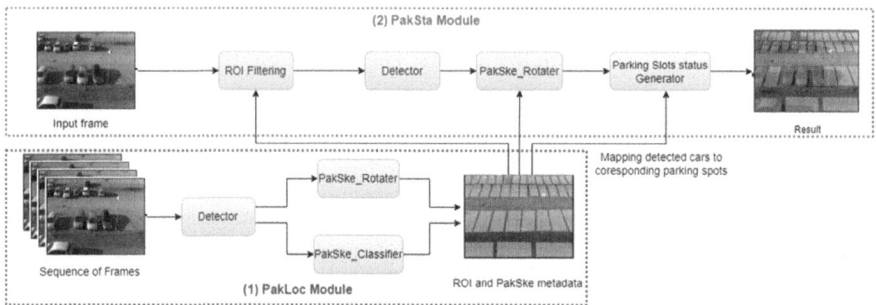

Fig. 2. Proposed Architecture. It includes two main modules: (1) PakLoc for automatic parking spots localization task and (2) PakSta for parking spots status identification task.

3.1   Vehicle Detector

There exist two methodologies for deep learning detection frameworks: the CNN-based approach and the Transformer-based model. The efficacy of transformer-based models in extracting various diversified discriminative parts of information and fine-grained features has been demonstrated in [12]. Furthermore, the PKlot dataset [10] includes variations in car scale and viewpoint. As a result, in this study, the we employ a transformer-based object detection model known as deformable DETR [33] that demonstrates effective performance in detecting objects of varying scales. This detector model has been pre-trained using the VeRi-776 [15] and CityFlow [26] datasets. The VeRi-776 dataset has a total of 49,357 images depicting 776 distinct vehicles. These images were captured using 20 different cameras. On the other hand, the CityFlow dataset consists of over 229,680 labeled bounding boxes of 666 distinct cars. These bounding boxes were obtained from 40 cameras positioned at 10 different junctions. The detector was pre-trained for 70 epochs using the same set of parameters as described in the original research. The detector model is fine-tuned on the training set of the PKlot dataset for a total of 40 epochs.

3.2   PakLoc Automatic Parking Spots Localization

PakLoc tackles the challenge of automatic parking spot localization through an innovative approach of vehicle movement tracking across consecutive frames. This method capitalizes on the temporal nature of parking lot surveillance to differentiate between moving vehicles and parked cars.

At its core, PakLoc operates by processing successive frames to detect cars and generate corresponding bounding boxes. These newly identified bounding boxes are then compared against an existing inventory of tracked vehicles using the Intersection over Union (IoU) metric. A vehicle is deemed stationary when its IoU value consistently exceeds a predefined threshold θ for a specified number of frames γ. Consequently, locations where vehicles remain stationary are designated as potential parking spots. The effectiveness of this approach is contingent upon the judicious selection of the IoU threshold θ and frame threshold γ. Our empirical studies, elaborated in Sect. 4.2, reveal optimal performance with $\theta = 0.75$ and $\gamma = 4$ for the PKLot dataset, which captures images at 5-minute intervals. These parameters, however, may require adjustment when applied to datasets with different temporal characteristics.

To address the prevalent issue of non-perpendicular parking spots (as illustrated in Fig. 1), we introduce the novel PakSke layers. During the training phase, the PakSke_Rotater layer determines optimal triplet hyperparameters [$angle, width\_scaling, height\_scaling$] to align detected spots with ground truth labels, as depicted in Fig. 3.

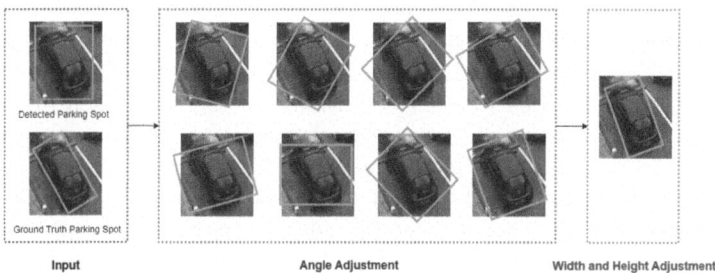

Fig. 3. PakSke_Rotater: Automatically adjusting parking slot angle, width, and height by maximizing IoU with ground truth.

For inference and deployment on new datasets where ground truth labels are unavailable, we employ the PakSke_classifier. This CNN-based classifier, derived from the CmAlexnet architecture [21], predicts the optimal triplet for newly detected parking spots. The PakSke_classifier is trained using the output produced by the PakSke_Rotater layer during the training phase, ensuring adaptability to various parking lot configurations. The culmination of PakLoc's processing is the generation of comprehensive Parking Spots metadata, encapsulated as [CameraId, x, y, w, h, angle, $width\_scaling$, $height\_scaling$]. This metadata provides a rich description of each identified parking spot, including its location, dimensions, and orientation adjustments. Through this multifaceted approach, PakLoc achieves accurate and adaptive parking spot localization across diverse parking lot layouts, overcoming the limitations of traditional perpendicular bounding box methods.

3.3   PakSta - Parking Spots Status Identification

The objective of PakSta is to forecast the state of parking spaces identified from the PakLoc. The PakSta system receives camera video input in a sequential manner, processing each frame individually. These frames are then sent via a region of interest (ROI) filtering layer. The ROI in question corresponds to the coordinates of parking spots, which are obtained from the parking spot metadata extracted in PakLoc. Subsequently, the filtered image is inputted into the detector in order to identify the presence of a car within the image. In the parking slots status generator layer, every identified vehicle is assigned a parking spot in the parking spot metadata and its bounding box is adjusted using the appropriate triplet parameter $[angle, width\_scaling, height\_scaling]$. Ultimately, the parking spaces that are mapped to any identified vehicle are categorized as occupied, whereas the remaining spaces are categorized as vacant.

4   Experiments and Results

4.1   Datasets, Evaluation Metrics and Baseline

Dataset:, Our study utilizes three datasets, each serving a distinct purpose in our model's development and evaluation. *PKlot Dataset* [10]: Our primary testing dataset, PKlot comprises 12,417 images, capturing 695,851 manually annotated spaces under various weather conditions. Images are taken at 5-minute intervals, providing a comprehensive view of parking dynamics across different times and environments. *CityFlow Dataset* [26]: Used for detector pre-training, CityFlow offers 229,680 labeled bounding boxes of 666 vehicles from 40 cameras across 10 intersections. This dataset provides diverse urban traffic scenarios, enhancing our model's ability to detect vehicles in complex environments. *Veri-776 Dataset* [15]: Also used in detector pre-training, VeRi-776 contains 49,357 images of 776 vehicles captured from multiple angles and under various conditions. Its real-world traffic scenarios and detailed vehicle annotations contribute to our model's robustness in diverse settings.

Evaluation Metric:, for PakLoc, we primarily use recall-based metrics: AR75 (average recall at IoU threshold 0.75), mAR40_90 (mean average recall for IoU thresholds 0.4 to 0.9), and AP50 (average precision at IoU threshold 0.5) for comparison with related work. For parking space status identification, we employ AP75 (average precision at IoU threshold 0.75) to assess accuracy.

Baseline: given the lack of standardized benchmarks for automatic parking spot localization, we selected representative works [6,13,18,20] with comparable metrics for evaluation. For parking spot status identification, PakSta is benchmarked against relevant studies [11,18,22] as discussed in Sect. 2.

4.2   Experimental Results

This section is divided into two subsections to evaluate the performance of Pak-Loc and PakSta.

PakLoc Performance. The evaluation of PakLoc is conducted using the test set of the PKLot dataset. Firstly, as discussed in Sect. 3.2, an ablation study is conducted to determine the best parameter IoU threshold θ. In this ablation study, we assess the results (AR) of PakLoc by varying the parameter θ throughout the range of 0.4 to 0.9, with a step size of 0.05. The experimental result indicates that the ideal value of θ is determined to be 0.75. Then, we select the θ as 0.75 for all next experiment. It is worth noting that in this ablation investigation, we adjusted the frame threshold γ to 4. To demonstrate the impact of the PakSke layer, we present outcomes obtained from both scenarios: one with the inclusion of the PakSke layer and the other without it using these three metrics. The results in Table 1 demonstrates the efficiency of our proposed PakSke layers. Using PakSke layers increases all three metrics by at least 6%.

Table 1. PakLoc result on testset of PKLot with and without PakSke layer

	AR75	mAR40_90	AP50
Without PakSke	88.31	74.38	80.23
With PakSke	**94.25**	**82.11**	**86.37**

Lastly, we compare PakLoc's performance to other baselines described above. The results presented in Table 2 indicate that our proposed method outperforms all prior work using the same dataset with 86.4% AP50. It even achieve a better result with 92.7% AP75.

Table 2. PakLoc result on testset of PKLot and other baselines method

Method/Ref	Backbone	Test Set	Metric	Result
Faster PSP [13]	Faster-RCNN	CNRPark-EXT	AP50	83.1
Auto PSP [20]	Yolo4	CNRPark-EXT	AP50	**97.6**
Realtime PSP [18]	Resnet and faster RCNN	PKLot	AP50	63.6
Cascade PSP [6]	Cascade Mask R-CNN	PKLot	AP50	59.1
PakLoc (ours)	Deformable DETR	PKLot	AP50	**86.4**
PakLoc (ours)	Deformable DETR	PKLot	AP75	**92.7**

PakSta Performance. A comparison is made between the results of PakStat and three additional baseline models [11,18,22]. The findings are presented in Table 3. The approach presented in [11] achieved the highest AP75 score of 98%. However, it should be noted that this system relied on the manual annotation of parking places during the training phase. This factor restricts the practical implementation of the paradigm in real-world scenarios. In contrast, our suggested solution, PakSta, does not necessitate human labeling of parking slots for implementation in a fresh dataset or a real parking lot environment. Furthermore, PakSta was able to obtain a notable outcome of 93.6% AP75, positioning it as the second best performer. This even surpasses the strategy employed in the study [22], where manually labeled parking spaces data was utilized. In addition, the results table further demonstrates the effectiveness of PakSke layers by indicating that they contribute a positive impact (improve 6%) on the ultimate outcome of PakSta.

Table 3. PakSta result on testset of PKLot and other baselines method

Method/Ref	Backbone	Data	Use Manually Label Of Parking Spots	AP75
POD [18]	RetinaNet	PKLot	No	61.8
Yolo SPS [22]	Yolo3	PKLot	Yes	93.3
OcpDept [11]	MBN-FPN	PKLot	Yes	**98.0**
PakSta without PakSke (ours)	Deformable DETR	PKLot	No	87.3
PakSta with PakSke (ours)	Deformable DETR	PKLot	No	**93.6**

5   Conclusion

In this paper, two new approaches, PakLoc and PakSta, are proposed to address the problems of automatic parking spot localization and parking spot status identification, respectively. Both of these methods demonstrate superior performance compared to the existing approaches in the given context. Additionally, we propose the incorporation of PakSke layers as a means to enhance the performance of these methods. The utilization of PakSke layers as a plug-in module is applicable in a comparable manner. In the forthcoming period, our objective is to construct a comprehensive smart parking system utilizing the way we have put forth.

Acknowledgement. This material is based upon work partially supported by the U.S. Department of Energy's Office of Energy Efficiency and Renewable Energy (EERE) under the Award Number DE-EE0009208. The views and opinions of authors expressed herein do not necessarily state or reflect those of the United States Government or any agency thereof.

References

1. Demand-responsive parking pricing. https://www.sfmta.com/demand-responsive-parking-pricing (2017). Accessed 31 July 2023
2. Smart parking – a silver bullet for parking pain. https://inrix.com/blog/parkingsurvey/ (2017). Accessed 31 July 2023
3. United nation. https://www.un.org/development/desa/en/news/population/2018-revision-of-world-urbanization-prospects.html (2018). Accessed 31 July 2023
4. Agrawal, T., Urolagin, S.: Multi-angle parking detection system using mask r-CNN. In: Proceedings of the 2020 2nd International Conference on Big Data Engineering and Technology, pp. 76–80 (2020)
5. Almeida, P., Oliveira, L.S., Silva, E., Britto, A., Koerich, A.: Parking space detection using textural descriptors. In: 2013 IEEE International Conference on Systems, Man, and Cybernetics, pp. 3603–3608. IEEE (2013)
6. de Almeida, P.R., Alves, J.H., Oliveira, L.S., Hochuli, A.G., Fröhlich, J.V., Krauel, R.A.: Vehicle occurrence-based parking space detection. arXiv preprint arXiv:2306.09940 (2023)
7. de Almeida, P.R.L., Alves, J.H., Parpinelli, R.S., Barddal, J.P.: A systematic review on computer vision-based parking lot management applied on public datasets. Expert Syst. Appl. **198**, 116731 (2022)
8. Bohush, R., Yarashevich, P., Ablameyko, S., Kalganova, T.: Extraction of image parking spaces in intelligent video surveillance systems. Mach. Graph. Vis. **27** (2018)
9. Coleiro, A., Scerri, D., Briffa, I.: Car parking detection in a typical village core street using public camera feeds. In: 2020 IEEE 10th International Conference on Consumer Electronics (ICCE-Berlin), pp. 1–6. IEEE (2020)
10. De Almeida, P.R., Oliveira, L.S., Britto, A.S., Jr., Silva, E.J., Jr., Koerich, A.L.: PKLot-a robust dataset for parking lot classification. Expert Syst. Appl. **42**, 4937–4949 (2015)
11. Duong, T.L., Le, V.D., Bui, T.C., To, H.T.: Towards an error-free deep occupancy detector for smart camera parking system. In: European Conference on Computer Vision, pp. 163–178. Springer (2022)
12. He, S., Luo, H., Wang, P., Wang, F., Li, H., Jiang, W.: TransReID: transformer-based object re-identification. In: Proceedings of the IEEE/CVF International Conference on Computer Vision, pp. 15013–15022 (2021)
13. Kirtibhai Patel, R., Meduri, P.: Faster r-CNN based automatic parking space detection. In: Proceedings of the 2020 3rd International Conference on Machine Learning and Machine Intelligence, pp. 105–109 (2020)
14. Lin, T., Rivano, H., Le Mouël, F.: A survey of smart parking solutions. IEEE Trans. Intell. Transp. Syst. **18**(12), 3229–3253 (2017)
15. Liu, X., Liu, W., Mei, T., Ma, H.: A deep learning-based approach to progressive vehicle re-identification for urban surveillance. In: Leibe, B., Matas, J., Sebe, N., Welling, M. (eds.) ECCV 2016. LNCS, vol. 9906, pp. 869–884. Springer, Cham (2016). https://doi.org/10.1007/978-3-319-46475-6_53
16. Nieto, R.M., Garcia-Martin, A., Hauptmann, A.G., Martinez, J.M.: Automatic vacant parking places management system using multicamera vehicle detection. IEEE Trans. Intell. Transp. Syst. **20**(3), 1069–1080 (2018)
17. Nyambal, J., Klein, R.: Automated parking space detection using convolutional neural networks. In: 2017 Pattern Recognition Association of South Africa and Robotics and Mechatronics (PRASA-RobMech), pp. 1–6. IEEE (2017)

18. Padmasiri, H., Madurawe, R., Abeysinghe, C., Meedeniya, D.: Automated vehicle parking occupancy detection in real-time. In: 2020 Moratuwa Engineering Research Conference (MERCon), pp. 1–6. IEEE (2020)

19. Pannerselvam, K.: Adaptive parking slot occupancy detection using vision transformer and LLIE. In: 2021 IEEE International Smart Cities Conference (ISC2), pp. 1–7. IEEE (2021)

20. Patel, R., Meduri, P.: Car detection based algorithm for automatic parking space detection. In: 2020 19th IEEE International Conference on Machine Learning and Applications (ICMLA), pp. 1418–1423. IEEE (2020)

21. Rahman, S., Ramli, M., Arnia, F., Sembiring, A., Muharar, R.: Convolutional neural network customization for parking occupancy detection. In: 2020 International Conference on Electrical Engineering and Informatics (ICELTICs), pp. 1–6. IEEE (2020)

22. Redmon, J., Farhadi, A.: YOLOv3: an incremental improvement. arXiv preprint arXiv:1804.02767 (2018)

23. Ren, S., He, K., Girshick, R., Sun, J.: Faster r-CNN: towards real-time object detection with region proposal networks. Adv. Neural Inf. Process. Syst. **28** (2015)

24. Sen, R., et al.: BTE-Sim: fast simulation environment for public transportation. In: 2022 IEEE International Conference on Big Data (Big Data), pp. 2886–2894. IEEE (2022)

25. Suwignyo, M.A., Setyawan, I., Yohanes, B.W.: Parking space detection using quaternionic local ranking binary pattern. In: 2018 International Seminar on Application for Technology of Information and Communication, pp. 351–355. IEEE (2018)

26. Tang, Z., et al.: CityFlow: a city-scale benchmark for multi-target multi-camera vehicle tracking and re-identification. In: Proceedings of the IEEE/CVF Conference on Computer Vision and Pattern Recognition, pp. 8797–8806 (2019)

27. Valipour, S., Siam, M., Stroulia, E., Jagersand, M.: Parking-stall vacancy indicator system, based on deep convolutional neural networks. In: 2016 IEEE 3rd World Forum on Internet of Things (WF-IoT), pp. 655–660. IEEE (2016)

28. Vítek, S., Melničuk, P.: A distributed wireless camera system for the management of parking spaces. Sensors **18**(1), 69 (2017)

29. Wang, L., et al.: Global perception-based robust parking space detection using a low-cost camera. IEEE Trans. Intell. Veh. **8**(2), 1439–1448 (2022)

30. Xu, D., Chen, Y., Ivanovic, B., Pavone, M.: BITS: bi-level imitation for traffic simulation. In: 2023 IEEE International Conference on Robotics and Automation (ICRA), pp. 2929–2936. IEEE (2023)

31. Zhang, W., Yan, J., Yu, C.: Smart parking system based on convolutional neural network models. In: 2019 6th International Conference on Information Science and Control Engineering (ICISCE), pp. 561–566. IEEE (2019)

32. Zhong, Z., et al.: Guided conditional diffusion for controllable traffic simulation. In: 2023 IEEE International Conference on Robotics and Automation (ICRA), pp. 3560–3566. IEEE (2023)

33. Zhu, X., Su, W., Lu, L., Li, B., Wang, X., Dai, J.: Deformable DETR: deformable transformers for end-to-end object detection. arXiv preprint arXiv:2010.04159 (2020)

LAVA

Questioning, Answering, and Captioning for Zero-Shot Detailed Image Caption

Duc-Tuan Luu[1,2,3,4], Viet-Tuan Le[5], and Duc Minh Vo[6(✉)]

[1] University of Information Technology, VNU-HCM, Ho Chi Minh City, Vietnam
tuanld@uit.edu.vn
[2] University of Science, VNU-HCM, Ho Chi Minh City, Vietnam
[3] John von Neumann Institute, VNU-HCM, Ho Chi Minh City, Vietnam
[4] Vietnam National University, Ho Chi Minh City, Vietnam
[5] Ho Chi Minh City Open University, Ho Chi Minh City, Vietnam
tuan.lv@ou.edu.vn
[6] The University of Tokyo, Bunkyō, Japan
vmduc@nlab.ci.i.u-tokyo.ac.jp

Abstract. End-to-end pre-trained large vision language models (VLMs) have made unprecedented progress in image captioning. Nonetheless, they struggle to generate detailed captions, which necessitate the models capturing spatial relations, counting, text rendering, world knowledge, and other presenting or not presenting aspects of the image. To overcome their inadequacies, we present a *Question – Answer – Caption* methodology, named QAC, that performs questioning and answering on many aspects of the given image, followed by captions based on the responses. Specifically, we use ChatGPT to produce a set of questions about the images' content. The questions are then answered using a pre-trained VLM. After gathering all answers, we prompt the pre-trained VLM to generate descriptive captions in a zero-shot setting. Our approach is plug-and-play and can be easily applied on any pre-trained VLM. We implement QAC on InstructBLIP and LLaVA, demonstrating comparable performance to fine-tuned models on a challenging DOCCI dataset.

1 Introduction

The field of image captioning has witnessed remarkable advancements with the emergence of VLMs [2,31,54,56,59]. These models have demonstrated impressive capabilities in generating detailed and informative captions based on visual input, making VLMs increasingly important across many downstream vision-language tasks. Usually, they are designed in a single-step paradigm, which returns a final caption directly from an image. This approach, however, hurts the performance when it comes to zero-shot reasoning [58] or visual entailment [52] tasks, which require in-depth captions from multi-step reasoning. Particularly, existing VLMs often struggle to capture the intricacies of complex visual scenes,

M. Cho et al. (Eds.): ACCV 2024 Workshops, LNCS 15482, pp. 249–266, 2025.
https://doi.org/10.1007/978-981-96-2641-0_17

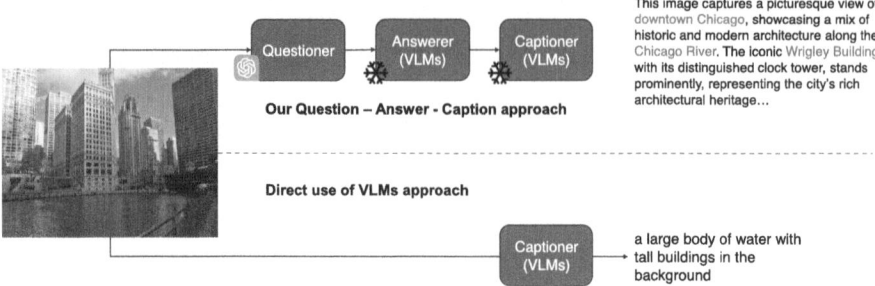

Fig. 1. Compared to traditional single-step image captioning VLMs, our multi-step QAC approach provides more insights and generates a higher level of descriptive captions, mimicking the process of human reasoning and understanding.

leading to captions that lack the specificity and richness necessary for accurate and informative descriptions. Hence, this drawback hinders VLMs applicability in various domains, such as content creation, image search, accessibility, etc.

In contrast, with a random photo, humans can provide a comprehensive fine-grained caption with ease and proficiency. Humans would intuitively break down the general image captioning task into different questions related to various aspects of the image. Moreover, the questions may even related one after another, acting like a human reasoning process from cumulative data. After answering all the questions, we can combine all the information and produce an insightful and descriptive caption, containing a deep understanding of multiple aspects, including objects, attributes, views, scenes, spatial relationships, text rendering, and world knowledge.

Inspired by this human nature behaviour, we propose a novel method that leverages the power of VLMs [1,4,12,46,48] to enhance the detail and accuracy of generated captions. Our method is reinforced by the observation that large language models excel at understanding and generating textual data, while deep learning visual models [10,22,25,26,35,38] are adept at processing and understanding visual information. We aim to create a robust and effective image captioning system by combining these strengths in a simple yet effective *Question – Answer – Caption* (QAC) paradigm. Figure 1 demonstrates the workflow of our proposed method, which contains three main components: the *Questioner*, *Answerer*, and *Captioner*. Based on the visual content of the input image, the *Questioner* generates a set of questions with different perspectives. The corresponding descriptive answers are obtained by utilizing the *Answerer*. Finally, the *Captioner* is responsible for synthesizing a fine-grained caption by analyzing and combining the information from the set of answers.

We carefully evaluate our proposed QAC image captioning method on the manually annotated image-caption pairs DOCCI dataset [32]. We implement our approach on the top of two open-source VLMs, including InstructBLIP (Vicuna-7B) [11,13] and LLaVA-1.5 7B [31] in a zero-shot manner, obtaining a marginal improvement on conventional metrics. We also notice that our generated captions contain information which does not present in the image such as world knowledge (words highlighted in red in Fig. 1, 2 and 7), revealing that *Question – Answer – Caption* is a good guidance for the image captioning task. Our contributions are summarized as follows:

- We introduce a straightforward yet effective approach called QAC, which is a *Question – Answer – Caption* framework designed to enhance the generation of detailed image captions in a zero-shot setting.
- Our method is designed to be plug-and-play, making it fully compatible with any pre-trained VLMs without requiring any additional re-training or fine-tuning.

2   Related Work

2.1   Multi-modality Models

Recent developments in multi-modality models [13,25,30,31,59] have notably enhanced the capabilities of comprehension and reasoning. These models often leverage the alignment of pre-trained large vision models [10,22,38,44] with large language models [4,11,12,48]. Early works, such as BLIP [25,26], LLaVA [23,30,31], and the Qwen series [30,31] bridged the modality gap through resampling or MLP projectors, demonstrating promising results. The Emu series [43,45] portrayed exceptional in-context learning ability for multi-modal content. Lately, there has been a growing trend towards developing high-resolution capability models in this research topic. Models like Monkey [28] and CogAgent [19] have adopted strategies to handle large images effectively, either by dividing them into patches or using separate low-resolution and high-resolution encoders. LLaVA-NEXT [23] and LLaVA-UHD [53] have introduced dynamic image aspect ratios, image partitioning and slicing techniques to capture more visual details. Moreover, Scaling on Scales [41] has demonstrated the ability to extract multi-scale features directly through image wrapping and rescaling without requiring an increase in image tokens.

2.2   Vision-Language Datasets

Early vision-language datasets were manually constructed using human annotations, such as Flickr30k [55] and COCO [29]. While these datasets offered high-quality annotations, they were limited in size and length. To address this, researchers turned to web-crawled datasets like YFCC100M [47] or RedCaps [14], which offered larger scales but faced challenges in terms of annotation quality. Many of these captions were only loosely related or unrelated to the corresponding images, impacting overall performance. To mitigate this issue, automatic filtering procedures were introduced to select higher-quality data samples, as seen in Localized Narratives [33] and Conceptual Captions [5,40]. These efforts have continued to improve, resulting in billion-scale datasets like LAION-5B

[39] and LAION-CAT [34], playing a crucial role in advancing vision-language pre-training. LaCLIP [15] and CapsFusion [57] have leveraged LLMs for caption rewriting and consolidation. Morover, recent studies have turned to GPT-4V or human-in-the-loop strategies to acquire detailed description datasets. ShareGPT4V [8] and ALLaVA [7] have generated large-scale synthetic datasets with detailed captions using GPT-4V. GLaMM [37] and all-seeing projects [50,51] have focused on region-level vision recognition and conversation generation. ImageInWords [16] and DOCCI [32] have introduced human-in-the-loop annotation frameworks for fine-grained detailed captions.

2.3 Dense Image Captioning

Dense image captioning has gained significant attention in recent years as a means to generate multiple descriptive captions for various objects and regions within a single image. This approach diverges from traditional image captioning, which typically produces a single sentence summarizing the entire image. Early works in this field, such as DenseCap [21], laid the groundwork by integrating visual features with textual descriptions, demonstrating the potential for more detailed image analysis. Subsequently, Anderson et al. [3] introduced the bottom-up and top-down attention mechanism, which effectively combined object detection with caption generation, allowing models to focus on salient areas of an image and generate context-aware descriptions. However, these pioneer approaches could not exploit all of the complementary nature of local and global visual cues. Hence, the generated captions are often concise, neglecting details about the intricate relation between objects. To alleviate this problem, recent studies [6,17,18,20] have focused on combining both LLMs and VLMs to boost the ability to generate high-fidelity dense captions. GBC [20] was proposed as a new vision-language data format that captions images with a graph-based structure akin to scene graphs while retaining the flexibility and intuitiveness of plain text description. VCB [17] introduced a blended mechanism that holistically captures various perspectives of the image while remaining anchored in human annotations. VFC [18] initiated a verification step after generating caption proposals by using tools such as object detection and visual question answering (VQA) models. This approach mitigates the challenge of hallucination in long captions. Furthermore, high-quality datasets (DenseFusion-1M [27], PixelProse [42], DCI [49], DOCCI [32]) are also proposed, having precise and reliable captions that can capture all of the aspects of the image (objects, attributes, spatial relations, scene, etc.). These datasets can benefit not only the image captioning task but also various vision-language tasks in general.

3 Question Answer Caption Method

We propose a plug-and-play novel method called *Question – Answer – Caption* (QAC) to enhance detailed image captioning in a zero-shot setting, leveraging pre-trained VLMs. Our method consists of three effective components that operate consecutively, namely the *Questioner, Answerer* and *Captioner*. The core idea behind QAC is to structure the image caption generation process as a sequence of questions and answers. Instead of solely relying on the model's direct image-captioning capabilities, we first generate a set of n relevant questions $\mathcal{Q} = \{q_1, q_2, \ldots, q_n\}$ about the image I and then use the pre-trained VLM to answer these questions, obtaining a set of n answers $\mathcal{A} = \{a_1, a_2, \ldots, a_n\}$. These answers are then used to construct a more comprehensive and enriched image caption c. Figure 2 illustrates a detailed example of each component in our proposed QAC approach for image captioning.

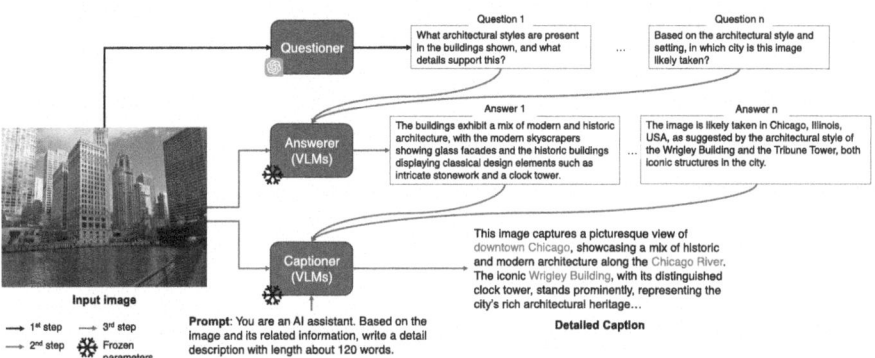

Fig. 2. The overall pipeline of our QAC approach. It consists of three sequential steps. The first step (*Questioner*) uses ChatGPT to generate a set of questions regarding the input image. The second step (*Answerer*) prompts a pre-trained VLM to answer the questions. The final step (*Captioner*) share the same VLM with a modified prompt to obtain the final detailed caption. We highlight the world knowledge extracted from the image that is successfully included in the caption.

3.1 Questioner

```
###Human: <Image><I></Image> Ask 10 questions about this image. The
question focuses on object and attributes, spatial relationships,
text rendering, world knowledge, view/scene.
###ChatGPT:
```

Fig. 3. ChatGPT prompt guidance for *Questioner*.

As mentioned earlier, the *Questioner* is responsible for creating a set of questions $\mathcal{Q}$ related to the visual information in the image. To ensure the generated captions capture multiple aspects of the image content, the *Questioner* must be capable of asking about a diverse range of characteristics, such as objects, attributes, spatial relationships, text rendering, world knowledge, and details about the view and scene. This comprehensive questioning ensures that all relevant information in the image is represented.

To achieve this level of detail, we select a powerful closed-source model as the *Questioner*, since open-source models do not produce satisfactory results, as discussed in our later analysis. Because we do not have direct access to this model on our machine, we upload the image I along with a prompt (shown in Fig. 3) through the API. Additionally, we specify the number n of questions q_i ($i = 1, \ldots, n$) to be generated, aiming for comprehensive coverage. As a result, we obtain a set $\mathcal{Q}$ consisting of n questions. Due to API costs, we conducted the questioning process only once without verifying whether all questions addressed every desired aspect.

3.2   Answerer

```
###Human: <Image><I></Image> <Questions><q1,q2,...,qn></Questions>
You are AI assistant with immersive knowledge about the world.
Based on that, answer the question related to the image.
###VLMs:
```

Fig. 4. Promt guidance for *Answerer*.

The second step of our proposed QAC method is the *Answerer*. This module is required to answer each question in $\mathcal{Q}$ generated from the *Questioner* based on the image content. Therefore, we employ a pre-trained VLM to serve as the *Answerer*.

Specifically, we first extract the image features $\mathbf{I}$ from the given image I using the image encoder in VLM. Then, we feed image features $\mathbf{I}$ along with all questions q_i in $\mathcal{Q}$, achieving corresponding answers $\mathcal{A} = \{a_1, a_2, \ldots, a_n\}$. The answer prompt is illustrated in Fig. 4, describing the VLM as an AI assistant with vast knowledge about the world. In fact, the content of this prompt can be manually changed to produce better answers. Moreover, these answers may contain details about world knowledge (words highlighted in red in Fig. 1, 2 and 7), giving people more insights about what information presents in the photo. After this step, a set $\mathcal{A}$ of n answers is ready to go through the *Captioner* to generate the final caption.

3.3 Captioner

```
###Human: <Image><I></Image> <Info><a₁,a₂,...,aₙ></Info> Given the
image and relevant information, describe this image in detail.
###VLMs:
```

Fig. 5. Prompt guidance for *Captioner*.

Captioner is the final stage of our proposed QAC approach. This step shares the same VLM with the *Answerer* but integrates with a different guidance prompt. The image features **I** from the previous step continue being the input, along with the relevant information from the set of answers $\mathcal{A}$. Figure 5 shows how we modify the guidance prompt to achieve the fine-grained caption c from the input image I. Similar to the previous step, the guidance caption prompt could also be altered to get a better image captioning result.

4 Experimental Results

4.1 Experimental Settings

Implementation. For the *Questioner*, we employ ChatGPT 4o to generate $n = 10$ questions for each image. Since the *Answerer* and *Captioner* in our QAC are flexible, we select two different methods, InstructBLIP [13] and LLaVA-1.5 7B [31], for each version. They are denoted as QAC-InstructBLIP and QAC-LLaVa, respectively. We note that all versions of our QAC are zero-shot settings. We tested our method on a single A100 80G GPU.

Evaluation Dataset. We evaluate QAC on DOCCI dataset [32] using its test set in a zero-shot manner without any training or finetuning. It contains 5,000 images with one caption for each. On average, each caption has 7 sentences with a length of 135 words. The caption covers multiple aspects, including objects, attributes, views, scenes, spatial relationships, text rendering, and world knowledge.

Compared Methods. We mainly compare our method with InstructBLIP [13] (Vicuna-7B), LLaVA-1.5 7B [31] since they are able to generate detailed captions. For other SOTAs on the image captioning task, such as MiniGPT4 [59], SmallCap [36], EVCap [24], since they focus on generating short captions, we only show their generated captions.

4.2   Case Study

In the case study, illustrated by Fig. 6, our method is demonstrated step-by-step. The process begins with the *Questioner*, which is responsible for generating detailed questions based on various aspects of the image content. As shown in Fig. 6, the questions cover a wide range of topics, including objects, attributes, views, scenes, spatial relationships, text rendering, and world knowledge. For instance, questions like "What text is prominently displayed on the object?" and "What colours are used in the circular sign?".

Next, the *Answerer* utilizes VLMs such as InstructBLIP [13] or LLaVA [31] to respond to the generated questions based on the image content. For instance, some answers are the object is a "sun-like shape", made out of "paper or cardboard" and located "inside a building, likely a library or bookstore", These answers, however, are not always entirely accurate, reflecting the models' varying ability to understand the image. For instance, while both InstructBLIP and LLaVA correctly identify the text "SUMMER READS", there are subtle differences in how they interpret elements like lighting and staircases. These discrepancies demonstrate the limitations of current VLMs in achieving perfect understanding.

Nevertheless, these answers serve their primary purpose, which is to guide the *Captioner* in generating detailed and accurate captions. The *Captioner* uses the information from the answers to produce captions that are both descriptive and informative. For example, the caption describes the scene as "a colorful sun-shaped sign reading 'SUMMER READS' in what appears to be a library or bookstore", incorporating specific details such as the object's material and the surrounding environment.

This case study provides strong evidence that our method effectively enhances the performance of pre-trained VLMs in a zero-shot setting. By using the *Questioner* and *Answerer* to extract relevant information and guide the *Captioner*, the approach enables the generation of more detailed and accurate image descriptions without requiring additional training data. Thus, it successfully improves the interpretive capabilities of VLMs.

Table 1. Quantitative comparison against SOTA methods on DOCCI dataset [32]. We report BLEU@4, ROUGE-L, METEOR, CIDEr, and average of caption length. The results from other methods are originally reported in [32].

Method	Eval mode	BLEU@4	ROUGE-L	METEOR	CIDEr	#Words
PaLI 5B [9]	Finetune on COCO	0.0	11.3	3.6	0.0	15.1
	Finetune on DOCCI	10.1	29.1	17.9	16.0	121.8
LLaVA-1.5 7B [31]	zero-shot	3.5	22.0	11.3	6.4	89.5
QAC-LLaVA (ours)		8.2	24.7	15.0	9.9	116.4
InstructBLIP (Vicuna-7B) [13]	zero-shot	3.5	20.5	10.6	5.9	84.4
QAC-InstructBLIP (ours)		7.6	24.6	13.5	9.4	115.8

4.3   Quantitative Comparison

Table 1 provides a comprehensive quantitative comparison of our method against other existing approaches. The results clearly demonstrate that QAC significantly enhances the performance of pre-trained models such as InstructBLIP [13] and LLaVA [31]. A particularly noteworthy observation is that our method improves all evaluation metrics, showing an overall boost in effectiveness. Additionally, our method allows the model to generate more detailed and longer image captions, with an average length of 115.8 and 116.4 words for each respective version. This substantial increase in caption length suggests that providing more detailed image information can greatly aid image-to-text generation tasks when leveraging pre-trained VLMs.

When we compare our method with other approaches that have been fine-tuned using datasets like COCO or DOCCI, the results become even more impressive. Our method outperforms PaLI [9], which has been fine-tuned on COCO, and closes the performance gap to PaLI fine-tuned on DOCCI to a

1. What text is prominently displayed on the object?
2. What colors are used in the circular sign?
3. What type of object is featured in the center?
4. What is the object made out of?
5. Where is the object located?
6. Are there any textual signs on the wall?
7. Are there any drawings or illustrated characters?
8. Is there any lighting visible in the image?
9. Are there any directional indications in the image?
10.Are there steps or a staircase present?

QAC-InstructBLIP

1. The text is "SUMMER READS."
2. Yellow, orange, red, pink, and white.
3. Sun-like shape.
4. Paper or cardboard.
5. Inside a building, likely a library or bookstore.
6. "Join Us Book Club" and "new stuff we love."
7. Illustrations of an owl and a red animal.
8. Visible lighting.
9. Indications to join a book club.
10. Yes, there is a staircase present.

The image captures a vibrant, sun-shaped decoration prominently displaying the text "SUMMER READS" in colorful, hand-drawn letters. The sunburst features a mix of yellow, orange, red, pink, and white hues crafted from paper or cardboard, suggesting a cheerful, welcoming theme. This decoration hangs inside a library or bookstore, as indicated by signs promoting a book club and new reading materials displayed below. An inviting staircase with railings leads to an upper level, illuminated by overhead lighting, which adds to the cozy atmosphere. Additionally, charming illustrations of an owl and another red animal adorn the walls, further enhancing the inviting and creative vibe of the space.

QAC-LLaVA

1. "SUMMER READS"
2. Yellow, orange, red, pink
3. Sun
4. Paper or cardboard
5. Library or bookstore
6. "Join Us BOOK CLUB", "our stuff we love"
7. Owl illustration
8. Ceiling light fixture
9. No arrows, invitation to Book Club
10. Yes, in background

A colorful sun-shaped sign reading "SUMMER READS" is prominently displayed in what appears to be a library or bookstore. The sign, made of paper or cardboard, is a vibrant mix of yellow, orange, red, and pink, symbolizing the warmth of summer. Behind the sign, a staircase leads to an upper level, suggesting a cozy, multi-level space. On the walls, additional signs are visible, including one inviting people to join a "BOOK CLUB" and another saying "our stuff we love." A small owl illustration adds a whimsical touch to the scene. A ceiling light fixture shines softly above the staircase, contributing to the warm, inviting atmosphere. The display serves as a cheerful reminder of summer reading and community engagement within the space.

Fig. 6. A case study shows step-by-step of our approach. We show the input image and 10 questions generated by ChatGPT. Different versions of our approach including QAC-InstructBLIP and QAC-LLaVA, first answers those question and then gathers all answers to generate the detailed caption.

greater extent than both InstructBLIP and LLaVA. While our method does not entirely surpass the performance of fine-tuned models, it is important to emphasize the considerable cost associated with fine-tuning. Finetuning requires extensive computational resources and high-quality datasets, which can be both expensive and time-consuming to acquire and process. In contrast, our method offers a more practical and efficient alternative, as it operates in a plug-and-play fashion, enabling it to be quickly integrated with any pre-trained VLMs without the need for extensive effort or resources.

4.4   Qualitative Comparison

Figure 7 presents examples of captions generated by different methods, offering a qualitative comparison of their performance. As shown, SmallCap [36] and EVCap [24] struggle to generate detailed captions, often failing to provide comprehensive descriptions of the image. In contrast, MiniGPT4 [59] manages to capture some level of detail, but its performance is still limited compared to the more sophisticated methods. InstructBLIP, LLaVA, QAC-InstructBLIP, and QAC-LLaVA perform noticeably better, generating captions that are richer in detail.

Although it may be challenging to definitively determine which method among InstructBLIP, LLaVA, QAC-InstructBLIP, and QAC-LLaVA performs best, especially since VLMs are generally capable of capturing basic image information, our methods show a clear advantage. Specifically, QAC-InstructBLIP and QAC-LLaVA enrich the captions with external knowledge, bringing them closer to the ground-truth descriptions than the other methods as highlighted in red. This ability to incorporate world knowledge demonstrates that our question-answer-caption (QAC) framework significantly enhances caption quality in a meaningful way. The benefits of our approach are further supported by the quantitative results discussed earlier, confirming that this simple yet effective method boosts the level of detail in image captions, particularly in zero-shot scenarios.

4.5   Plug-and-Play Ability

Our method can be implemented on pre-trained VLMs in a zero-shot manner, enabling it to function as a plug-and-play module compatible with any pre-trained VLMs. To verify this capability, we implement it on InstructBLIP and LLaVA. As demonstrated in Figs. 6 and 7, as well as Table 1, our method consistently boosts the performance of both InstructBLIP and LLaVA.

However, we observe variations in performance between QAC-InstructBLIP and QAC-LLaVA models, which could be attributed to differences in the original capabilities of each pre-trained VLM and the prompts used. Firstly, as shown in Fig. 6, the two models respond differently to the same questions, indicating variations in their ability to comprehend and interpret image content. For instance, InstructBLIP might identify the owl illustration more effectively, while LLaVA might provide a more detailed description of the staircase. Secondly,

Ground-truth: A medium-long shot view of downtown Chicago, Illinois, with a view of the Chicago River. Along the near side of the river, a metal railway can be seen running horizontally. On the other side of the river, there are large buildings. Along the middle, there is a large rectangular building, and the same one is behind it. This building is made of light brown cement, and along this building there are squared windows. To the left of these buildings is one that is made up of glass, the bottom portion of this glass is covered in shade while the top portion of it is being lit up by the sun. In front of this building, there are trees that have orange leaves. To the right of the buildings in the middle, there is a skinny concrete building, and along the top portion of the building there are small cylindrical pillars that surround the building. To the right of this building are two glass-squared buildings that are made of glass and are reflecting the baby blue sky. Along the right side of the river there is a large square building that is submerged in shade, and in front of this building there is a bridge that runs horizontally, and is also covered in shade.

InstructBLIP: The image presents a tranquil cityscape, where nature and architecture coexist in harmony. The river, calm and reflective, serves as a mirror to the city's skyline, its surface undisturbed except for the occasional ripple. On the left side of the image, a large, ornate building stands tall. Its architecture is intricate, with a clock tower reaching towards the sky, punctuated by a pointed spire that adds to its grandeur. The building's facade is a mix of stone and glass, reflecting the sunlight and adding a touch of elegance to the scene. Opposite the ornate building, on the right side of the image, a modern skyscraper reaches for the heavens. Its glass exterior is a testament to contemporary architecture, reflecting the city's bustling life.

LLaVA: The image captures a serene urban scene, dominated by the presence of a river that cuts through the cityscape. The river, calm and reflective, serves as a mirror to the city's skyline, its surface undisturbed except for the occasional ripple.On the left side of the image, a large, ornate building stands tall. Its architecture is intricate, with a clock tower reaching towards the sky, punctuated by a pointed spire that adds to its grandeur. The building's facade is a mix of stone and glass, reflecting the sunlight and adding a touch of elegance to the scene.Opposite the ornate building, on the right side of the image, a modern skyscraper reaches for the heavens. Its glass exterior is a testament to contemporary architecture, reflecting the city's bustling life. The skyscraper's sleek lines contrast with the ornate details of the building across the river, creating a harmonious blend of old and new.

QAC-InstructBLIP: This image captures a picturesque view of downtown Chicago, showcasing a mix of historic and modern architecture along the Chicago River. The iconic Wrigley Building, with its distinguished clock tower, stands prominently, representing the city's rich architectural heritage. Adjacent skyscrapers with sleek glass facades rise against the blue sky, creating a striking contrast between old and new. A walkway runs parallel to the river, offering a scenic path for pedestrians, while the river itself reflects the towering structures, emphasizing the harmonious integration of urban design and natural elements. The late afternoon light casts a warm glow on the stone buildings, highlighting their intricate details and adding to the vibrant atmosphere of this bustling urban landscape.

QAC-LLaVA: This image showcases a prominent view of Chicago's iconic architecture along the Chicago River. In the foreground, the Wrigley Building stands tall with its historic clock tower, a symbol of the city's architectural heritage since 1920. Flanking it are modern glass skyscrapers, reflecting Chicago's balance between its storied past and cutting-edge present. The Chicago Riverwalk is visible along the right, a popular public space offering stunning views of the city's skyline. The Tribune Tower, another landmark with gothic architecture, adds to the blend of old and new. The calm waterway mirrors the buildings, highlighting Chicago's reputation as a global center for architecture and design. The scene captures the city's dynamic urban landscape, blending history, innovation, and the beauty of its riverside location.

MiniGPT4: The image shows a view of a river running through the city, with tall buildings on either side. The buildings are made of concrete and steel, with large windows and pointed roofs. The river is clear and blue, with boats and other watercraft moving along it. The sky is a bright blue, with a few clouds in the distance. The sun is shining, casting a warm glow over the scene. The overall atmosphere is peaceful and serene.

SmallCap: A large body of water with tall buildings in the background.

EVCap: A large body of water next to tall buildings.

Ground-truth: A view of a black Dodge Ram truck parked on pavement with two black cars lined up behind it. There are diagonal white lines on the pavement to the right of them. The cars are facing the front, and all have reflections on their windshields. There is a stoplight overhead on the left. There is a strip of buildings on the left, alongside the road. The first one has flowers around the top of the striped awning. The tall building in the middle with red brick has ladders on the front of it and graffiti sprayed on the side of it. There is part of a tree visible on the right, next to the third parked car. There are more tall buildings visible in the background.

InstructBLIP: The image shows a city street scene with a row of parked vehicles. Behind these vehicles, there are more cars parked, and the street is lined with buildings that have storefronts and awnings. The architecture suggests an urban environment, and there are traffic lights visible, indicating a regulated traffic area. The sky is partly cloudy, suggesting it might be a cool or overcast day.

LLaVA: The image depicts a city street scene with a row of parked vehicles. The most prominent vehicle is a large, black pickup truck with a distinctive front grille design, which suggests it might be a luxury or high-performance model. The truck is parked alongside other vehicles, including a smaller car and another truck. The street is lined with buildings, and there are traffic lights visible, indicating an urban setting. The presence of a fire escape on one of the buildings and the architecture of the buildings suggest this could be a city in the United States.

QAC-InstructBLIP: This image captures a street scene in an urban area, possibly in New York City, recognizable by its characteristic architecture and graffiti-covered walls. In the foreground, there's a lineup of black vehicles, led by a customized Ram 1500 truck. The truck features an aggressive, armored front bumper, designed for off-road use and heavy-duty applications, giving it a formidable presence. Following the truck are other black SUVs, adding to the tough, unified look. The background reveals classic New York City buildings with red brick facades, fire escapes, and storefronts, while overhead traffic lights and street signs emphasize the city's dynamic, fast-paced environment.

QAC-LLaVA: The image captures an urban street scene with three black vehicles parked in a row along the curb. These vehicles include a Ram truck at the front and two SUVs trailing behind. The street is lined with a variety of buildings, indicating a mixed-use neighborhood possibly in a city. Some buildings have notable fire escapes on their facades, while others display commercial signage. Visible graffiti adds to the urban aesthetic. A traffic signal hangs above the street, and there are also trees providing some greenery in the scene. Sidewalks are bordered with flowers and decorative elements, enhancing the urban environment. The overall atmosphere is a blend of modern urban life with hints of historical architecture.

MiniGPT4: The image depicts a row of three dark-colored vehicles parked in a line on a city street. The vehicles are parked along the curb in a clearly marked parking area. The parked vehicles seem to be part of a bustling, well-visited area.

SmallCap: A black truck is driving down the street.

EVCap: A row of black trucks parked on a city street.

Fig. 7. Generated captions by compared methods. Captions generated by QAC-InstructBLIP and QAC-LLaVA (both are ours) are more detailed than those by other methods. They also contain world knowledge (highlighted in red) which is closed to the ground-truth captions.

our method employed a simple prompt to interact with the VLMs, which might not be robust enough to fully guide the models. This remains a room where further refinement could lead to improved performance, suggesting that prompt engineering could enhance the overall effectiveness of our approach. The potential impact of more sophisticated prompts remains an avenue for future research. Additionally, exploring our method's compatibility with other pre-trained VLMs is crucial to further elaborate on its efficacy. Despite these considerations, current observations suggest that our method is beneficial in enhancing both world knowledge and generating more detailed captions.

4.6 Impact of Questioner

To understand the impact of different *Questioners* on our approach, we conduct an ablation study by comparing the effectiveness of using InstructBLIP, LLaVA, and ChatGPT as the *Questioner* module (see Table 2). When using InstructBLIP and LLaVA, the generated questions are simpler and lack of insights, resulting in less detailed answers and ultimately reducing the quality of the captions. In contrast, using ChatGPT as the *Questioner* leads to more contextually rich and comprehensive questions, which will provide the *Captioner* with more information to create detailed and accurate captions. Our ablation study shows that the choice of *Questioner* significantly affects the performance of our proposed method.

Table 2. Ablation study on the effect of different generators.

Method	Questioner	BLEU@4	ROUGE-L	METEOR	CIDEr	#Words
QAC-InstructBLIP	ChatGPT	7.6	24.6	13.5	9.4	115.8
QAC-LLaVA		8.2	24.7	15.0	9.9	116.4
QAC-InstructBLIP	InstructBLIP [13]	5.2	21.1	11.7	6.7	92.8
QAC-LLaVA		5.4	22.3	12.0	6.9	91.3
QAC-InstructBLIP	LLaVA [31]	5.6	22.4	11.9	6.4	93.7
QAC-LLaVA		6.2	22.7	12.5	7.6	95.1

We also conduct an ablation study to examine the impact of the number of questions generated by the *Questioner* on the overall performance of our method. In this experiment, we use ChatGPT as the *Questioner* and generate a number of 1, 5, 10, 15, and 20 questions, respectively. The BLEU@4 and CIDEr scores are plotted in Fig. 8. We see that when a smaller number of questions is used (i.e., 1 or 5), the answers are less comprehensive, leading to captions that missed certain details and nuances of the image content. On the other hand, as the number of questions increases (i.e., 10), the *Questioner* is able to extract more detailed information, enabling the *Captioner* to produce more informative and accurate captions. However, we notice that beyond a certain threshold (i.e., 15, 20), the additional questions provide diminishing returns, as

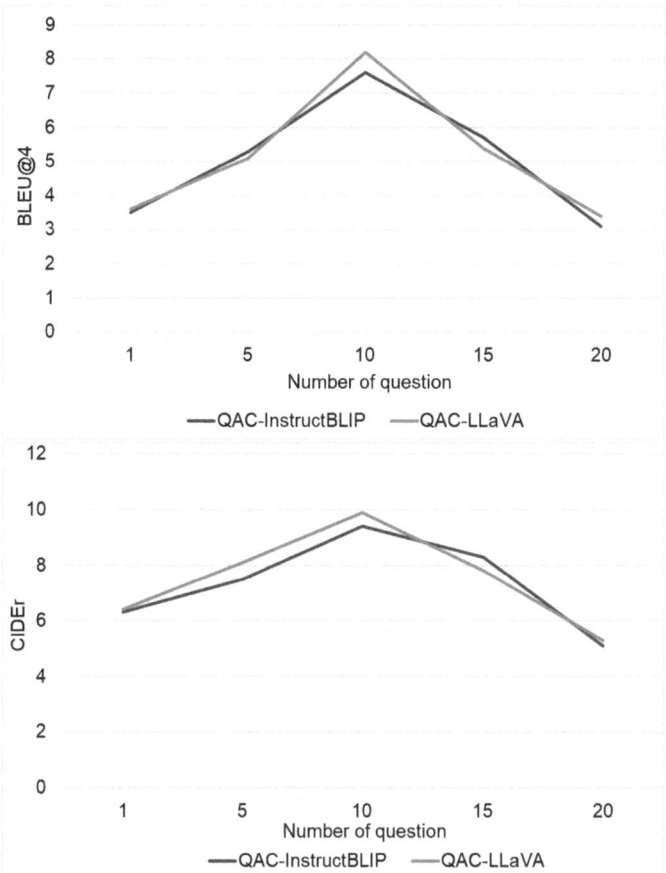

Fig. 8. Ablation study on the impact of the number of question. When fewer questions or more questions are asked, the performance degrades in both BLEU@4 (above) and CIDEr (below) scores.

the extra information becomes repetitive and does not significantly improve the final captions. Therefore, finding an optimal number of questions is crucial for balancing detail and efficiency, ensuring the most effective enhancement of VLM performance.

5   Discussion

Our method, QAC, delivers several advantages. (1) **Adaptability:** Due to the continuing advancement of both vision and language models, our proposed approach can be adopted by any state-of-the-art VLMs effortlessly in order to enhance the overall performance. (2) **Interpretability:** The generated questions set are easy to understand and portray multiple perspectives of the visual content. With our own knowledge, human can comfortably verify the relevance of the questions related to the input photo as well as how accurate of the corresponding answers. (3) **Robustness:** By leveraging a wide range of descriptive information, our approach produces accurate and detailed descriptions that are comparable to human annotations.

Nonetheless, we recognize some failure cases that reveal limitations of our method. Firstly, our method heavily relies on the performance of the *Questioner*. As highlighted in Table 2 above, low-quality *Questioner* leads to diminished performance. Specifically, when the *Questioner* generates questions that are either irrelevant or overly specific, especially in complex images with intricate details, the answers do not contribute meaningfully to the overall understanding. This results in captions that are cluttered with unnecessary information, reducing their clarity and coherence.

Secondly, the performance of the *Answerer* significantly influences the quality of the final captions. When the *Answerer* struggles to provide accurate responses, the generated captions are often incomplete or misleading. In fact, when we used ChatGPT as the *Answerer* on samples with low quantitative scores from QAC-InstructBLIP and QAC-LLaVA, the performance improved noticeably, indicating the importance of the interpretive ability of VLMs. However, since our goal is to develop a plug-and-play method adaptable to any pre-trained VLMs, we do not delve deeply into this issue.

Thirdly, our method does not intentionally control the hallucination tendencies inherent in VLMs. As a result, the generated captions sometimes contain irrelevant information that does not correspond to the image. Addressing this limitation would require improving the fairness and accuracy of LLMs, which falls beyond the scope of our current work.

6   Conclusion

In this paper, we present a novel method that leverages a *Question – Answer – Caption* framework to enhance the performance of pre-trained VLMs in zero-shot detailed image captioning task. Our approach effectively utilizes the *Questioner* to generate questions to extract detailed information from the image. Then, the answers by the *Answerer* guide the *Captioner* to produce rich and informative captions. Through extensive experiments with InstructBLIP and LLaVA, we demonstrate that our method improves their captioning capabilities. Moreover, our method serves as a plug-and-play module that is compatible with any pre-trained VLMs. While we identified certain limitations, they highlight room for future refinement, providing a pathway for further advancements in image captioning.

Acknowledgements. This work was supported by JSPS/MEXT KAKENHI Grant Numbers JP24K20830, and ROIS NII Open Collaborative Research 2024-24S1201.

References

1. Achiam, J., Adler, S., Agarwal, S., Ahmad, L., Akkaya, I., Aleman, F.L., Almeida, D., Altenschmidt, J., Altman, S., Anadkat, S., et al.: Gpt-4 technical report. arXiv preprint arXiv:2303.08774 (2023)
2. Alayrac, J.B., Donahue, J., Luc, P., Miech, A., Barr, I., Hasson, Y., Lenc, K., Mensch, A., Millican, K., Reynolds, M., et al.: Flamingo: a visual language model for few-shot learning. Adv. Neural. Inf. Process. Syst. **35**, 23716–23736 (2022)
3. Anderson, P., He, X., Buehler, C., Teney, D., Johnson, M., Gould, S., Zhang, L.: Bottom-up and top-down attention for image captioning and visual question answering. In: Proceedings of the IEEE conference on computer vision and pattern recognition. pp. 6077–6086 (2018)
4. Anil, R., Dai, A.M., Firat, O., Johnson, M., Lepikhin, D., Passos, A., Shakeri, S., Taropa, E., Bailey, P., Chen, Z., et al.: Palm 2 technical report. arXiv preprint arXiv:2305.10403 (2023)
5. Changpinyo, S., Sharma, P., Ding, N., Soricut, R.: Conceptual 12m: Pushing web-scale image-text pre-training to recognize long-tail visual concepts. In: Proceedings of the IEEE/CVF conference on computer vision and pattern recognition. pp. 3558–3568 (2021)
6. Chen, D., Cahyawijaya, S., Ishii, E., Chan, H.S., Bang, Y., Fung, P.: The pyramid of captions. arXiv preprint arXiv:2405.00485 (2024)
7. Chen, G.H., Chen, S., Zhang, R., Chen, J., Wu, X., Zhang, Z., Chen, Z., Li, J., Wan, X., Wang, B.: Allava: Harnessing gpt4v-synthesized data for a lite vision-language model. arXiv preprint arXiv:2402.11684 (2024)
8. Chen, L., Li, J., Dong, X., Zhang, P., He, C., Wang, J., Zhao, F., Lin, D.: Sharegpt4v: Improving large multi-modal models with better captions. arXiv preprint arXiv:2311.12793 (2023)
9. Chen, X., Wang, X., Changpinyo, S., Piergiovanni, A., Padlewski, P., Salz, D., Goodman, S., Grycner, A., Mustafa, B., Beyer, L., et al.: Pali: A jointly-scaled multilingual language-image model. arXiv preprint arXiv:2209.06794 (2022)
10. Cherti, M., Beaumont, R., Wightman, R., Wortsman, M., Ilharco, G., Gordon, C., Schuhmann, C., Schmidt, L., Jitsev, J.: Reproducible scaling laws for contrastive language-image learning. In: Proceedings of the IEEE/CVF Conference on Computer Vision and Pattern Recognition. pp. 2818–2829 (2023)
11. Chiang, W.L., Li, Z., Lin, Z., Sheng, Y., Wu, Z., Zhang, H., Zheng, L., Zhuang, S., Zhuang, Y., Gonzalez, J.E., et al.: Vicuna: An open-source chatbot impressing gpt-4 with 90%* chatgpt quality. See https://vicuna.lmsys.org (accessed 14 April 2023) **2**(3), 6 (2023)
12. Chowdhery, A., Narang, S., Devlin, J., Bosma, M., Mishra, G., Roberts, A., Barham, P., Chung, H.W., Sutton, C., Gehrmann, S., et al.: Palm: Scaling language modeling with pathways. J. Mach. Learn. Res. **24**(240), 1–113 (2023)
13. Dai, W., Li, J., Li, D., Tiong, A.M.H., Zhao, J., Wang, W., Li, B., Fung, P., Hoi, S.: Instructblip: Towards general-purpose vision-language models with instruction tuning. In: Advances in Neural Information Processing Systems (NeurIPS) (2023)
14. Dai, W., Li, J., Li, D., Tiong, A.M.H., Zhao, J., Wang, W., Li, B., Fung, P., Hoi, S.: Instructblip: Towards general-purpose vision-language models with instruction tuning (2023)
15. Desai, K., Kaul, G., Aysola, Z., Johnson, J.: Redcaps: Web-curated image-text data created by the people, for the people. arXiv preprint arXiv:2111.11431 (2021)

16. Fan, L., Krishnan, D., Isola, P., Katabi, D., Tian, Y.: Improving clip training with language rewrites. Advances in Neural Information Processing Systems **36** (2024)
17. Garg, R., Burns, A., Ayan, B.K., Bitton, Y., Montgomery, C., Onoe, Y., Bunner, A., Krishna, R., Baldridge, J., Soricut, R.: Imageinwords: Unlocking hyper-detailed image descriptions. arXiv preprint arXiv:2405.02793 (2024)
18. Gaur, M., Tapaswi, M., et al.: No detail left behind: Revisiting self-retrieval for fine-grained image captioning. arXiv preprint arXiv:2409.03025 (2024)
19. Ge, Y., Zeng, X., Huffman, J.S., Lin, T.Y., Liu, M.Y., Cui, Y.: Visual fact checker: Enabling high-fidelity detailed caption generation. In: Proceedings of the IEEE/CVF Conference on Computer Vision and Pattern Recognition. pp. 14033–14042 (2024)
20. Hong, W., Wang, W., Lv, Q., Xu, J., Yu, W., Ji, J., Wang, Y., Wang, Z., Dong, Y., Ding, M., et al.: Cogagent: A visual language model for gui agents. In: Proceedings of the IEEE/CVF Conference on Computer Vision and Pattern Recognition. pp. 14281–14290 (2024)
21. Hsieh, Y.G., Hsieh, C.Y., Yeh, S.Y., Béthune, L., Ansari, H.P., Vasu, P.K.A., Li, C.L., Krishna, R., Tuzel, O., Cuturi, M.: Graph-based captioning: Enhancing visual descriptions by interconnecting region captions. arXiv preprint arXiv:2407.06723 (2024)
22. Johnson, J., Karpathy, A., Fei-Fei, L.: Densecap: Fully convolutional localization networks for dense captioning. In: Proceedings of the IEEE conference on computer vision and pattern recognition. pp. 4565–4574 (2016)
23. Kirillov, A., Mintun, E., Ravi, N., Mao, H., Rolland, C., Gustafson, L., Xiao, T., Whitehead, S., Berg, A.C., Lo, W.Y., et al.: Segment anything. In: Proceedings of the IEEE/CVF International Conference on Computer Vision. pp. 4015–4026 (2023)
24. Li, F., Zhang, R., Zhang, H., Zhang, Y., Li, B., Li, W., Ma, Z., Li, C.: Llava-next-interleave: Tackling multi-image, video, and 3d in large multimodal models. arXiv preprint arXiv:2407.07895 (2024)
25. Li, J., Vo, D.M., Sugimoto, A., Nakayama, H.: Evcap: Retrieval-augmented image captioning with external visual-name memory for open-world comprehension. In: Proceedings of the IEEE/CVF Conference on Computer Vision and Pattern Recognition. pp. 13733–13742 (2024)
26. Li, J., Li, D., Savarese, S., Hoi, S.: Blip-2: Bootstrapping language-image pre-training with frozen image encoders and large language models. In: International conference on machine learning. pp. 19730–19742. PMLR (2023)
27. Li, J., Li, D., Xiong, C., Hoi, S.: Blip: Bootstrapping language-image pre-training for unified vision-language understanding and generation. In: International conference on machine learning. pp. 12888–12900. PMLR (2022)
28. Li, X., Zhang, F., Diao, H., Wang, Y., Wang, X., Duan, L.Y.: Densefusion-1m: Merging vision experts for comprehensive multimodal perception. arXiv preprint arXiv:2407.08303 (2024)
29. Li, Z., Yang, B., Liu, Q., Ma, Z., Zhang, S., Yang, J., Sun, Y., Liu, Y., Bai, X.: Monkey: Image resolution and text label are important things for large multi-modal models. In: Proceedings of the IEEE/CVF Conference on Computer Vision and Pattern Recognition. pp. 26763–26773 (2024)
30. Lin, T.-Y., Maire, M., Belongie, S., Hays, J., Perona, P., Ramanan, D., Dollár, P., Zitnick, C.L.: Microsoft COCO: Common Objects in Context. In: Fleet, D., Pajdla, T., Schiele, B., Tuytelaars, T. (eds.) ECCV 2014. LNCS, vol. 8693, pp. 740–755. Springer, Cham (2014). https://doi.org/10.1007/978-3-319-10602-1_48

31. Liu, H., Li, C., Li, Y., Lee, Y.J.: Improved baselines with visual instruction tuning. In: Proceedings of the IEEE/CVF Conference on Computer Vision and Pattern Recognition. pp. 26296–26306 (2024)
32. Liu, H., Li, C., Wu, Q., Lee, Y.J.: Visual instruction tuning. Advances in neural information processing systems **36** (2024)
33. Onoe, Y., Rane, S., Berger, Z., Bitton, Y., Cho, J., Garg, R., Ku, A., Parekh, Z., Pont-Tuset, J., Tanzer, G., et al.: Docci: Descriptions of connected and contrasting images. arXiv preprint arXiv:2404.19753 (2024)
34. Pont-Tuset, J., Uijlings, J., Changpinyo, S., Soricut, R., Ferrari, V.: Connecting Vision and Language with Localized Narratives. In: Vedaldi, A., Bischof, H., Brox, T., Frahm, J.-M. (eds.) ECCV 2020. LNCS, vol. 12350, pp. 647–664. Springer, Cham (2020). https://doi.org/10.1007/978-3-030-58558-7_38
35. Radenovic, F., Dubey, A., Kadian, A., Mihaylov, T., Vandenhende, S., Patel, Y., Wen, Y., Ramanathan, V., Mahajan, D.: Filtering, distillation, and hard negatives for vision-language pre-training. In: Proceedings of the IEEE/CVF conference on computer vision and pattern recognition. pp. 6967–6977 (2023)
36. Radford, A., Kim, J.W., Hallacy, C., Ramesh, A., Goh, G., Agarwal, S., Sastry, G., Askell, A., Mishkin, P., Clark, J., et al.: Learning transferable visual models from natural language supervision. In: International conference on machine learning. pp. 8748–8763. PMLR (2021)
37. Ramos, R., Martins, B., Elliott, D., Kementchedjhieva, Y.: Smallcap: lightweight image captioning prompted with retrieval augmentation. In: Proceedings of the IEEE/CVF Conference on Computer Vision and Pattern Recognition. pp. 2840–2849 (2023)
38. Rasheed, H., Maaz, M., Shaji, S., Shaker, A., Khan, S., Cholakkal, H., Anwer, R.M., Xing, E., Yang, M.H., Khan, F.S.: Glamm: Pixel grounding large multimodal model. In: Proceedings of the IEEE/CVF Conference on Computer Vision and Pattern Recognition. pp. 13009–13018 (2024)
39. Ravi, N., Gabeur, V., Hu, Y.T., Hu, R., Ryali, C., Ma, T., Khedr, H., Rädle, R., Rolland, C., Gustafson, L., et al.: Sam 2: Segment anything in images and videos. arXiv preprint arXiv:2408.00714 (2024)
40. Schuhmann, C., Beaumont, R., Vencu, R., Gordon, C., Wightman, R., Cherti, M., Coombes, T., Katta, A., Mullis, C., Wortsman, M., et al.: Laion-5b: An open large-scale dataset for training next generation image-text models. Adv. Neural. Inf. Process. Syst. **35**, 25278–25294 (2022)
41. Sharma, P., Ding, N., Goodman, S., Soricut, R.: Conceptual captions: A cleaned, hypernymed, image alt-text dataset for automatic image captioning. In: Proceedings of the 56th Annual Meeting of the Association for Computational Linguistics (Volume 1: Long Papers). pp. 2556–2565 (2018)
42. Shi, B., Wu, Z., Mao, M., Wang, X., Darrell, T.: When do we not need larger vision models? arXiv preprint arXiv:2403.13043 (2024)
43. Singla, V., Yue, K., Paul, S., Shirkavand, R., Jayawardhana, M., Ganjdanesh, A., Huang, H., Bhatele, A., Somepalli, G., Goldstein, T.: From pixels to prose: A large dataset of dense image captions. arXiv preprint arXiv:2406.10328 (2024)
44. Sun, Q., Cui, Y., Zhang, X., Zhang, F., Yu, Q., Wang, Y., Rao, Y., Liu, J., Huang, T., Wang, X.: Generative multimodal models are in-context learners. In: Proceedings of the IEEE/CVF Conference on Computer Vision and Pattern Recognition. pp. 14398–14409 (2024)
45. Sun, Q., Fang, Y., Wu, L., Wang, X., Cao, Y.: Eva-clip: Improved training techniques for clip at scale. arXiv preprint arXiv:2303.15389 (2023)

46. Sun, Q., Yu, Q., Cui, Y., Zhang, F., Zhang, X., Wang, Y., Gao, H., Liu, J., Huang, T., Wang, X.: Emu: Generative pretraining in multimodality. In: The Twelfth International Conference on Learning Representations (2023)
47. Team, G., Anil, R., Borgeaud, S., Wu, Y., Alayrac, J.B., Yu, J., Soricut, R., Schalkwyk, J., Dai, A.M., Hauth, A., et al.: Gemini: a family of highly capable multimodal models. arXiv preprint arXiv:2312.11805 (2023)
48. Thomee, B., Shamma, D.A., Friedland, G., Elizalde, B., Ni, K., Poland, D., Borth, D., Li, L.J.: Yfcc100m: The new data in multimedia research. Commun. ACM **59**(2), 64–73 (2016)
49. Touvron, H., Martin, L., Stone, K., Albert, P., Almahairi, A., Babaei, Y., Bashlykov, N., Batra, S., Bhargava, P., Bhosale, S., et al.: Llama 2: Open foundation and fine-tuned chat models. arXiv preprint arXiv:2307.09288 (2023)
50. Urbanek, J., Bordes, F., Astolfi, P., Williamson, M., Sharma, V., Romero-Soriano, A.: A picture is worth more than 77 text tokens: Evaluating clip-style models on dense captions. In: Proceedings of the IEEE/CVF Conference on Computer Vision and Pattern Recognition. pp. 26700–26709 (2024)
51. Wang, W., Ren, Y., Luo, H., Li, T., Yan, C., Chen, Z., Wang, W., Li, Q., Lu, L., Zhu, X., et al.: The all-seeing project v2: Towards general relation comprehension of the open world. arXiv preprint arXiv:2402.19474 (2024)
52. Wang, W., Shi, M., Li, Q., Wang, W., Huang, Z., Xing, L., Chen, Z., Li, H., Zhu, X., Cao, Z., et al.: The all-seeing project: Towards panoptic visual recognition and understanding of the open world. arXiv preprint arXiv:2308.01907 (2023)
53. Xie, N., Lai, F., Doran, D., Kadav, A.: Visual entailment: A novel task for fine-grained image understanding. arXiv preprint arXiv:1901.06706 (2019)
54. Xu, R., Yao, Y., Guo, Z., Cui, J., Ni, Z., Ge, C., Chua, T.S., Liu, Z., Sun, M., Huang, G.: Llava-uhd: an lmm perceiving any aspect ratio and high-resolution images. arXiv preprint arXiv:2403.11703 (2024)
55. You, H., Guo, M., Wang, Z., Chang, K.W., Baldridge, J., Yu, J.: Cobit: A contrastive bi-directional image-text generation model. arXiv preprint arXiv:2303.13455 (2023)
56. Young, P., Lai, A., Hodosh, M., Hockenmaier, J.: From image descriptions to visual denotations: New similarity metrics for semantic inference over event descriptions. Transactions of the Association for Computational Linguistics **2**, 67–78 (2014)
57. Yu, J., Wang, Z., Vasudevan, V., Yeung, L., Seyedhosseini, M., Wu, Y.: Coca: Contrastive captioners are image-text foundation models. arXiv preprint arXiv:2205.01917 (2022)
58. Yu, Q., Sun, Q., Zhang, X., Cui, Y., Zhang, F., Cao, Y., Wang, X., Liu, J.: Capsfusion: Rethinking image-text data at scale. In: Proceedings of the IEEE/CVF Conference on Computer Vision and Pattern Recognition. pp. 14022–14032 (2024)
59. Zellers, R., Bisk, Y., Farhadi, A., Choi, Y.: From recognition to cognition: Visual commonsense reasoning. In: Proceedings of the IEEE/CVF conference on computer vision and pattern recognition. pp. 6720–6731 (2019)

Exploring Cross-Attention Maps in Multi-modal Diffusion Transformers for Training-Free Semantic Segmentation

Rento Yamaguchi[✉] [iD] and Keiji Yanai[iD]

The University of Electro-Communications, Tokyo, Japan
{yamaguchi-r,yanai}@mm.inf.uec.ac.jp

Abstract. This paper presents a novel training-free semantic segmentation method that leverages a pre-trained large-scale image generation model incorporating the Multi-modal Diffusion Transformer (MM-DiT) architecture. Inspired by training-free segmentation techniques using the U-Net-based noise removal model in the Stable Diffusion framework, our approach extracts cross-attention maps between textual and visual features during the inference stages of the MM-DiT to generate mask images. Experimental results demonstrate that our method achieves segmentation accuracy comparable to CLIP-based and U-Net-based stable diffusion methods. While the direct segmentation scores are relatively modest, the significance of our work lies in the exploration of cross-attention maps within the DiT. This investigation provides critical insights that could advance training-free segmentation methodologies and enhance the interpretability of diffusion-based models.

Keywords: Training-Free Semantic Segmentation · Multi-modal Diffusion Transformer · Stable Diffusion · Cross-Attention Maps

1 Introduction

Semantic segmentation in computer vision involves assigning a class label to each pixel in an image, which is a task that holds significant importance across a myriad of domains such as image editing, autonomous driving, and medical image analysis. Conventional supervised learning approaches to semantic segmentation demand extensive labeled annotation datasets, which are costly and labor-intensive to generate. Additionally, these models typically exhibit poor generalization to unseen classes. This limitation hinders their practical applicability.

To alleviate these challenges, unsupervised and zero-shot segmentation techniques have surfaced as promising alternatives. Zero-shot semantic segmentation, in particular, aims to generalize to new categories without requiring explicit training on those categories, thereby addressing the data scarcity and labeling cost issues. Noteworthy advancements in zero-shot segmentation leverage large pre-trained models like CLIP (Contrastive Language-Image Pretraining) [12] and U-Net-based architectures integrated within the Stable Diffusion framework [14], which exploit the synergy of textual and visual features.

Despite the efficacy of these approaches, limitations remain, especially when adapting to the latest architectures. For instance, while U-Net-based noise reduction models within Stable Diffusion have demonstrated some success, the emergence of the Multi-modal Diffusion Transformer (MM-DiT) in Stable Diffusion 3 [4] introduces a new paradigm that existing methods cannot directly apply to. This necessitates innovative methodologies to harness the potential of MM-DiT effectively.

In this paper, we propose a novel zero-shot training-free semantic segmentation method that utilizes the MM-DiT architecture from the pre-trained large-scale image generation model Stable Diffusion. Drawing inspiration from U-Net-based training-free segmentation techniques, our approach extracts cross-attention maps during the inference stages of MM-DiT to produce segmentation masks. Our experiments reveal that this method attains segmentation accuracy on par with established CLIP-based and U-Net-based approaches, although direct segmentation metrics are still moderate. Crucially, our work emphasizes the exploration of cross-attention mechanisms within the DiT architecture, offering vital insights that could drive the future development of training-free segmentation techniques and enhance the interpretability of diffusion-based models.

2   Related Work

2.1   Training-Free Semantic Segmentation

Zero-shot learning is a paradigm wherein models classify and segment data into categories that are not explicitly encountered during their training phase. This methodology utilizes knowledge from known categories to infer and segment unseen categories, significantly alleviating the challenges associated with collecting and annotating large volumes of labeled data. Specifically, in the realm of semantic segmentation, zero-shot learning is invaluable, offering practical solutions in areas like autonomous driving, healthcare, and remote sensing.

One significant advancement in this domain is the Segment Anything model by Kirillov et al. [7]. The model serves as a foundation for segmentation, being pre-trained on extensive annotated datasets. It provides exceptional performance across diverse images and annotations, establishing a benchmark in the use of large-scale pre-trained models for semantic segmentation.

CLIP models have also been paramount in zero-shot semantic segmentation. Works by Rao et al. [13] and Luddecke et al. [9] have combined textual descriptions with visual features. The key idea lies in mapping both text and images into a unified embedding space, aligning similar concepts closely. This shared space allows models to infer segmentation masks for unseen classes based on textual descriptions, circumventing the need for annotated examples for every possible class. Recently, diffusion models have drawn attention for their precision in image generation. Studies by Wu et al. [16] and Tian et al. [15] have proposed leveraging these models to generate accurate mask images for segmentation tasks.

Moreover, training-free semantic segmentation has emerged as a promising avenue for improving model performance. This approach leverages pre-trained models and existing knowledge to achieve high-accuracy segmentation on new datasets or tasks without additional training, which is particularly advantageous in scenarios with time constraints or limited resources. For example, MaskCLIP by Zhou et al. [17] and FreeDA by Barsellotti et al. [1] employ pre-trained large vision models to perform segmentation of unseen classes without any training data, significantly reducing data collection and annotation time, thus enabling rapid deployment. Additionally, the StableSeg model introduced by Honbu et al. [6] utilizes Cross Attention Maps and Self Attention Maps within the U-Net architecture of the Stable Diffusion model to achieve segmentation without further supervision. This innovative approach highlights the potential of diffusion models to enhance segmentation accuracy through inherent attention mechanisms.

In conclusion, zero-shot learning and training-free semantic segmentation represent crucial methodologies for minimizing the cost and time associated with labeled data acquisition, allowing models to swiftly and efficiently adapt to new tasks and environments.

2.2   Diffusion Transformers

Diffusion models have emerged as powerful tools for generating high-quality visual content through a sequential noise removal process. These models initiate the process with a noisy version of an image, progressively denoising it to produce a clear and detailed final output. Diffusion models have demonstrated significant potential in various tasks such as image generation, editing, and semantic segmentation. Stable Diffusion, based on the Latent Diffusion Model [14], was trained on hundreds of millions of pairs of text and image data. This model employs an extended U-Net with integrated attention mechanisms as the primary noise reduction component, enabling the generation of high-quality images. The widespread availability and success of Stable Diffusion have underscored the effectiveness of this approach.

On the other hand, Peebles et al. [10] introduced the Diffusion Transformer (DiT), a Transformer-based noise reduction model. The DiT model has been adapted to create even higher quality image generation models, such as Stable Diffusion 3 [4], and video generation models like OpenAI Sora [2], which generate videos indistinguishable from real footage. Esser et al. [4] proposed the Multimodal Diffusion Transformer by integrating the Flow Matching technique [8] with the robust noise reduction capabilities of DiT. This multi-modal approach enables the generation of high-quality, coherent images aligned with free-form text prompts, pushing the boundaries of what is achievable in image generation and demonstrating the versatility and power of diffusion models and transformer-based architectures.

2.3   Attention Mechanisms in Diffusion Transformers

The evolution of diffusion models, particularly with the integration of Transformer architectures, has marked significant advancements in their ability to handle complex and multi-modal data. A crucial element driving this evolution is the cross-attention mechanism, which facilitates the interaction of diverse types of information, such as textual and visual data, during the generative process. Cross-attention mechanisms play a pivotal role in enhancing the understanding of contextual relationships within data, thereby contributing significantly to the interpretability and performance improvements of these models.

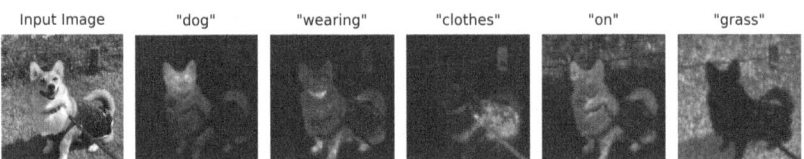

Fig. 1. Visualization of cross-attention maps between text and images. The input image is shown on the left, followed by attention maps for different text prompts. Each attention map highlights the regions of the image that correspond to the given text prompt.

Within the framework of the Multi-modal Diffusion Transformer (MM-DiT) in Stable Diffusion 3, Joint Attention layers promote the alignment and fusion of multi-modal information, enabling the generation of high-quality and coherent images that are consistent with meaningful text prompts. This alignment is achieved by mapping textual embeddings to visual features, effectively guiding the noise reduction process in accordance with the semantic content of the text prompts. Joint-Attention Maps, which capture these interactions, serve as a foundational element in our proposed methodology, providing invaluable insights for extracting meaningful segmentation masks even in zero-shot contexts. Existing research, such as that by Honbu *et al.* [6], has demonstrated the promise of utilizing attention maps within U-Net-based architectures for zero-shot training-free segmentation. However, the integration of attention mechanisms within the DiT architecture remains an underexplored area with substantial potential. Leveraging these mechanisms offers the prospect of enhancing segmentation accuracy and interpretability without the need for extensive annotated datasets.

In the following section, we describe our proposed method, which utilizes cross-attention maps extracted during the inference stages of MM-DiT to generate segmentation masks. This technique aligns with the principles established by prior research while pioneering new possibilities achievable with state-of-the-art diffusion transformers.

3 Methodology

3.1 Overview

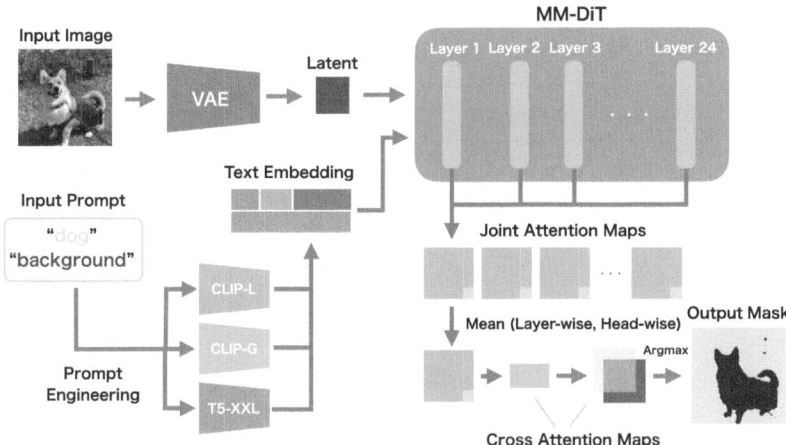

Fig. 2. Overview of the Proposed Method

Inspired by the concept of utilizing Cross-Attention Maps as demonstrated by Honbu *et al.* [6], in this paper, we propose a novel approach for training-free semantic segmentation by leveraging MM-DiT within the Stable Diffusion 3 framework. Unlike its predecessors in the v1 and v2 series, Stable Diffusion 3 employs three separate text encoders to precisely capture the features of input prompts and generate text-aligned images. The architecture of Stable Diffusion 3 repeatedly applies MM-DiT through Joint-Attention mechanisms that synergize image and text embeddings.

Our proposed method encompasses two primary steps: (1) generation of text embeddings using three distinct text encoders, and (2) usage of Joint-Attention for segmenting regions within the image. A single inference step within MM-DiT facilitates training-free semantic segmentation.

Figure 2 provides an overview of the proposed method's workflow. This diagram illustrates the sequential steps involved in the methodology, highlighting key processes such as data acquisition, preprocessing, feature extraction, model training, and evaluation. Each step is critical for the successful implementation and validation of the proposed approach.

3.2 Generation of Text Embeddings

In Stable Diffusion 3, three models—CLIP/L-14, CLIP/G-14, and T5-XXL—are utilized for text encoding. Here, we describe the generation of text embeddings needed for DiT inference. Given a set of k class labels $\{c_0, c_1, \ldots, c_{k-1}\}$ representing target segmentation classes, initial preprocessing involves appending a prefix prompt like "a photo of" to each class label, resulting in modified prompts $\{p_0 + c_0, p_1 + c_1, \ldots, p_{k-1} + c_{k-1}\}$. Each modified prompt is fed into the three respective text models (CLIP/L-14, CLIP/G-14, T5-XXL) to generate corresponding text embeddings.

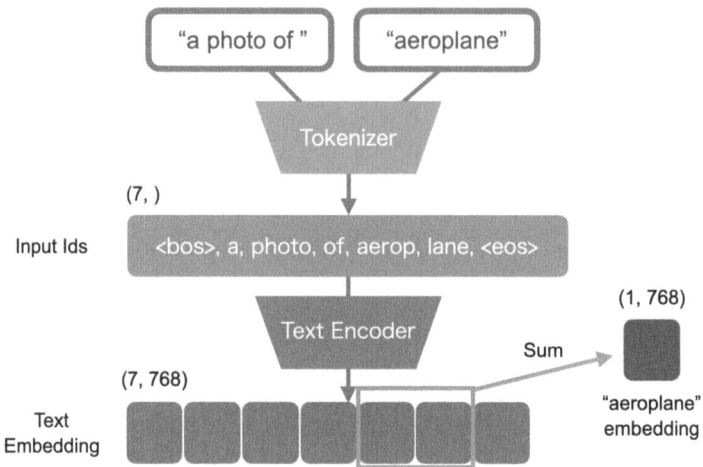

Fig. 3. Overview of the method for generating text embeddings for classes. A prefix prompt like 'a photo of' is added to the class category, and the sum of the portions corresponding to the class category from the generated embeddings is utilized as the embedding for that category.

From these generated embeddings, as illustrated in Fig. 3, we extract text embeddings that correspond to each class label c_i. Taking into account the presence of special tokens such as <bos> and <eos> in the CLIP tokenizer, and defining n_i as the number of tokens in each prefix prompt p_i', the following operations are performed to obtain the class-specific text embeddings:

$$\mathcal{E}_{\text{CLIP-L},i} = \text{Sum}\left(\text{CLIP-L}(p_i')[n_i + 1 : -1]\right) \tag{1}$$

$$\mathcal{E}_{\text{CLIP-G},i} = \text{Sum}\left(\text{CLIP-G}(p_i')[n_i + 1 : -1]\right) \tag{2}$$

$$\mathcal{E}_{\text{T5},i} = \text{Sum}\left(\text{T5}(p_i')[n_i : -1]\right) \tag{3}$$

For all three models, the embeddings for class tokens 0 to $k - 1$ are summed according to the above operations, while the pad token embeddings are inserted for positions from k to 76. This results in the final embeddings $\mathcal{E}_{\text{CLIP-L}}$, $\mathcal{E}_{\text{CLIP-G}}$, $\mathcal{E}_{\text{T5}}$.

Subsequently, the two CLIP embeddings are concatenated along the embedding dimension and padded to match the dimensions of the T5 embeddings. These are then concatenated along the token dimension, resulting in the final text embedding $\mathcal{E}'$ for DiT inference.

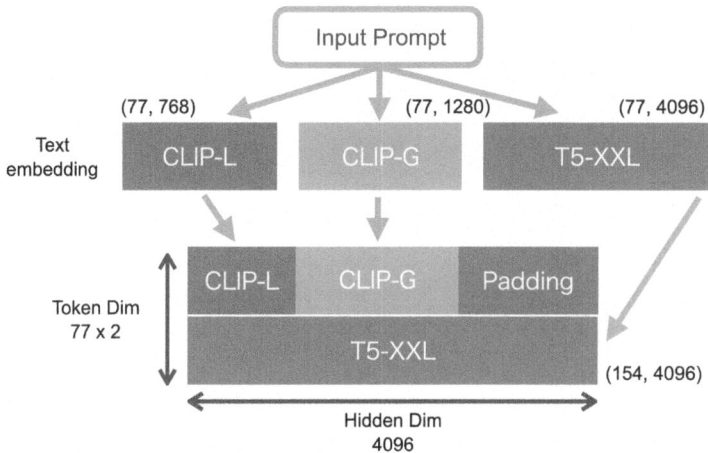

Fig. 4. Procedure for creating a text embedding from the three text encoders

3.3   Segmentation Using Joint-Attention

For an input image dimensioned at 1024×1024, the VAE encoder in Stable Diffusion 3 generates a 16-channel latent variable z. The latent variable z is patched into a 2×2 grid and positional embeddings are added to form image features x. Assuming the image contains noise corresponding to predefined timestep t, the MM-DiT model predicts the next step of noise.

Stable Diffusion 3 does not separately compute Self-Attention for image and text features or Cross-Attention between them. Instead, it combines them into a unified Query, Key, and Value for computing Joint-Attention. If the linear transformation layers for image and text features are l_{IQ}, l_{IK}, l_{IV}, and l_{TQ}, l_{TK}, l_{TV} respectively, Joint-Attention Map (JAMap) is computed as follows:

$$Q = \text{Concat}(l_{IQ}(x),\ l_{TQ}(\mathcal{E}')) \tag{4}$$

$$K = \text{Concat}(l_{IK}(x),\ l_{TK}(\mathcal{E}')) \tag{5}$$

$$V = \text{Concat}(l_{IV}(x),\ l_{TV}(\mathcal{E}')) \tag{6}$$

$$\text{JAMap} = \text{softmax}\left(\frac{QK^T}{\sqrt{d_k}}\right) \tag{7}$$

The JAMap, as shown in Fig. 5, can be divided into four quadrants representing Self-Attention for image features, Self-Attention for text features, and two Cross-Attention components between them. The regions in JAMap corresponding to Cross-Attention are extracted, transposed, and averaged to yield the Cross-Attention Map (CAMap) for the i-th DiT layer.

$$\text{CAMap}_i = \frac{\text{JAMap}[:, d_i :, : d_i] + \text{JAMap}[:, : d_i, d_i :]^T}{2}, \tag{8}$$

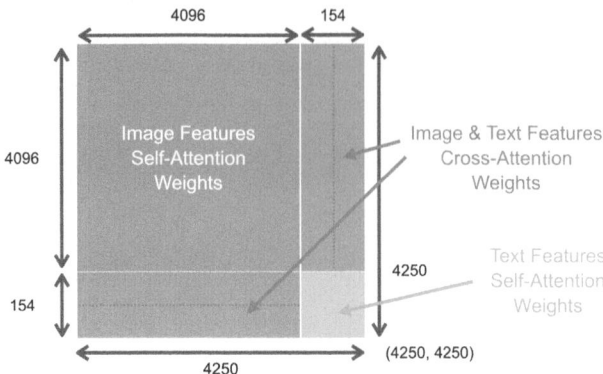

Fig. 5. Visualization of the Joint-Attention Map. The JAMap consists of four regions corresponding to the Self Attention and Cross Attention between image and text features.

where d_i denotes the dimension of image features. Each CAMap is calculated using 24 Multi-Head Attention layers. Given the token dimensions and image feature dimensions d_t and d_i, respectively, the Head-dimension-averaged CAMap reshapes into $(2, d_t, d_i, d_i)$, representing CAMaps from CLIP and T5.

If there are k class labels, the final Cross-Attention-based segmentation map M is computed as follows, considering token indices impacted by special tokens.

$$\text{CAMap}_{\text{CLIP},i} = \text{CAMap}[0, 1 : k + 1] \qquad (9)$$

$$\text{CAMap}_{\text{T5},i} = \text{CAMap}[1, : k] \qquad (10)$$

After extracting CAMaps for CLIP and T5, and averaging them over 24 heads, the segmentation mask is generated by applying Argmax across the class dimension, C. The mean map from all CAMaps sourced from the three text encoders is referred to as the Cross Attention Probability Map (CAPM).

$$\text{CAPM}_i = \frac{\text{CAMap}_{\text{CLIP}} + \text{CAMap}_{\text{T5}}}{2} \qquad (11)$$

To obtain the final mask image M, the cross-attention probability maps (CAPM) from selected layers k are summed and then the Argmax function is applied across the class dimension. Here, the index set k can be any subset of the layers from 1 to 24.

$$M = \text{Argmax}_C \left(\sum_{j \in k} \text{CAPM}_j \right) \qquad (12)$$

Here, k_j represents the index of each selected layer from which the cross-attention probability maps are obtained.

4   Experiments

4.1   Experimental Setup

In this study, we utilize Stable Diffusion 3, which employs a DiT-based architecture, as our model. We used the 'stabilityai/stable-diffusion-3-medium-diffusers' checkpoint from the diffusers library by Hugging Face [11]. Unless otherwise specified, all input images are resized to a uniform size of 1024 × 1024. For the class labels, which are considered to be known, we prepend the prefix prompt "a photo of" when inputting into the text encoders. Quantitative evaluation is performed using the standard segmentation datasets PASCAL VOC [5] and Cityscapes [3].

In this paper, we perform segmentation using the MM-DiT model of Stable Diffusion 3 by specifying a single timestep out of a total of 1000 image generation steps. The appropriate timestep value is experimentally determined and will be discussed in detail in Sect. 4.3. Additionally, the choice of layers from which to extract attention maps significantly impacts segmentation accuracy. This choice will be explored in detail in Sect. 4.2.

4.2   Layer-Wise Segmentation Differences in MM-DiT

Visualization of the cross-attention maps between text and image features across the 24 layers of MM-DiT is presented in Fig. 6.

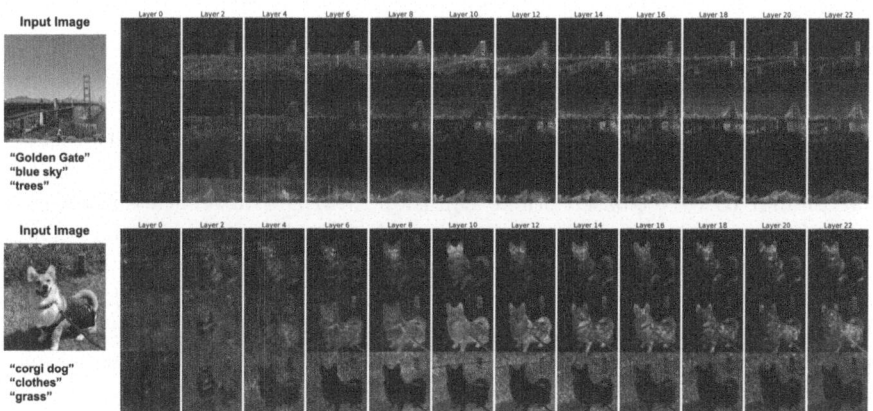

Fig. 6. Qualitative segmentation results from different layers of MM-DiT.

Upon examining each layer, we observe that attention maps closer to the initial or final layers show weaker responses and appear noisier concerning the regions corresponding to the text. In contrast, the maps from the middle layers exhibit strong responses to image keypoints that correspond to the text tokens. This leads us to hypothesize that the choice of attention maps significantly affects segmentation accuracy. Accordingly, we evaluate the performance on the PASCAL VOC dataset using different layers' attention maps, as shown in Fig. 7.

Starting with layers 11 and 12, we incrementally included more layers around the center to determine the optimal range. We expanded the range to include layers 10 and 13, and ultimately found that the highest scores were obtained when using layers 9 through 14. This confirms that the choice of layers significantly impacts segmentation performance.

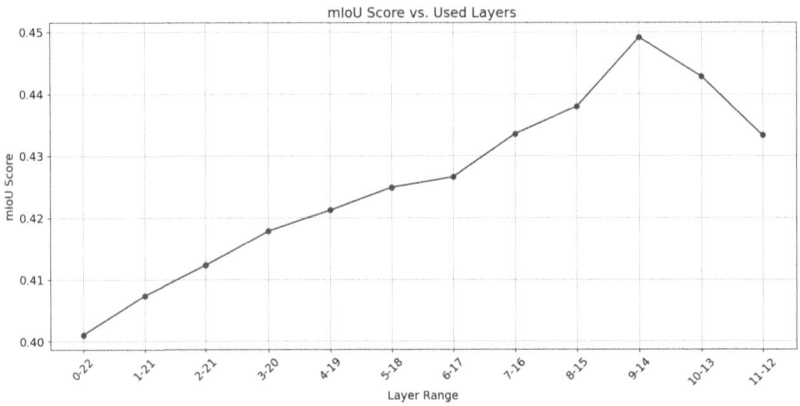

Fig. 7. Quantitative segmentation results on PASCAL VOC by utilizing different layers of MM-DiT, showing how the choice of layers from 0 to 23 affects the mIoU score.

As observed in Fig. 7, the evaluation scores are highest when selecting multiple layers from the central part of MM-DiT. This can be attributed to the fact that, similar to U-Net-based Stable Diffusion, MM-DiT also encapsulates more semantic information of the image features in layers closer to the model's center. The incremental inclusion of layers around the center—starting with layers 11 and 12, and expanding to include layers 9 to 14—showed a noticeable improvement in segmentation accuracy, supporting our hypothesis. As the experiments progressed, we decided to use this optimal range of layers (9 to 14) for subsequent evaluations.

4.3 Segmentation Variability with Different Timesteps

We evaluated the impact of different timesteps on segmentation accuracy. Specifically, we conducted experiments using comparable training-free segmentation methods, such as MaskCLIP and StableSeg [6,17]. Notably, for a fair comparison, we assessed StableSeg—which employs U-Net-based Stable Diffusion—under identical conditions by utilizing only the Cross-Attention Maps as we do.

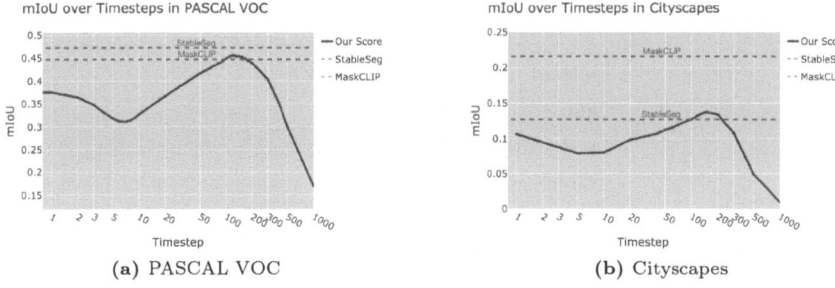

Fig. 8. Comparison of mIoU values for different methods with varying timesteps.

The evaluation was carried out on benchmarks including PASCAL VOC and Cityscapes, with results illustrated in Figs. 8a and 8b.

These results indicate that the optimal segmentation accuracy is attained when assuming noise is applied around $t = 150$. We hypothesize that at timesteps approximating this value, the model accentuates depicting objects according to the textual description rather than merely denoising images, reflecting a different interpretation phase of the diffusion model's learning process. As the experiments progressed, we decided to use this optimal timestep (around $t = 150$) for subsequent evaluations. Additionally, the results of region segmentation at different time steps are shown in Fig. 9.

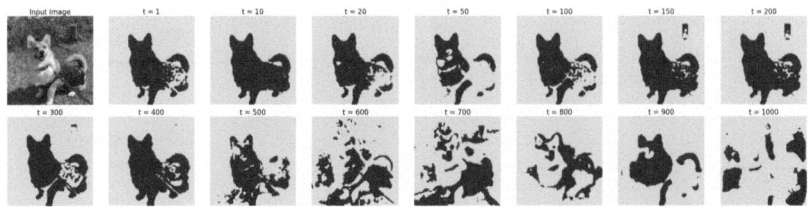

Fig. 9. Results of region segmentation for 'clothed corgi' and 'background' at different time steps.

4.4   Comparison with Existing Training-Free Methods

Using the optimal timestep and hyperparameters for the layer of the MM-DiT determined from the above experiments, we conducted a quantitative comparison with the CLIP-based method MaskCLIP [17] and the Stable Diffusion v1 based method StableSeg [6]. The results are presented in Table 1. The evaluation settings are consistent with those described in Sect. 4.3. These results indicate that our method achieves comparable segmentation accuracy to the two existing methods.

Table 1. Comparison of mIoU scores for different methods across two datasets

Method	PascalVOC	Cityscapes
Ours	0.452	0.137
MaskCLIP	0.447	0.216
StableSeg	0.472	0.127

4.5   Qualitative Comparison of Different Text Encoders

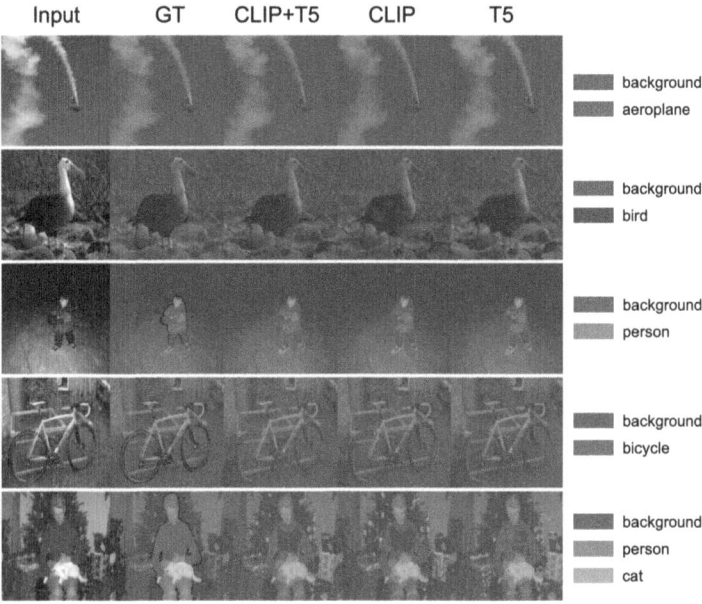

Fig. 10. Qualitative comparison showing how different text encoders affect segmentation outcomes on PASCAL VOC. The upper four images exhibit accurate segmentation, while the bottom image shows a failure case where the cat's position is misaligned.

Stable Diffusion 3 leverages text embeddings generated by three text encoders: two instances of CLIP and one T5. To assess the impact of these distinct text encoders on the resultant segmentation, we qualitatively analyzed their cross-attention maps as demonstrated in Fig. 10. While foreground extraction of images containing a single foreground object displayed high accuracy, segmentation performance deteriorated in scenarios where multiple foreground objects were present, occasionally leading to segmentation failures.

4.6   Open Vocabulary Segmentation

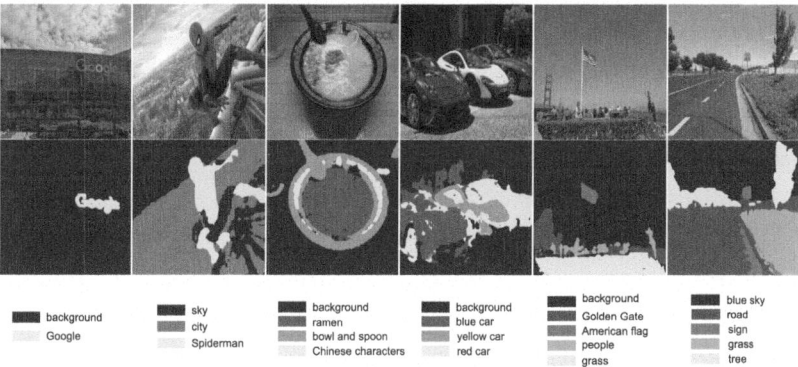

Fig. 11. Open vocabulary segmentation results, demonstrating capability beyond traditional class labels, including segmentation of classes such as proper nouns.

Leveraging the Stable Diffusion 3 model trained on extensive web-based data, our method enables open vocabulary segmentation, transcending the limitations of conventional class labels typically found in standard segmentation datasets. The segmentation results are illustrated in Fig. 11. This approach allows for flexible segmentation of classes, including sentences containing adjectives and proper nouns.

5   Conclusion

In this paper, we proposed a novel training-free semantic segmentation method leveraging the Multi-modal Diffusion Transformer (MM-DiT) architecture within the Stable Diffusion 3 framework. Through extensive experiments, we demonstrated that our approach achieved the comparable performance to existing CLIP-based and U-Net-based methods within the Stable Diffusion framework, albeit with relatively modest direct segmentation scores. Our contribution lies in the exploration and utilization of cross-attention maps in image generation diffusion models, representing a significant step toward enhancing the interpretability and accuracy of zero-shot segmentation methods.

Our findings indicate that selecting attention maps from the middle layers of the MM-DiT model significantly improves segmentation results, emphasizing the importance of these layers in capturing semantic information. Additionally, our experiments revealed the impact of varying timesteps on segmentation accuracy, with optimal results achieved around the timesteps where the model transitions from merely denoising to aligning with textual descriptions.

Qualitative comparisons further validated the influence of different text encoders on the segmentation process. While the model exhibited high performance in single-object scenarios, its performance deteriorated in complex scenes with multiple foreground objects, highlighting areas for future improvement.

In conclusion, this research provides foundational insights into the utilization of cross-attention mechanisms within diffusion-based models for training-free segmentation. These insights pave the way for future research to refine and enhance the performance and applicability of training-free techniques across diverse and practical domains.

References

1. Barsellotti, L., Amoroso, R., Cornia, M., Baraldi, L., Cucchiara, R.: Training-free open-vocabulary segmentation with offline diffusion-augmented prototype generation. In: CVPR (2024)
2. Brooks, T., et al.: Video generation models as world simulators (2024). https://openai.com/research/video-generation-models-as-world-simulators
3. Cordts, M., et al.: The cityscapes dataset for semantic urban scene understanding. In: CVPR (2016)
4. Esser, P., et al.: Scaling rectified flow transformers for high-resolution image synthesis. In: ICML (2024)
5. Everingham, M., Van Gool, L., Williams, C.K.I., Winn, J., Zisserman, A.: The Pascal visual object classes (VOC) challenge. Int. J. Comput. Vis. **88**(2), 303–338 (2010)
6. Honbu, Y., Yanai, K.: Training-free region prediction with stable diffusion. In: ACM MM (2024)
7. Kirillov, A., et al.: Segment anything. In: ICCV, pp. 4015–4026 (2023)
8. Lipman, Y., Chen, R.T., Ben-Hamu, H., Nickel, M., Le, M.: Flow matching for generative modeling. In: ICLR (2022)
9. Lüddecke, T., Ecker, A.: Image segmentation using text and image prompts. In: CVPR, pp. 7086–7096 (2022)
10. Peebles, W., Xie, S.: Scalable diffusion models with transformers. In: ICCV, pp. 4195–4205 (2023)
11. von Platen, P., et al.: Diffusers: state-of-the-art diffusion models (2022). https://github.com/huggingface/diffusers
12. Radford, A., et al.: Learning transferable visual models from natural language supervision. In: ICML, pp. 8748–8763 (2021)
13. Rao, Y., et al.: DenseCLIP: language-guided dense prediction with context-aware prompting. In: CVPR (2022)
14. Rombach, R., Blattmann, A., Lorenz, D., Esser, P., Ommer, B.: High-resolution image synthesis with latent diffusion models. In: CVPR, pp. 10684–10695 (2022)
15. Tian, J., Aggarwal, L., Colaco, A., Kira, Z., Gonzalez-Franco, M.: Diffuse, attend, and segment: unsupervised zero-shot segmentation using stable diffusion. In: CVPR (2024)
16. Wu, W., Zhao, Y., Shou, M.Z., Zhou, H., Shen, C.: DiffuMask: synthesizing images with pixel-level annotations for semantic segmentation using diffusion models. In: ICCV, pp. 1206–1217 (2023)
17. Zhou, C., Loy, C.C., Dai, B.: Extract free dense labels from clip. In: Avidan, S., Brostow, G., Cissé, M., Farinella, G.M., Hassner, T. (eds.) ECCV, pp. 696–712. Springer, Cham (2022)

Enhancing Visual Question Answering with Pre-trained Vision-Language Models: An Ensemble Approach at the LAVA Challenge 2024

Trong-Hieu Nguyen-Mau[1,2(✉)] 📵, Nhu-Binh Nguyen Truc[1,2] 📵,
Nhu-Vinh Hoang[1,2] 📵, Minh-Triet Tran[1,2] 📵, and Hai-Dang Nguyen[1,2] 📵

[1] University of Science, VNU-HCM, Ho Chi Minh City, Vietnam
{nmthieu,nhdang}@selab.hcmus.edu.vn,
{ntnbinh21,hnvinh21}@apcs.fitus.edu.vn, tmtriet@fit.hcmus.edu.vn
[2] Viet Nam National University, Ho Chi Minh City, Vietnam

Abstract. The LAVA challenge presents complex visual question answering tasks involving intricate diagrams, each accompanied by multiple-choice questions in English or Japanese. Addressing this challenge, we - the team v1olet - explore the capabilities of pre-trained Large Vision-Language Models to interpret and reason over such sophisticated visual data. We utilize models including Qwen2-VL, InternVL2, MiniCPM, and Llama-3.2-Vision-Instruct, employing a structured prompt template designed to standardize response generation and facilitate step-by-step reasoning. To enhance accuracy and robustness, we implement an ensemble method using majority voting to combine outputs from different models and configurations. Our experimental results demonstrate that the ensemble approach significantly improves performance, achieving a higher public score on the LAVA challenge dataset compared to individual models. Specifically, the ensemble of Qwen2-VL, InternVL2, and Llama-3.2 models attained the highest public score of 82, outperforming the best single model. This study highlights the effectiveness of combining multiple Large Vision-Language Models through ensemble methods and underscores the potential of prompt-based inference in enhancing model reasoning capabilities for complex VQA tasks. The provided code is here.

Keywords: Large Vision-Language Models · Visual Question Answering · Ensemble Methods

1 Introduction

Large Vision-Language Models (LVLMs) have emerged as a transformative force in artificial intelligence by seamlessly integrating visual and textual data at scale [12]. These generative models are engineered to process and understand multiple

M. Cho et al. (Eds.): ACCV 2024 Workshops, LNCS 15482, pp. 281–292, 2025.
https://doi.org/10.1007/978-981-96-2641-0_19

modalities simultaneously, enabling them to interpret visual content and generate coherent textual responses or outputs [17]. This multimodal capability is particularly valuable in tasks that require the fusion of visual and textual information, making LVLMs critical tools in addressing complex challenges across various real-world applications [3, 21].

The advent of LVLMs has significantly advanced fields such as image captioning [15], visual question answering (VQA) [4], and multimodal dialogue systems [25]. In VQA, for instance, models are required not only to recognize objects within an image but also to comprehend contextual cues and infer relationships between entities to answer questions accurately [27]. This necessitates a deep understanding of both visual content and natural language, highlighting the importance of sophisticated LVLMs in achieving high performance in such tasks.

Despite these advancements, current LVLMs face significant challenges when dealing with complex visual data that go beyond straightforward photographic images. Real-world applications often involve intricate diagrams, technical schematics, and multilingual texts, which are common in fields like engineering, architecture, and data analysis. These complex visual representations require models to perform advanced reasoning and interpretation, pushing the boundaries of current multimodal understanding capabilities.

Addressing this gap, the LAVA Challenge 2024 introduces a dataset specifically designed to evaluate and enhance the capabilities of LVLMs in interpreting complex visual information [2]. The dataset consists of two parts: a public dataset with approximately 3,000 samples sourced from the internet, and a private dataset provided by the TASUKI team (SoftBank), containing around 1,100 samples. The visual data encompass a wide range of complex diagrams such as Data Flow Diagrams, Class Diagrams, Gantt Charts, and Building Design Drawings. Each image is accompanied by a multiple-choice question in either English or Japanese, with four possible answers derived from the visual content.

The LAVA Challenge presents several unique difficulties:

- **Complex Visual Structures:** The images contain detailed and abstract representations that require models to understand not just objects but also relationships, hierarchies, and flows of information.
- **Multilingual Text Understanding:** Questions are provided in both English and Japanese, necessitating models to possess or integrate cross-lingual comprehension capabilities.
- **Advanced Reasoning:** Answering the questions correctly often involves multi-step reasoning processes, including deduction, inference, and sometimes even external knowledge.

These challenges make the LAVA dataset a rigorous benchmark for testing the limits of current LVLMs and exploring new methodologies to enhance their performance.

In this paper, we introduce an ensemble approach leveraging pre-trained LVLMs to enhance visual question answering performance on the LAVA

Challenge 2024. Our framework integrates multiple state-of-the-art LVLMs, including Qwen2-VL, InternVL2, MiniCPM, and Llama-3.2-Vision-Instruct, utilizing a structured prompt design to standardize responses and facilitate step-by-step reasoning. Our contributions are multifaceted and are presented as follows:

- **Structured Prompt Design for Enhanced Reasoning:** We design a structured prompt template that standardizes response generation across different models. This template guides the models through a step-by-step reasoning process, improving their ability to interpret intricate visual data and generate consistent answers.
- **Majority Voting Ensemble Method:** We implement a majority voting scheme to aggregate outputs from various models and configurations. This method leverages the strengths of individual models while mitigating their weaknesses, resulting in improved accuracy and robustness in the final predictions.
- **Significant Performance Improvement on LAVA Challenge Dataset:** Our experimental results demonstrate that the ensemble approach significantly enhances performance, achieving higher scores on the LAVA Challenge 2024 dataset compared to individual models. This highlights the effectiveness of our method in advancing the state-of-the-art in visual question answering.

This paper is structured as follows. In Sect. 2, we provide a concise overview of existing methods relevant to our research. Section 3 introduces our proposed approach in detail. We then discuss our experimental findings in Sect. 4. In Sect. 5, we discuss and outline open problems and future investigations. Finally, Sect. 6 concludes the paper.

2   Related Works

2.1   Large Vision-Language Models

Large vision-language models (LVLMs) represent the intersection of machine vision and natural language processing, combining visual interpretation with linguistic capabilities to answer questions, perform visual reasoning, and generate textual descriptions [30]. These models leverage advancements in both machine vision—traditionally used for tasks like image classification, object detection, and counting—and the robust inference capabilities of large language models (LLMs) [8]. With access to vast amounts of data from the internet, pre-trained LVLMs are particularly adept at domain-specific adaptation [31], even when faced with previously unseen images. This flexibility is largely attributed to contrastive learning techniques [6], which enhance the model's ability to align visual and textual modalities.

In recent advancements, the alignment of modalities allows models to answer questions about images, generate captions, and even engage in multimodal dialogues, as seen in models like InternVl [23], and Qwen-VL [9]. These models connect vision encoders with LLMs using advanced architectures like vision transformer (ViT) [29], multilayer perceptrons (MLP) [23].

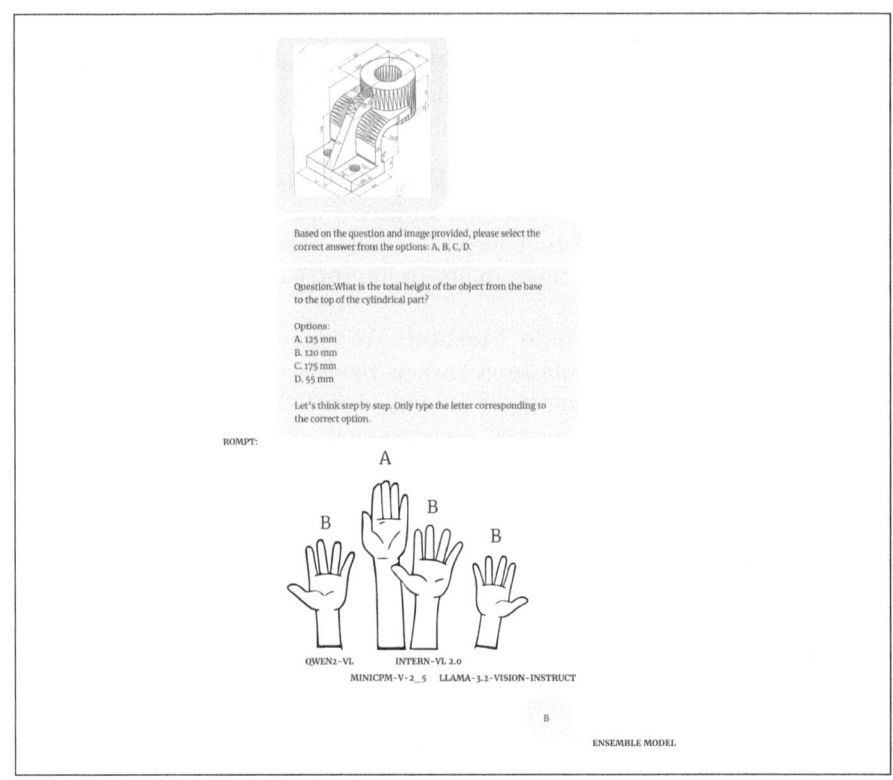

Fig. 1. Overall process of our approaches.

2.2   Visual Question Answering

The task of Visual Question Answering (VQA) involves answering natural language questions based on a given image. It represents the intersection of computer vision and natural language processing. VQA requires models to perform multimodal reasoning and integrate visual with text understanding.

Early approaches used simple CNN-LSTM architectures [5] to extract visual features from images and encode questions using recurrent neural networks. Later, Stacked Attention Networks (SAN) [28] enabled models to focus on specific regions of the image based on the questions. Furthermore, multimodal fusion techniques such as Multimodal Compact Bilinear Pooling (MCB) [10] and Bilinear Attention Networks (BAN) [13] improved the interaction between visual and textual features, leading to enhanced performance. In 2019, the advent of transformer-based models like VilBERT [18] and LXMERT [22] extended the success of transformers to vision-language tasks.

2.3   Ensemble Approaches for Visual Question Answering

Ensemble methods in VQA leverage the strengths of multiple models to enhance overall performance, reducing biases and increasing the robustness of predictions. One such method, the Greedy Gradient Ensemble [11], employs a strategy to sequentially fit bias models to data, thus enabling a base model to focus more effectively on unbiased data distribution. This method helps improve model generalization by minimizing biases inherent in the training data, which often skew predictions.

Additionally, the use of explicit attention models has been shown to improve VQA accuracy by allowing models to focus on the most relevant parts of an image [16]. This approach leverages separate word embedding models for textual and visual inputs, enhancing the expressive power of the attention mechanism and improving the accuracy of answering visually grounded questions.

Recently, some papers have utilized simple aggregation methods yet achieved impressive results [14]. These methods, which simply select the most frequent output from various models, have surprisingly performed well and demonstrated good generalization.

These ensemble and attention methods illustrate advanced strategies in handling the complex interplay of visual and textual information essential for robust VQA systems. They address the critical challenges of bias and focus on model responses, contributing to more accurate and reliable VQA outcomes.

3   Proposed Method

We conduct experiments with pre-trained LVLMs through prompt-based inference, utilizing the following models: Qwen2-VL [24], InternVL2 [23], MiniCPM [29], and Llama-3.2-Vision-Instruct [20]. We chose these models because they are the top-performing models on the leaderboard [1] that we could access and apply to our experiments.

We design a structured prompt template to standardize the process of generating responses, guiding them through a structured, step-by-step reasoning framework. The overall approach is shown in Fig. 1.

We obtain answers to each model's input questions using the prompt template. To enhance accuracy and robustness, we employ an ensemble approach to combine the outputs from different models, known as majority voting, where the most commonly selected answer by the models is chosen as the final prediction. This ensemble method increases accuracy by leveraging the collective wisdom of several models, mitigating the risks associated with individual model errors and capitalizing on their diverse strengths for more reliable and robust results.

Specifically, we generate results from multiple inference runs with different configurations, including variations in step-by-step reasoning, or different model sizes (e.g., InternVL 26B, 40B, 76B [23] or Llama-3.2-Vision-Instruct with 11B or 90B versions [20]). The outputs from these models are then aggregated by selecting the most frequent response for each question, based on a majority voting scheme. We gather the answers from all models for each image-question pair and select the most commonly occurring answer as the final prediction.

We will detail how we designed the prompt and the specifics of each model family in subsequent sections.

3.1   Prompt Design

In our experiments, we utilized a structured prompt template to standardize the input provided to the models and guide them toward generating accurate answers. The prompt facilitates step-by-step reasoning and ensures that the models focus on selecting the correct option from the given choices. The prompt template is as follows:

This prompt begins by instructing the model to select the correct answer based on the provided question and image, explicitly mentioning that the options are labeled A, B, C, and D. This sets clear expectations for the model's response format.

The placeholders `question` and `option1` to `option4` are filled with the actual question and options from the dataset. Presenting the options in a labeled list helps the model to reference them easily during reasoning.

The instruction "Let's think step by step." encourages the model to engage in a chain-of-thought reasoning process [26], which has been shown to improve performance on complex tasks. By prompting the model to consider the problem methodically, we aim to enhance its ability to arrive at the correct answer.

Finally, we include the directive "Only type the letter corresponding to the correct option." to constrain the model's output to the required format. This minimizes the chance of generating extraneous text and ensures that the answer can be easily parsed and evaluated.

3.2   Qwen2-VL

Qwen2-VL [24] series represents an advanced upgrade from the original Qwen Large Vision Language Model (Qwen-VL) which is proposed by Alibaba Cloud. It is a powerful tool for a variety of applications, including visual analysis, multilingual support, and autonomous agent capabilities [9].

One of the key innovations in Qwen2-VL is its Naive Dynamic Resolution mechanism [24], allowing the model to dynamically adjust the resolution of input images and closely mirroring the way humans perceive visual information. Additionally, Qwen2-VL incorporates Multimodal Rotary Position Embedding (M-RoPE), enabling effective fusion of positional data across text, images, and videos, hence improving overall comprehension of complex, multimodal inputs.

The Qwen2-VL series also scales across different parameter sizes—2 billion, 7 billion, and 72 billion parameters—making it adaptable to a range of tasks, from efficient on-device performance to tackling more complex visual reasoning tasks. The largest model, Qwen2-VL-72B, achieves results comparable to state-of-the-art systems like GPT-4 and Claude 3.5 [24].

3.3 InternVL2

InternVL represents a significant leap in vision-language foundation models by scaling up the vision foundation model to 6 billion parameters [7]. The architecture of InternVL2, consistent with the ViT-MLP-LLM configuration of previous version InternVL 1.5, has been further enhanced in version 2.0 with a variety of instruction-tuned models, ranging from 1 billion to 108 billion parameters [23].

Through these advanced techniques, InternVL2 establishes itself as a competitive alternative to other large-scale vision models, delivering robust performance in chart comprehension and infographics QA. Its ability to seamlessly integrate with LLMs positions it as an essential tool for developing comprehensive multimodal AI applications.

3.4 MiniCPM

MiniCPM-Llama3-V 2.5 [29], the latest in the MiniCPM-V series, builds on SigLip-400M and Llama3-8B-Instruct, offering significant performance enhancements over its predecessor, MiniCPM-V 2.0 [29]. This model excels in chart comprehension and infographics QA, achieving an impressive 65.1 average score on OpenCompass across 11 benchmarks, and outperforms leading models like GPT-4V-1106 and Gemini Pro with its 8 billion parameters. Additionally, its advanced OCR capabilities allow for adept processing of images up to 1.8 million pixels in any aspect ratio, excelling in data extraction from charts and infographics with a score over 700 on OCRBench [29]. Enhanced instruction-following and reasoning abilities improve multimodal interactions and utility in applications requiring nuanced visual content understanding.

3.5 Llama-3.2-Vision-Instruct

Llama-3.2-Vision-Instruct [20] includes multimodal large language models for both image and text processing in 11B and 90B sizes, as well as lightweight, text-only models in 1B and 3B sizes that support a context length of 128K tokens and are state-of-the-art in their class for on-device use cases. These models are pre-trained and fine-tuned for visual recognition, image reasoning, captioning, and answering general questions about images.

The model's evaluation shows that the Llama-3.2-Vision model performs similarly to leading models like Claude 3 Haiku and GPT4o-mini in tasks of image recognition and visual understanding. This 3B model performs better than the Gemma 2 2.6B and Phi 3.5-mini models in tasks involving following instructions, summarization, and prompt rewriting, while the 1B model is comparable to Gemma [20].

3.6 Post-processing Approach

In our experiments, we observed that individual models sometimes produced inconsistent or incorrect answers due to the complexity of the visual questions and the diversity of the data. To address this issue and enhance the overall accuracy of our system, we implemented a post-processing step that aggregates the outputs from multiple models using a majority voting scheme.

Our post-processing approach operates as follows:

1. For each image-question pair in the dataset, we obtain the predicted answers from multiple models (e.g., Qwen2-VL-7B-Instruct, InternVL2 models, Llama-3.2-Vision-Instruct models).
2. We collect these predictions into a list of candidate answers.
3. We apply majority voting to determine the final answer, selecting the option that appears most frequently among the models' predictions.
4. In cases where there is a tie (i.e., multiple options receive the same highest number of votes), we apply a predefined tie-breaking strategy by selecting the answer from the model with the highest individual performance, which is the Qwen2-VL model.

This ensemble method utilizes the strengths of individual models while mitigating their weaknesses. By aggregating the predictions, it diminishes the effect of any single model's errors and enhances the robustness of the overall result. This method is straightforward, efficient, and scalable, enabling us to process the entire dataset with minimal computational overhead.

4   Experiments

4.1   Datasets and Evaluation Metrics

The datasets employed in our study are integral components of the LAVA challenge [2], designed to evaluate the effectiveness of LVLMs in understanding and interpreting complex visual data accompanied by textual queries. The public dataset comprises around 3,000 visual samples, each paired with a multiple-choice question in either English or Japanese. These samples predominantly include intricate diagrams such as Data Flow Diagrams, Gantt Charts, and Building Designs, which demand a high level of visual-textual comprehension. On the other hand, the private dataset contains approximately 1,100 samples provided by the TASUKI team at SoftBank, featuring a similar composition and complexity but used exclusively for private leaderboard evaluation.

The evaluation of this competition will be based on the MMMU metric. The final score will be calculated as follows: Final score = $0.3 \times$ Public dataset score $+ 0.7 \times$ Private dataset score. We can only access the Public dataset score.

4.2   Implementation Details

We utilize PyTorch and the `transformers` library from Hugging Face for our experiments. Our study explores the range of image dimensions within our dataset, noting a maximum height of 7041 pixels and a maximum width of 9600 pixels, with minimums at 60 pixels in height and 130 pixels in width. We retain the original image sizes, which are processed according to each model's specific requirements, as detailed in our code. The inference pipeline is constructed using the prompt template described in 3.1. To minimize randomness in our experiments, we try to set `do_sample=False`, and if not, we try to adjust the `temperature=0.000001` and `top_k=1`. These experiments were conducted on a single NVIDIA H100 80GB graphics card, with a batch size set to 1.

4.3 Results

Table 1. Comparison results on Public dataset.

Index	Method	Public Score
(1)	Qwen2-VL-7B-Instruct	**80**
(2)	MiniCPM-Llama3-V-2_5	64
(3)	InternVL2-26B	75
(4)	InternVL2-40B	73
(5)	InternVL2-76B	79
(6)	Llama-3.2-11B-Vision-Instruct	70
(7)	Llama-3.2-90B-Vision-Instruct	77
(8)	(3) + (4) + (5)	78
(9)	(1) + (3) + (4) + (5)	80
(10)	(1) + (3) + (4) + (5) + (6) + (7)	**82**

Table 1 presents the performance of various LVLMs and their ensembles on the public dataset, evaluated using the Public Score metric. Among the individual models, Qwen2-VL-7B-Instruct (Index 1) achieves the highest score of 80, outperforming larger models like InternVL2-76B (Index 5) and Llama-3.2-90B-Vision-Instruct (Index 7), which score 79 and 77 respectively. This indicates that model performance is not solely dependent on the number of parameters; factors such as model architecture and training data play significant roles. Notably, MiniCPM-Llama3-V-2_5 (Index 2) scores the lowest at 64, suggesting limitations in handling the complexity of the dataset.

Ensemble methods show improved performance over individual models. The ensemble of all InternVL2 models (Index 8) increases the score to 78, while adding Qwen2-VL-7B-Instruct to the ensemble (Index 9) maintains the score at 80. The highest score of 82 is achieved by the comprehensive ensemble (Index 10) combining Qwen2-VL, InternVL2 models, and Llama-3.2 models, excluding the lowest-performing MiniCPM-Llama3-V-2_5. This demonstrates that combining models with diverse strengths through ensemble methods like majority voting can enhance predictive accuracy, mitigating individual model errors and leveraging complementary capabilities.

5 Discussion

Although our model performed well, there is still room for improvement. The challenge dataset does not provide the correct answers, so we cannot confirm if the model's predictions are accurate. We noticed that the model's performance varies across different versions and configurations. This observation led us to use an ensemble approach, which helps to stabilize the model's performance by combining predictions from multiple models. This method aims to reduce errors and increase the likelihood of selecting the most accurate answer, especially in difficult or ambiguous cases.

In future work, we plan to develop more advanced ensemble methods beyond simple majority voting. We will explore options like decision trees or assigning specific models to particular tasks or languages, which could improve the model's accuracy. Another promising method is the Mixture of Experts (MoE) [19], which divides a model into specialized parts that each handle a subset of the data. We believe that incorporating MoE could greatly enhance our model's effectiveness. Additionally, we intend to fine-tune our models specifically for the tasks in the LAVA challenge if we could find the dataset with the actual label. This should lead to better performance, particularly for processing both English and Japanese, as current models may not handle different languages equally well.

6 Conclusion

In summary, our study demonstrates the effectiveness of pre-trained Large Vision-Language Models (LVLMs) in tackling complex visual question answering tasks using the LAVA challenge dataset. By employing a structured prompt template that encourages step-by-step reasoning, we enhanced the models' ability to interpret intricate visual data. Among individual models, Qwen2-VL-7B-Instruct achieved the highest public score, highlighting its capability despite a smaller parameter size. Moreover, ensemble methods using majority voting further improved performance, with the comprehensive ensemble attaining a public score of 82. These results underscore the benefits of combining different LVLMs to leverage their diverse strengths and improve predictive accuracy. Our work highlights the potential of ensemble approaches and prompt-based inference in advancing multimodal AI applications, particularly for tasks requiring sophisticated visual and textual understanding.

References

1. OpenVLM leaderboard. https://huggingface.co/spaces/opencompass/open_vlm_leaderboard
2. ACCV workshop on large vision – language model learning and applications (2024). https://lava-workshop.github.io/
3. Gopalkrishnan, A., Greer, R., Trivedi, M.: Multi-frame, lightweight & efficient vision-language models for question answering in autonomous driving. arXiv preprint arXiv:2403.19838 (2024)

4. Antol, S., et al.: VQA: visual question answering. In: Proceedings of the IEEE International Conference on Computer Vision, pp. 2425–2433 (2015)

5. Antol, S., et al.: VQA: visual question answering. arXiv preprint arXiv:1505.00468 (2015)

6. Chen, L., et al.: Are we on the right way for evaluating large vision-language models? arXiv preprint arXiv:2403.20330 (2024)

7. Chen, Z., Wang, W., Tian, H., et al.: How far are we to GPT-4v? Closing the gap to commercial multimodal models with open-source suites. arXiv preprint arXiv:2404.16821 (2024)

8. Cheng, S., et al.: EgoThink: evaluating first-person perspective thinking capability of vision-language models. In: Proceedings of the IEEE/CVF Conference on Computer Vision and Pattern Recognition, pp. 14291–14302 (2024)

9. Emanuilov, S.: Qwen2-VL — a new milestone in vision-language AI. https://unfoldai.com/qwen2-vl/

10. Fukui, A., Park, D.H., Yang, D., Rohrbach, A., Darrell, T., Rohrbach, M.: Multimodal compact bilinear pooling for visual question answering and visual grounding. arXiv preprint arXiv:1606.01847 (2016)

11. Han, X., Wang, S., Su, C., Huang, Q., Tian, Q.: Greedy gradient ensemble for robust visual question answering. In: Proceedings of the IEEE/CVF International Conference on Computer Vision, pp. 1584–1593 (2021)

12. Jiang, Y., et al.: Effectiveness assessment of recent large vision-language models. Vis. Intell. **2**(1), 17 (2024)

13. Kim, J., Jun, J., Zhang, B.: Bilinear attention networks. arXiv preprint arXiv:1805.07932 (2018)

14. Le, B.H., Nguyen-Mau, T.H., Nguyen-Vu, D.K., Ho-Ngoc, V.P., Nguyen, H.D., Tran, M.T.: Leveraging large vision-language models for visual question answering in VizWiz grand challenge (2024)

15. Li, J., Li, D., Xiong, C., Hoi, S.: BLIP: bootstrapping language-image pre-training for unified vision-language understanding and generation. In: International Conference on Machine Learning, pp. 12888–12900. PMLR (2022)

16. Lioutas, V., Passalis, N., Tefas, A.: Explicit ensemble attention learning for improving visual question answering. Pattern Recogn. Lett. **111**, 51–57 (2018)

17. Liu, H., Li, C., Wu, Q., Lee, Y.J.: Visual instruction tuning. In: NeurIPS (2023)

18. Lu, J., Batra, D., Parikh, D., Lee, S.: ViLBERT: pretraining task-agnostic Visiolinguistic representations for vision-and-language tasks. arXiv preprint arXiv:1908.02265 (2019)

19. Masoudnia, S., Ebrahimpour, R.: Mixture of experts: a literature survey. Artif. Intell. Rev. **42**, 275–293 (2014)

20. Meta: Llama 3.2: revolutionizing edge AI and vision with open, customizable models. https://ai.meta.com/blog/llama-3-2-connect-2024-vision-edge-mobile-devices/

21. Saxena, S., Sharma, M., Kroemer, O.: MResT: multi-resolution sensing for real-time control with vision-language models. arXiv preprint arXiv:2401.14502 (2024)

22. Tan, H., Bansal, M.: LXMERT: learning cross-modality encoder representations from transformers. arXiv preprint arXiv:1908.07490 (2019)

23. Team, O.: InternVL2: better than the best—expanding performance boundaries of open-source multimodal models with the progressive scaling strategy (2024). https://internvl.github.io/blog/2024-07-02-InternVL-2.0/

24. Wang, P., et al.: Qwen2-VL: enhancing vision-language model's perception of the world at any resolution. arXiv preprint arXiv:2409.12191 (2024)

25. Wang, W., et al.: VisionLLM: large language model is also an open-ended decoder for vision-centric tasks. Adv. Neural Inf. Process. Syst. **36** (2024)
26. Wei, J., et al.: Chain-of-thought prompting elicits reasoning in large language models. Adv. Neural. Inf. Process. Syst. **35**, 24824–24837 (2022)
27. Yang, A., Miech, A., Sivic, J., Laptev, I., Schmid, C.: Just ask: learning to answer questions from millions of narrated videos. In: Proceedings of the IEEE/CVF International Conference on Computer Vision, pp. 1686–1697 (2021)
28. Yang, Z., He, X., Gao, J., Deng, L., Smola, A.J.: Stacked attention networks for image question answering. arXiv preprint arXiv:1511.02274 (2015)
29. Yao, Y., Yu, T., Zhang, et al.: MiniCPM-v: a GPT-4v level MLLM on your phone. arXiv preprint arXiv:2408.01800 (2024)
30. Zhang, J., Huang, J., Jin, S., Lu, S.: Vision-language models for vision tasks: a survey. IEEE Trans. Pattern Anal. Mach. Intell. (2024)
31. Zhou, L., Palangi, H., Zhang, L., Hu, H., Corso, J., Gao, J.: Unified vision-language pre-training for image captioning and VQA. In: Proceedings of the AAAI Conference on Artificial Intelligence, vol. 34, pp. 13041–13049 (2020)

DermAI: A Chatbot Assistant for Skin Lesion Diagnosis Using Vision and Large Language Models

Viet-Tham Huynh[1,2], Trong-Thuan Nguyen[1,2,3],
Thao Thi-Phuong Dao[1,2,4], Tam V. Nguyen[5], and Minh-Triet Tran[1,2(✉)]

[1] Software Engineering Laboratory and Faculty of Information Technology,
University of Science VNU-HCM, Ho Chi Minh City, Vietnam
{hvtham,ntthuan,tmtriet}@selab.hcmus.edu.vn
[2] Vietnam National University, Ho Chi Minh City, Vietnam
[3] Department of Electrical Engineering and Computer Science, University of
Arkansas, Fayetteville, USA
[4] Department of Otolaryngology, Thong Nhat Hospital, Ho Chi Minh City, Vietnam
thao.dao2020@ict.jvn.edu.vn
[5] Department of Computer Science, University of Dayton, Dayton, USA
tamnguyen@udayton.edu

Abstract. In dermatology, the demand for accurate skin lesion diagnoses is critical, especially during peak times like summer when skin cancer screenings surge. The need for efficient processing of large volumes of medical images and the risk of human error highlights the importance of innovative diagnostic tools. In this paper, we propose DermAI, an advanced AI-driven framework to improve diagnostic accuracy and efficiency in skin lesion analysis. Our DermAI framework combines a state-of-the-art segmentation model and a large language model to assist clinicians in interpreting medical images swiftly and precisely. Our framework isolates and analyzes key lesion features using advanced segmentation models and vision encoders, while a large language model provides contextual insights to understand lesion characteristics and potential malignancies. By integrating visual and linguistic analysis, our DermAI framework reduces diagnostic errors, alleviates clinician workloads, and enhances patient care with faster, more accurate results, supporting dermatologists in making informed decisions and advancing AI-assisted diagnostics.

Keywords: Large Language Model · Medical Vision-Language Model · Skin lesion Segmentation

1 Introduction

In today's healthcare landscape, dermatologists and clinicians face increasing pressure to interpret complex medical images, such as skin lesion scans, within limited timeframes. During peak seasons, particularly in summer, when skin cancer screenings surge, clinicians are tasked with reviewing hundreds of lesion images daily, where even minor delays in diagnosis can significantly impact patient outcomes [13,22]. These high-stakes situations increase the risk of human error or oversight, placing additional pressure on healthcare professionals.

The growing demand for enhanced diagnostic accuracy and efficiency calls for advanced tools in medical imaging. Artificial Intelligence (AI) [23,28] offers a transformative solution, reshaping how medical images are analyzed. To this end, we address this need through an AI-driven chatbot that integrates advanced vision and language models to assist in skin lesion diagnosis. The objective is to ease the workload on healthcare professionals, minimize diagnostic errors, and deliver faster, more accurate insights, ultimately improving patient care.

In this paper, we present **DermAI** (Dermatology AI Assistant), a novel framework that utilizes a segmentation model to isolate key areas of skin lesions, focusing on clinically relevant features. The vision component, powered by a state-of-the-art medical image segmentation model [6] and vision encoder [26,31], processes visual data, while a linear transformation enhances image quality for more accurate interpretation. In addition, based on Generative Pre-trained Transformer 4 (GPT-4) [17], the language component contextualizes the visual data, providing deeper insights into lesion characteristics and potential malignancy. Our approach delivers comprehensive analyses, supporting healthcare professionals in making well-informed decisions.

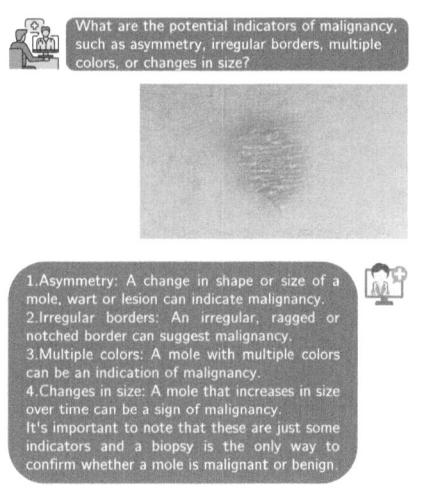

Fig. 1. Our DermAI with an interactive chat assistant. The user enters a prompt and a skin photo. The assistant responds with text feedback.

The Contributions of this Work.
We first investigate state-of-the-art skin lesion segmentation methods. Then, we introduce a novel framework integrating vision with a Large Language Model (LLM) trained on large-scale medical data, namely **DermAI**, illustrated in Fig. 1. Our experiments show exceptional performance in skin lesion segmentation and practical interpretation, earning recognition from medical professionals.

2   Related Work

2.1   Medical Chatbot

The introduction of large language models, particularly ChatGPT, has ignited increasing interest in developing medical chatbots, especially for their capacity to automate X-ray image analysis. These technologies serve as valuable tools for patients and healthcare professionals by facilitating a more comprehensive understanding of diagnostic findings from X-ray images. Recent advancements in Large Language Models (LLMs) and Multi-Modal Learning have highlighted the potential of these systems in medical applications. Several notable works have emerged in this domain, including Chatdoctor [15], LLaMA [27], MedAlpaca [11], PMC-LLaMA [33], and DoctorGLM [35]. For instance, Chatdoctor [15], built upon the LLaMA [27] model, provides reliable interpretations of X-ray images for both patients and clinicians, offering personalized medical advice. Similarly, MedAlpaca [11], PMC-LLaMA [33], and DoctorGLM [35] have fine-tuned open-source LLMs on medical data to develop chatbots tailored to healthcare contexts. These advancements emphasize the growing potential of integrating LLMs and multi-modal learning into medical applications, paving the way for more personalized, accurate, and accessible diagnostic tools in healthcare.

2.2   Skin Lesion Segmentation

Recently, medical image segmentation methods inspired by Transformer [16] and CNN architectures [20]. In particular for skin lesion segmentation methods (e.g., MobileUNetR [19] and DuAT [25]), which balance efficiency and accuracy by addressing the challenges of preserving global context and local detail. Polar transformations [5] and boundary-aware mechanisms [30] also enhance segmentation performance and data efficiency by capturing crucial boundary information. In addition, the integrating diffusion models (e.g., DermoSegDiff [6] and MedSegDiff [34]) with attention mechanisms further improve boundary delineation, which improving overall segmentation quality. To tackle data imbalance, particularly in small lesion segmentation, the Focal Tversky loss function [1] improves the precision-recall trade-off. At the same time, Inconsistency Masks (IM) [29] enable strong results with minimal labeled data. Moreover, enhanced U-Net variants such as DoubleU-Net [12] and BCDU-Net [4], which utilize pre-trained encoders, dense connections, and bi-directional ConvLSTM for feature extraction and fusion. Multi-scale approaches like MSRF-Net [24] and context-gating mechanisms [3] address variable object sizes and complex anatomical variations.

2.3   Discussion

As presented in Sects. 2.2 and 2.1, while advancements in skin lesion segmentation and medical chatbots have progressed, a noticeable gap remains in integrating chatbots with skin lesion segmentation capabilities. Although chatbots have proven valuable in tasks like X-ray analysis, none currently harness the power of segmentation for skin lesions, a crucial tool in dermatology. Developing a DermAI chatbot, which combines skin lesion segmentation with interactive conversational capabilities, offers transformative potential. Our system assists healthcare professionals in making precise diagnoses by delivering real-time segmented imagery with expert guidance while empowering patients with

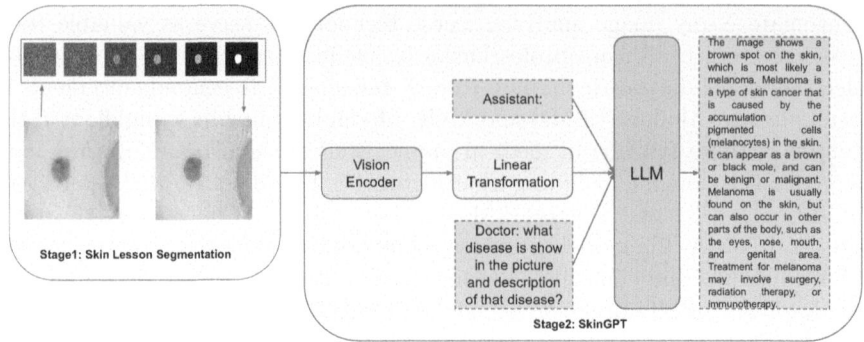

Fig. 2. Overview of the proposed framework, namely DermAI. Our approach consists of two stages: skin lesion segmentation and SkinGPT. In the first stage, the segmentation model [6] isolates the lesion from the input image. In the second stage, SkinGPT [37] leverages a pre-trained Vision Transformer and LLM (Llama-2 [27]) to provide context-aware diagnostic insights based on the segmented image.

personalized, easily understandable feedback. This innovation has the potential to streamline dermatological workflows, boost diagnostic accuracy, improve patient outcomes, and bridge a gap in healthcare technology for dermatology.

3   Our Proposed Framework

Framework Overview. Our proposed framework, illustrated in Fig. 2, introduces a two-stage process designed to revolutionize dermatological diagnostics. In the first stage, we leverage DermoDiff [6] to segment the input image, isolating key features for analysis. This segmented output is fed into SkinGPT [37], an innovative, multimodal diagnostic system powered by large language models. By aligning a pre-trained Vision Transformer with the Llama-2 LLM, SkinGPT leverages an extensive dataset of skin disease images enriched with clinical concepts and doctors' notes to generate highly insightful diagnostic outputs.

3.1   Skin Lesion Segmentation

Skin lesion segmentation plays a critical role in medical imaging and dermatology, serving as a cornerstone in diagnosing and analyzing various skin conditions, including life-threatening cancers such as melanoma. With the global incidence of skin cancer rising and melanoma being one of the most aggressive forms, the demand for accurate, automated diagnostic tools has grown significantly. Segmentation techniques are pivotal in this context, enabling healthcare professionals to identify, assess, and monitor skin lesions from dermoscopic or clinical images, supporting more timely and informed medical decisions. In our approach, we levage DermoSegDif [6] including *an encoder*, *a bottleneck*, and *a decoder*.

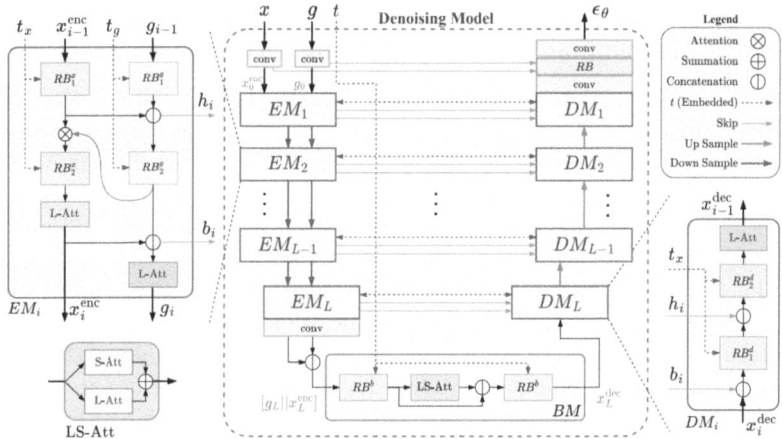

Fig. 3. Overview of the DermoSegDif [6] method.

Encoder. The Encoder consists of a series of stacked Encoder Modules (EM), followed by a convolution layer that reduces the spatial dimensions to a four-by-four tensor. Instead of the conventional approach of concatenating ϵ_θ and g_{i-1} before feeding them into the network, as proposed in prior work [32], the authors introduce a two-path feature extraction strategy within each EM. This method emphasizes the mutual influence between the noisy segmentation mask and the guidance image. Each path in the encoder includes two ResNet Blocks (RB) and a Linear Attention (L-Att) mechanism, providing computational efficiency and non-redundant feature extraction. Time embeddings are incorporated into each RB using sinusoidal positional embeddings, processed through linear layers and GeLU activation functions. Separate time embeddings are used for the guidance image (t_g) and the noisy segmentation mask (t_x), allowing the model to capture the temporal dynamics of both inputs. To enhance feature extraction, knowledge from the noise path (RB_1^x) is transferred and concatenated with the guidance path, creating an intermediate feature (h_i) that captures complementary representations. The guidance path processes the output through RB_2^g, and a feedback mechanism applies a convolution to the output, reconnecting it to RB_2^x. This feedback loop ensures boundary and noise information integration, allowing the model to emphasize key features while suppressing irrelevant details.

Bottleneck. The final outputs of the encoder, x_L^{enc} and g_L, are concatenated and passed through the Bottleneck Module (BM). This module includes a ResNet Block (RB), a Linear Self-Attention (LS-Att) mechanism, and another ResNet Block. The LS-Att module enhances feature representation by combining the spatial relationships captured by Self-Attention (S-Att) and the semantic context captured by Linear Attention (L-Att). These two attention mechanisms operate in parallel, allowing the model to integrate spatial and contextual information effectively. The output from the Bottleneck Module is then passed to the decoder.

Decoder. The Decoder comprises stacked Decoder Modules (DM) that match the number of Encoder Modules (EM). Each DM operates as a single-path module, consisting of two consecutive ResNet Blocks (RB) and one Linear Attention (L-Att) module, followed by a convolutional block that outputs the estimated noise ϵ_θ. The decoder integrates information from both the noise and guidance paths by concatenating the encoder outputs, b_i and h_i, before and after applying RB_1^d. This enables the decoder to effectively utilize the refined features from the encoder, improving its ability to estimate the added noise and recover missing segmentation details. A skip connection is also introduced, linking the original input x to the final decoder layer. This skip connection concatenates the output of the first decoder module (DM_1) with x_1, and the combined features are processed through a final convolutional block to produce the estimated noise ϵ_θ.

Figure 4 illustrates the diffusion process, a key component in generating and refining noise for reconstructing segmentation masks. In this process, noise is incrementally added to the input image to simulate data uncertainty, which the model progressively estimates and reduces to recover the lesion's clear boundaries. The model refines its predictions as noise is removed, producing an accurate segmentation mask that outlines the lesion's exact contours. A skip connection linking the original input to the final decoder layer preserves critical details, combining them with intermediate features to enhance segmentation. The model effectively handles complex, irregular lesion boundaries by predicting the noise (ϵ_θ) at each step, resulting in more precise and robust segmentation.

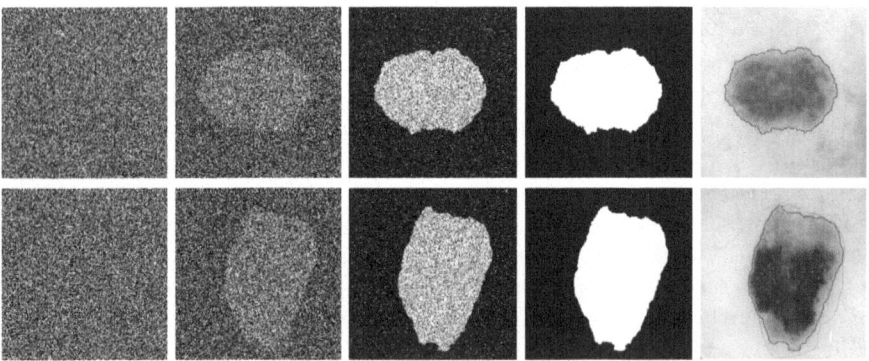

Fig. 4. Illustration of the diffusion process: noise is progressively added to the input image and then estimated and reduced by the model to reconstruct an accurate segmentation mask, delineating the boundaries of the lesion.

3.2   SkinGPT

SkinGPT, built on ChatGPT [17] and fine-tuned for dermatology, has gained attention for its ability to assist healthcare professionals in analyzing and interpreting medical images. A key application involves integrating a skin lesion segmentation model with ChatGPT to enhance diagnostic accuracy when processing medical images and responding to prompts, such as doctors' inquiries.

In the first stage, MiniGPT-4 [38] is trained to understand the alignment between visual information and language by learning from a large dataset of image-text pairs. It uses Vicuna [18], a language decoder based on LLaMA [27], and adopts the Vision Transformer (ViT) [2] from BLIP-2 [14], extracting features with Q-Former [36]. In addition, a linear projection layer is introduced to align the features from the visual encoder with the LLM, transforming them into soft prompts for generating textual descriptions. During pretraining, only the projection layer is trained while the vision encoder and LLM remain frozen.

In the second stage, the model undergoes fine-tuning to address issues from the first stage, such as incoherent outputs and repetitive phrases. Since high-quality vision-language datasets are scarce, the authors created their own by generating detailed image descriptions using the pre-trained model. Prompts were designed to encourage exhaustive descriptions in a conversational format, such as:

```
###Human:<Img><ImgFeature></Img> Describe this image in detail.
###Assistant:
```

If descriptions were too short, additional prompts were used to elicit more comprehensive responses. Despite generating a large number of image-text pairs, many descriptions still contained errors like redundancy or irrelevant content. ChatGPT was used to clean the data, automate refinement, and remove errors and redundant information. Moreover, MiniGPT-4 is fine-tuned using the curated dataset. The predefined prompt template follows the below format:

```
###Human: <Img><ImageFeature></Img <Instruction>
###Assistant:
```

where <Instruction> refers to a randomly selected prompt, such as *"Describe this image in detail"* or *"Could you describe the contents of this image for me?"*. The fine-tuning process aims to enhance the model's ability to generate natural, coherent language that aligns contextually with the visual input. Importantly, no regression loss is calculated for these text-image prompts, as the primary objective is to improve the fluency and reliability of the model's output.

3.3   DermAI: Dermatology AI Assistant

In *Stage 1*, the system processes an input image to isolate the region of interest (ROI) using a segmentation technique that generates a mask M, which highlights relevant areas such as skin lesions or other significant dermatological features. This mask is applied to the image, producing a masked version IM, where only the important regions remain visible. By removing irrelevant parts of the image, the model can concentrate solely on the segmented region that contains the most critical diagnostic information. In *Stage 2*, the masked image IM is fed into SkinGPT, a large language model fine-tuned specifically for dermatology. By focusing exclusively on the segmented region, SkinGPT analyzes key features such as the lesions shape, color, borders, and texture. This targeted approach ensures that the model remains focused on clinically significant areas, allowing it to generate detailed, context-aware explanations. The focus on these critical features enhances the accuracy of diagnostic insights, ensuring more relevant and precise outcomes.

Discussion. DermAI's integration of segmentation in *Stagre 2* delivers significant advantages by focusing the model's attention on the region of interest (ROI), allowing for a more precise and detailed analysis of the lesion. This targeted approach enhances the ability of SkinGPT to identify critical lesion characteristics, such as asymmetry, border irregularity, and color variation, which are essential for diagnosing conditions like melanoma. Our framework minimizes distractions by filtering out irrelevant areas, leading to more specific and clinically relevant insights, ultimately improving the reliability of DermAI's diagnostic outputs.

Segmentation and SkinGPT work together within our DermAI framework to maximize precision and contextual relevance in dermatological analysis. Segmentation ensures the model concentrates on the most important regions of the image, while SkinGPT, fine-tuned for medical language tasks, generates detailed, context-aware explanations. Our approach enables more accurate diagnoses by leveraging the strengths of both components, ultimately providing healthcare professionals with reliable and actionable information for evaluating skin lesions.

4   Experiment Results

4.1   Implementation Details

Training. We first train a segmentation model on the ISIC [8,10] datasets. The segmented images are then analyzed with the pre-trained SkinGPT [10], which was initially trained on SKINCON [9], a dataset densely annotated by dermatologists across multiple skin disease concepts, and Dermnet, which includes a diverse range of images features skin disease classes by board-certified dermatologists.

Inference. During inference, DermAI operates as a unified framework by processing the skin image and the user-provided prompt. Our system uses the segmentation model to isolate the skin lesion, generating a masked image. This segmented image and the prompt are fed into the SkinGPT model, where the visual data and the prompt are jointly analyzed to produce a response.

Datasets. The *ISIC 2016* [10] dataset includes 900 training images and 379 test images, each with expert-annotated binary masks delineating lesion boundaries. The *ISIC 2017* [8] dataset includes 2,000 training images and 150 validation images. Additionally, the SkinGPT model is trained on two large-scale datasets.

While the SKINCON [9] dataset features diverse dermatological annotations across 48 clinical concepts, the Dermnet[1] dataset covers 15 skin diseases.

Evaluation Metrics. We evaluate the segmentation model using the *Dice* and Intersection over Union (*IoU*) as previous work for fair comparison. Higher Dice values reflect better region overlap, indicating improved segmentation accuracy. Higher IoU values denote closer alignment between predictions and ground truth.

4.2   Quantitative Analysis

The results in Table 1 compare Dice scores across the ISIC 2016 and ISIC 2017 datasets, highlighting the performance variation of DermoDiff against baseline methods. On the ISIC 2016 dataset, DermoDiff achieves a competitive Dice score of 90.37%, closely following the highest scorer, UNet,

Table 1. Comparison (%) on ISIC 2016 and ISIC 2017 against baseline methods at *Dice*.

Methods	ISIC'16	ISIC'17
Swin-UNet [7]	85.68	**79.14**
UNet [21]	89.84	77.08
DermoDiff [6]	**90.37**	74.63

which reaches 89.84%. Our experimental results suggest that DermoDiff is well-suited for handling relatively straightforward lesion segmentation tasks, where the boundaries between lesions and healthy tissue are well-defined, making segmentation less complex.

However, the performance gap becomes more pronounced when analyzing the ISIC 2017 dataset. The Dicde score of DermoDiff drops to 74.63%, notably behind Swin-UNet at 79.14%. This decrease indicates that DermoDiff struggles with the more challenging images in the ISIC 2017 dataset, where lesions are often more irregular in shape, size, and texture and may present higher variability in appearance. For example, cases require advanced feature extraction and spatial context understanding, which DermoDiff appears to have limitations in addressing effectively. The lower score of DermoDiff points to a need for enhancing the ability to generalize across diverse datasets, particularly when dealing with more nuanced clinical cases where lesion characteristics are less homogeneous.

Table 2 further reinforces these observations through IoU scores, which offer an additional perspective on model accuracy in identifying lesion boundaries. DermoDiff performs admirably on the ISIC 2016 dataset, securing an IoU of 82.43%, slightly outperforming UNet (82.09%) and significantly ahead of Swin-UNet (75.59%). This strong performance on ISIC 2016 underscores the capability of DermoDiff in environments where segmentation challenges are moderate and lesions are relatively distinguishable. However, the drop in IoU to 59.53% on the ISIC 2017 dataset is significant, highlighting the limitations of DermoDiff in more intricate segmentation tasks. This decline suggests that DermoDiff still struggles with distinguishing between lesion boundaries and surrounding tissues when faced with more significant variability in lesion presentations.

---

[1] https://www.kaggle.com/datasets/shubhamgoel27/dermnet

Additionally, the disparity in results between the ISIC 2016 and ISIC 2017 datasets suggests that DermoDiff is optimized for cases where lesion appearances are more consistent and well-defined, such as those on ISIC 2016, but performs less effectively with the more heterogeneous cases on the ISIC 2017. This underperformance on ISIC 2017 may stem from its reduced ability to capture complex spatial relationships or its reliance on prominent features in simpler images. In contrast, the superior performance of Swin-UNet on ISIC 2017 highlights that models using advanced transformer-based architectures are better equipped to handle complex skin lesion segmentation, as they can capture long-range dependencies and contextual information more effectively.

Table 2. Comparison (%) on ISIC 2016 and ISIC 2017 against baseline methods at IoU.

Methods	ISIC'16	ISIC'17
Swin-UNet [7]	75.59	**66.76**
UNet [21]	82.09	64.10
DermoDiff [6]	**82.43**	59.53

4.3 Qualitative Analysis

Figure 5 provides a qualitative evaluation of a dermatological segmentation model, illustrating the precision of the predictions (blue outlines) compared to the ground truth (green outlines) across a spectrum of skin lesions. The model exhibits commendable accuracy in lesions with homogeneous boundaries, such as in the top left and bottom center images, where the blue and green outlines align closely. However, it encounters challenges with lesions that exhibit more complex features (e.g., irregular borders or heterogeneous pigmentation). For instance, the image in the top second from left shows the model overextending the lesion boundary, a common issue in cases where the lesion fades gradually into surrounding tissue, leading to over-segmentation. In contrast, the bottom second from the left image highlights under-segmentation, where the model fails to capture the entire lesion, likely due to its inability to detect subtle differences in color and texture that a trained dermatologist would note.

Figure 6 underscores the critical importance of segmentation in medical image analysis, particularly when leveraged by the advanced capabilities of DermAI. In addition, Fig. 1 illustrates the interaction of the DermAI chat assistant with an internet-sourced image (not from the ISIC 2016 or 2017 datasets), demonstrating the generalization of our DermAI framework. By isolating regions of interest, DermAI can focus its pattern recognition algorithms on specific areas, identifying subtle dermatological markers such as asymmetry, irregular borders, color variation, and changes in size, which are factors crucial for distinguishing malignant lesions, like melanoma, from benign growths such as warts and skin tags. This segmentation-driven approach closely mimics the systematic evaluation process employed by dermatologists, ensuring that the model's analysis adheres to clinical best practices. DermAI effectively reduces image noise, enhances diagnostic accuracy, and delivers clinically relevant insights by isolating and analyzing

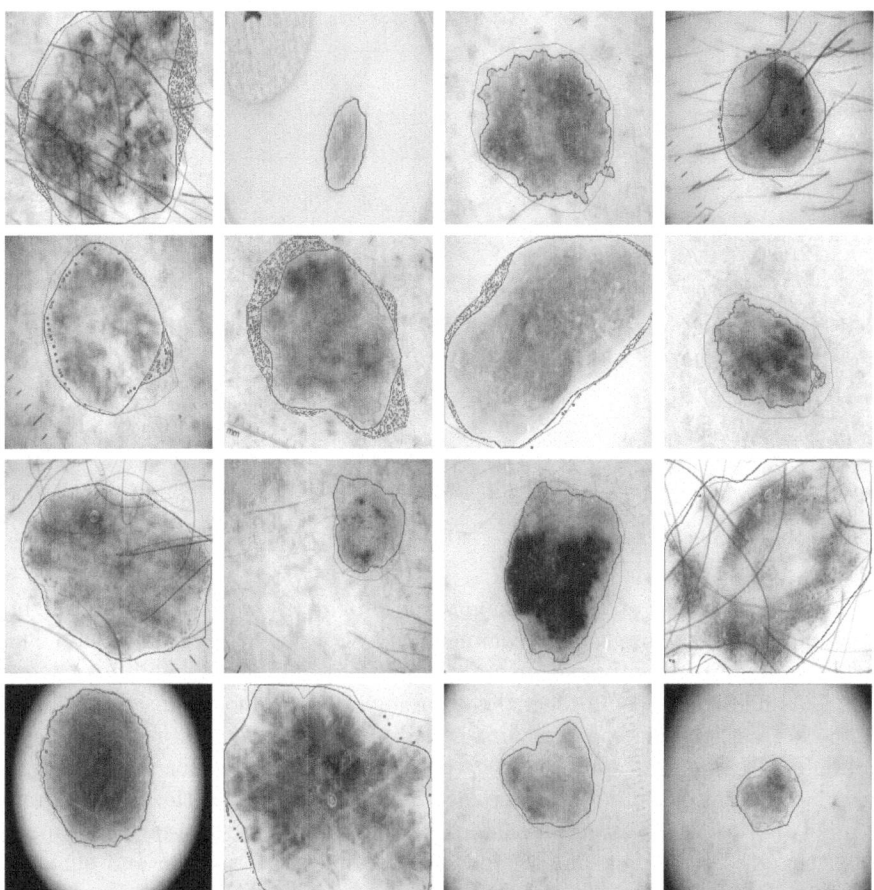

Fig. 5. Comparison of skin lesion segmentation results: Blue outlines show model predictions, while green outlines represent the ground truth. (Color figure online)

distinct, well-defined regions. The refined focus of our proposed approach significantly enhances the precision of lesion identification by concentrating on the most relevant areas of interest. Therefore, DermAI leads to more accurate and reliable diagnoses, enabling healthcare professionals to make informed, actionable decisions. By reducing false positives and increasing diagnostic clarity, our approach contributes to higher-quality assessments in dermatological care, ultimately supporting improved patient outcomes.

Expert Medical Evaluation. In a comprehensive evaluation, we presented our software to three medical experts, who compared two versions of DermAI: one with segmentation and one without. All three unanimously agreed that the version with segmentation significantly outperformed the one without. The segmentation-enhanced model consistently identified critical lesion features such

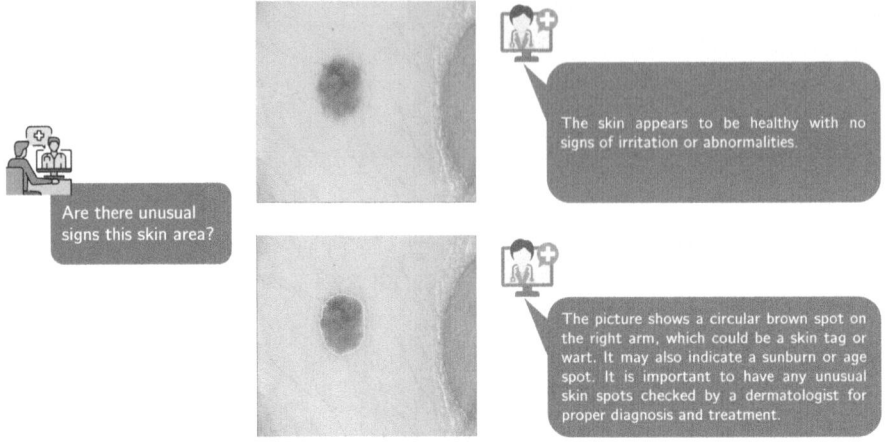

Fig. 6. Example of how segmentation (shown below) enhances dermatological analysis by highlighting a skin lesion for focused and precise assessment by DermAI.

as shape, color, borders, and location while providing more actionable insights regarding potential causes and recommendations for diagnosis or treatment. This feedback demonstrates that segmentation enhances the model's ability to detect key visual features and improves the accuracy of its diagnostic insights.

In contrast, the version without segmentation struggled to produce focused and detailed responses, often lacking the specificity required for accurate diagnosis. Segmentation empowered the model to concentrate on the most relevant areas of the lesion, resulting in a more thorough and precise analysis. Experts noted that the segmented model delivered far more informative and clinically useful descriptions. Based on this feedback, we will include additional lesion attributes such as size, surface texture, and fluid presence, which will further refine diagnostic accuracy. Importantly, these evaluations underscore the importance of segmentation in boosting diagnostic precision and clinical effectiveness.

5   Conclusion

In this paper, we have introduced DermAI, a novel framework designed to enhance the efficiency of skin lesion diagnosis by integrating advanced vision and language models. Our system combines state-of-the-art segmentation models with LLM, allowing clinicians to swiftly and accurately interpret medical images while reducing the risk of diagnostic errors. Integrating vision encoders with language models provides a comprehensive understanding of skin lesions, helping dermatologists make well-informed decisions. The results demonstrate the effectiveness of DermAI in streamlining diagnostic workflows and improving patient care, particularly during peak times, such as in summer when skin cancer screenings surge. Despite its promising capabilities, DermAI also highlights the importance of continuous refinement to better generalize across diverse datasets, as shown by performance variations between ISIC 2016 and ISIC 2017 datasets.

In the future, we aim to enhance the robustness and clinical utility of DermAI. First, we will refine the segmentation model to handle complex and diverse datasets, improving its generalization across different lesion types and skin conditions. Second, we will incorporate active learning strategies, enabling the model to continuously learn and improve from real-world data with minimal human intervention. Third, we plan to expand DermAI's capabilities to support real-time image processing and provide clinicians with interactive, interpretable AI-driven insights. Lastly, future iterations will explore integrating multimodal data inputs, such as patient history or genetic information, to create a more holistic diagnostic tool, further increasing precision and clinical relevance.

References

1. Abraham, N., Khan, N.M.: A novel focal Tversky loss function with improved attention U-Net for lesion segmentation. In: 2019 IEEE 16th International Symposium on Biomedical Imaging (ISBI 2019), pp. 683–687. IEEE (2019)
2. Alexey, D.: An image is worth 16×16 words: transformers for image recognition at scale. arXiv preprint arXiv: 2010.11929 (2020)
3. Asadi-Aghbolaghi, M., Azad, R., Fathy, M., Escalera, S.: Multi-level context gating of embedded collective knowledge for medical image segmentation. arXiv preprint arXiv:2003.05056 (2020)
4. Azad, R., Asadi-Aghbolaghi, M., Fathy, M., Escalera, S.: Bi-directional ConvLSTM U-Net with densley connected convolutions. In: Proceedings of the IEEE/CVF International Conference on Computer Vision Workshops, pp. 0–0 (2019)
5. Benčević, M., Galić, I., Habijan, M., Babin, D.: Training on polar image transformations improves biomedical image segmentation. IEEE access **9**, 133365–133375 (2021)
6. Bozorgpour, A., Sadegheih, Y., Kazerouni, A., Azad, R., Merhof, D.: DermoSegDiff: a boundary-aware segmentation diffusion model for skin lesion delineation. In: International Workshop on PRedictive Intelligence In MEdicine, pp. 146–158. Springer (2023). https://doi.org/10.1007/978-3-031-46005-0_13
7. Chen, J., et al.: TransUet: transformers make strong encoders for medical image segmentation. arXiv preprint arXiv:2102.04306 (2021)
8. Codella, N.C., et al.: Skin lesion analysis toward melanoma detection: a challenge at the 2017 international symposium on biomedical imaging (ISBI), hosted by the international skin imaging collaboration (ISIC). In: 2018 IEEE 15th International Symposium on Biomedical Imaging (ISBI 2018), pp. 168–172. IEEE (2018)
9. Daneshjou, R., Yuksekgonul, M., Cai, Z.R., Novoa, R., Zou, J.Y.: SkinCon: a skin disease dataset densely annotated by domain experts for fine-grained debugging and analysis. Adv. Neural. Inf. Process. Syst. **35**, 18157–18167 (2022)
10. Gutman, D., et al.: Skin lesion analysis toward melanoma detection: a challenge at the international symposium on biomedical imaging (ISBI) 2016, hosted by the international skin imaging collaboration (ISIC). arXiv preprint arXiv:1605.01397 (2016)
11. Han, T., et al.: MedAlpaca–An open-source collection of medical conversational AI models and training data. arXiv preprint arXiv:2304.08247 (2023)

12. Jha, D., Riegler, M.A., Johansen, D., Halvorsen, P., Johansen, H.D.: DoubleU-Net: a deep convolutional neural network for medical image segmentation. In: 2020 IEEE 33rd International Symposium on Computer-based Medical Systems (CBMS), pp. 558–564. IEEE (2020)
13. Jones, O.T., Ranmuthu, C.K., Hall, P.N., Funston, G., Walter, F.M.: Recognising skin cancer in primary care. Adv. Ther. **37**(1), 603–616 (2020)
14. Li, J., Li, D., Savarese, S., Hoi, S.: Blip-2: bootstrapping language-image pre-training with frozen image encoders and large language models. In: International Conference on Machine Learning, pp. 19730–19742. PMLR (2023)
15. Li, Y., Li, Z., Zhang, K., Dan, R., Jiang, S., Zhang, Y.: ChatDoctor: a medical chat model fine-tuned on a large language model meta-AI (LLAMA) using medical domain knowledge. Cureus **15**(6), e40895 (2023)
16. Nguyen, T.T., Nguyen, T.V., Tran, M.T.: Collaborative consultation doctors model: unifying CNN and VIT for Covid-19 diagnostic. IEEE Access **11**, 95346–95357 (2023)
17. OpenAI: GPT-4 technical report (2023)
18. Peng, B., Li, C., He, P., Galley, M., Gao, J.: Instruction tuning with GPT-4. arXiv preprint arXiv:2304.03277 (2023)
19. Perera, S., Erzurumlu, Y., Gulati, D., Yilmaz, A.: MobileUNETR: a lightweight end-to-end hybrid vision transformer for efficient medical image segmentation. arXiv preprint arXiv:2409.03062 (2024)
20. Phung, K.A., Nguyen, T.T., Wangad, N., Baraheem, S., Vo, N.D., Nguyen, K.: Disease recognition in x-ray images with doctor consultation-inspired model. J. Imaging **8**(12), 323 (2022)
21. Ronneberger, O., Fischer, P., Brox, T.: U-Net: convolutional networks for biomedical image segmentation. In: Navab, N., Hornegger, J., Wells, W.M., Frangi, A.F. (eds.) MICCAI 2015. LNCS, vol. 9351, pp. 234–241. Springer, Cham (2015). https://doi.org/10.1007/978-3-319-24574-4_28
22. Soenksen, L.R., et al.: Using deep learning for dermatologist-level detection of suspicious pigmented skin lesions from wide-field images. Sci. Transl. Med. **13**(581), eabb3652 (2021)
23. Srivastav, S., et al.: ChatGPT in radiology: the advantages and limitations of artificial intelligence for medical imaging diagnosis. Cureus **15**(7), e41435 (2023)
24. Srivastava, A., et al.: MSRF-NET: a multi-scale residual fusion network for biomedical image segmentation. IEEE J. Biomed. Health Inform. **26**(5), 2252–2263 (2021)
25. Tang, F., et al.: DUAT: dual-aggregation transformer network for medical image segmentation. In: Chinese Conference on Pattern Recognition and Computer Vision (PRCV), pp. 343–356. Springer, Singapore (2023). https://doi.org/10.1007/978-981-99-8469-5_27
26. Thawkar, O., et al.: XrayGPT: chest radiographs summarization using medical vision-language models. arXiv preprint arXiv:2306.07971 (2023)
27. Touvron, H., et al.: Llama: open and efficient foundation language models. arXiv preprint arXiv:2302.13971 (2023)
28. Tu, T., et al.: Towards conversational diagnostic AI. arXiv preprint arXiv:2401.05654 (2024)
29. Vorndran, M.R., Roeck, B.F.: Inconsistency masks: removing the uncertainty from input-pseudo-label pairs. arXiv preprint arXiv:2401.14387 (2024)
30. Wang, J., Wei, L., Wang, L., Zhou, Q., Zhu, L., Qin, J.: Boundary-aware transformers for skin lesion segmentation. In: de Bruijne, M., et al. (eds.) MICCAI 2021. LNCS, vol. 12901, pp. 206–216. Springer, Cham (2021). https://doi.org/10.1007/978-3-030-87193-2_20

31. Wang, Z., Wu, Z., Agarwal, D., Sun, J.: MedClip: contrastive learning from unpaired medical images and text. arXiv preprint arXiv:2210.10163 (2022)
32. Wolleb, J., Sandkühler, R., Bieder, F., Valmaggia, P., Cattin, P.C.: Diffusion models for implicit image segmentation ensembles. In: Konukoglu, E., Menze, B., Venkataraman, A., Baumgartner, C., Dou, Q., Albarqouni, S. (eds.) Proceedings of The 5th International Conference on Medical Imaging with Deep Learning. Proceedings of Machine Learning Research, vol. 172, pp. 1336–1348. PMLR 2022). https://proceedings.mlr.press/v172/wolleb22a.html
33. Wu, C., Zhang, X., Zhang, Y., Wang, Y., Xie, W.: PMC-LLaMA: further finetuning llama on medical papers. arXiv preprint arXiv:2304.14454 (2023)
34. Wu, J., et al.: MedSegDiff: Medical image segmentation with diffusion probabilistic model. In: Medical Imaging with Deep Learning, pp. 1623–1639. PMLR (2024)
35. Xiong, H., et al.: DoctorGLM: fine-tuning your chinese doctor is not a herculean task. arXiv preprint arXiv:2304.01097 (2023)
36. Zhang, Q., Zhang, J., Xu, Y., Tao, D.: Vision transformer with quadrangle attention. IEEE Trans. Pattern Anal. Mach. Intell. **46**, 3608–3624 (2024)
37. Zhou, J., et al.: Pre-trained multimodal large language model enhances dermatological diagnosis using SkinGPT-4. Nat. Commun. **15**(1), 5649 (2024)
38. Zhu, D., Chen, J., Shen, X., Li, X., Elhoseiny, M.: MiniGPT-4: enhancing vision-language understanding with advanced large language models. In: The Twelfth International Conference on Learning Representations (2024). https://openreview.net/forum?id=1tZbq88f27

Mitigating Backdoor Attacks Using Activation-Guided Model Editing

Felix Hsieh[1,2]([✉]), Huy H. Nguyen[1], AprilPyone MaungMaung[1],
Dmitrii Usynin[2,3], and Isao Echizen[1,4]

[1] National Institute of Informatics, Tokyo, Japan
[2] Technical University of Munich, Munich, Germany
Felix.Hsieh98@googlemail.com
[3] Imperial College London, London, UK
[4] The University of Tokyo, Tokyo, Japan

Abstract. Backdoor attacks compromise the integrity and reliability of machine learning models by embedding a hidden trigger during the training process, which can later be activated to cause unintended misbehavior. We propose a novel backdoor mitigation approach via machine unlearning to counter such backdoor attacks. The proposed method utilizes model activation of domain-equivalent unseen data to guide the editing of the model's weights. Unlike the previous unlearning-based mitigation methods, ours is computationally inexpensive and achieves state-of-the-art performance while only requiring a handful of unseen samples for unlearning. In addition, we also point out that unlearning the backdoor may cause the whole targeted class to be unlearned, thus introducing an additional repair step to preserve the model's utility after editing the model. Experiment results show that the proposed method is effective in unlearning the backdoor on different datasets and trigger patterns.

Keywords: Backdoor Mitigation · Machine Unlearning · Model Editing

1 Introduction

Machine learning models highly depend on the quality and quantity of data available during training. As the demand for more powerful models increases, so does the need for vast data collections and significant computational resources for model training. Except for major corporations, most entities rely on uncurated data, such as publicly available data online and third-party services that run learning protocols. The loss of control of the training opens up an attack vector for a malicious actor to use backdoor attacks to poison the training data [29].

Supplementary Information The online version contains supplementary material available at https://doi.org/10.1007/978-981-96-2641-0_21.

Gu et al. [14] proposed BadNets, the first backdoor attack. BadNets overlays a small subset of training samples with a square of fixed size and position, and it changes the labels to a target class, thus poisoning the samples. During training, the victim model learns to associate the trigger pattern with the target class, creating a hidden backdoor reactive to the trigger. During inference, the model behaves as usual on clean data. Still, when a malicious actor forwards a sample with a specific trigger, the backdoor in the neural network is activated, leading to model misbehavior, such as misclassification. A survey from Microsoft stated that data poisoning is one of the top attacks on machine learning systems [24].

The security risk backdoor attacks impose creates the need for contrary defense methods. In this work, we focus on one type of defense where one mitigates the influence of a backdoor attack on an adversarial-modified model. Retraining the model from scratch with clean training data is the most straightforward approach for obtaining an adversarial-free model. Retraining is computationally expensive and requires access to clean training data. Filtering out poisoned samples in a training dataset is often unfeasible because the dataset is too large. Backdoor mitigation with machine unlearning has emerged as a promising approach to overcoming the limitations of retraining and efficiently removing a backdoor in a poisoned model. Many methods omit the need for original training data.

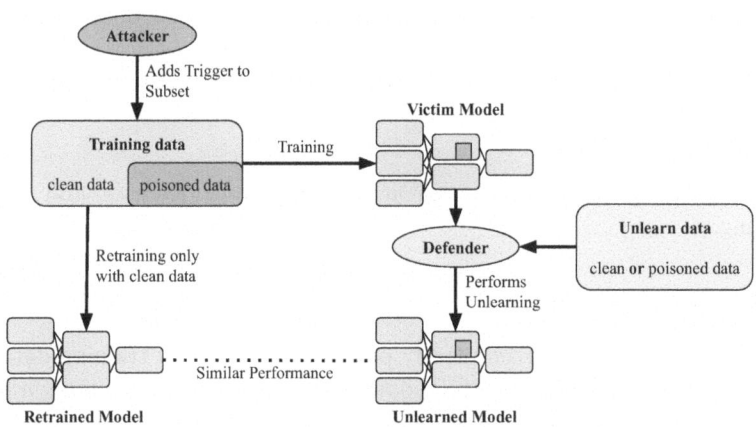

Fig. 1. Summary of backdoor unlearning setting.

This work focuses on a realistic scenario where the defender cannot access the dataset used to train the victim model. Figure 1 shows an overview of our unlearning setting. Unlearning aims to obtain a model that performs similarly to a model retrained on a clean subset of training data for clean and poisoned data. For unlearning the backdoor, we have access only to a limited domain-equivalent unseen dataset, the length of which is an order of magnitude smaller than the original training dataset. In practice, collecting a fitting dataset with annotations is expensive and time-consuming. Methods that are effective with an even smaller data count available are of particular interest for this scenario because they allow secure hand-picking of clean data for unlearning.

We propose a novel backdoor-unlearning approach that uses information from an extracted activation to guide the editing of model weights. This process aims to mitigate the influence of backdoor samples in the training dataset. Editing the weights is beneficial because it allows us to selectively target and repair the parts compromised by the attack. In addition to directly editing the weights, unlearning benefits from optionally allowing the parameters of the Batch Normalization (BN) layer to be changed during activation extraction.

The contributions of this work are as follows:

- We propose a novel model-editing method for unlearning samples with backdoor triggers by utilizing the activation of clean or poisoned samples extracted for a backdoored model. The proposed method is time- and sample-efficient.
- We point out that the proposed unlearning might unlearn the targeted class, thus introducing an optional repair process to preserve utility while forgetting only the backdoor trigger.
- We conduct experiments under two scenarios (with or without knowledge of the backdoor trigger) with three state-of-the-art backdoor attacks on different models and datasets. We present the results with an analysis.

In the experiments, the proposed method can consistently outperform other baseline methods.

2   Related Work

This section briefly reviews backdoor attacks, backdoor defenses, and machine unlearning.

2.1   Backdoor Attacks

Backdoor attacks involve preemptively poisoning a subset of training data with a specific backdoor trigger pattern and a target label. During training, a neural network learns that images with a specified trigger correspond to a target class, thus introducing an additional adversarial task. During inference, the network works as usual on benign data. A malicious actor can activate the backdoor to manipulate model response, causing misbehavior, such as misclassifications.

There exist various types of backdoor attacks [29]. BadNets [14], the first backdoor attack, uses noticeable square patches as triggers. In contrast to visible triggers, for invisible triggers, poisoned images are indistinguishable from clean ones, as in [27,30]. Backdoor attacks with optimized triggers [34,50] are designed to be more effective and thus usually require fewer poisoned training samples. Moreover, a shared semantic part of the images can be used as a trigger [1,31] without manipulating the images and only changing the labels. In addition, instead of using a single trigger pattern, certain methods allow for varying sample-specific triggers [36]. Although the targeted label is usually for a single class, there are all-to-all attacks [15] that use different target labels. In this work, we use visible, invisible, and optimized triggers for our experiments, and examples of such triggers are shown in Fig. 2.

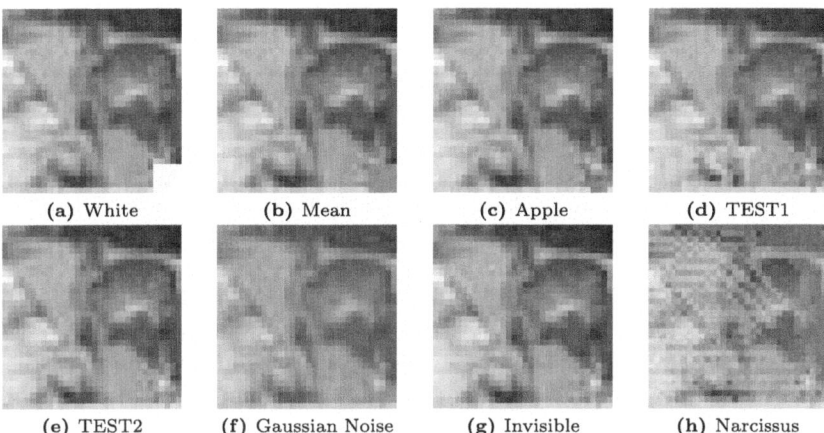

    (a) White   (b) Mean   (c) Apple   (d) TEST1

    (e) TEST2   (f) Gaussian Noise   (g) Invisible   (h) Narcissus

Fig. 2. Examples of eight backdoor triggers on CIFAR10. Images (a)–(f) are poisoned by BadNets [14] with different patches, (g) is with Steganography [27], and (h) is with Narcissus [50].

2.2 Backdoor Defenses

With the development of backdoor attacks, researchers have proposed various backdoor defenses as countermeasure [29]. Most defense methods should instead be considered mitigation methods because, in most cases, they cannot entirely erase the influence of the attacks. One such mitigation is data pre-processing prior to model inference, which aims to perturb the trigger to not activate the backdoor [7,35]. Another form of defense, typically used to aid other defense methods, is trigger synthesis [45]. With reverse engineering, trigger synthesis approximates the trigger, which can be used for trigger-guided defense or to retrieve the target class. Model diagnosis [22,49] is another type of defense to detect a backdoor and prevent model deployment. Moreover, poison suppression defenses modify the training process to be robust against backdoor creation [9, 19]. Other methods utilize sample filtering to detect trigger images and remove them from the training set or decline them during model inference [5,10,44]. Some defense methods aim to remove the backdoor from an infected model by directly modifying the model [32,35]. In this work, we propose a method that directly edits model weights to erase the backdoor in a poisoned model. The editing uses the activation of clean or poisoned samples as a guide.

2.3 Machine Unlearning

The influence of specific training samples on a model can be mitigated with machine unlearning. This influence can be entirely removed by retraining the model from scratch without the data we want to forget. One limitation of this approach is the requirement for training data, which can be inaccessible or too big to filter out the data we want to forget. Another issue is the high computational and time resource expense associated with retraining [48].

Different machine unlearning methods try to evade some of those limitations [48]. One approach is data obfuscation [13,42], where the model is fine-tuned with additional obfuscated data that disturbs the functionality of the data we want to forget. Certain approaches require design choices prior to training, like multi-model-aggregation [3,17] or a transformation layer inserted between data and model [4]. For specific model manipulation methods, model weights can be shifted by an update value [11,16], replaced by new values [38,47], or pruned [2,46] and usually repaired with a subsequent fine-tuning step. The scope of the information targeted for unlearning can range from whole classes [39,42] to individual samples [13]. This work focuses on backdoor attacks and aims to unlearn the features of a backdoor trigger pattern learned by the victim model.

3   Methodology

We consider an adversarial-modified (backdoored) image classifier f_θ parameterized by θ, which is trained with a dataset D that is comprised of clean and backdoored data ($D = D_C \cup D_B$). Samples in D_B contain the backdoor trigger δ and have the target label y_t. D_B is usually a small fraction of a clean training set D_T with a budget ρ such that $|D_B| \leq \rho|D_T|$. f_θ takes an input image $x \in \mathcal{X}$ and $f_\theta(x)_i$ represents the probability that x corresponds to label $i \in \mathcal{Y}$. $\mathcal{X}$ is the input space, and $\mathcal{Y}$ is the label space. The predicted label $\hat{y}$ is obtained by using the arg max operation ($\arg\max_i f_\theta(x)_i$). Since f_θ is backdoored, f_θ works as normal on a clean input x_c (i.e. predicting $\hat{y}$) and predicts y_t for input x_b embedded with the backdoor trigger δ. We aim to unlearn D_B that f_θ does not predict y_t when given x_b. Here, we slightly abuse the notation and imply that f_θ is a deep neural network with multiple layers. Specifically, we consider a neural network with multiple blocks of convolutional layers with or without BN.

Given f_θ without having access to the training dataset D, we propose an activation-guided model editing approach to unlearn D_B under two assumptions: (1) we have Backdoor Knowledge (BDK), and (2) we do not have it (¬BDK). For both assumptions, we split the total weights of f_θ into two halves and add those layers corresponding to the weights of the second half to a layer list L, for which we want to edit the weights. We target those later layers because they have the highest proximity to the classification output. We do not want to edit the early layers associated with general low-level feature extraction [12]. The authors of other backdoor mitigation approaches also suggest that focusing the unlearning on the later layers improves performance [32,45]. First, we prepare an unlearning dataset D_U with the same distribution as the training dataset D.

However, D_U is not used in training f_θ. Our empirical experiments suggest that D_U can be as small as four samples.

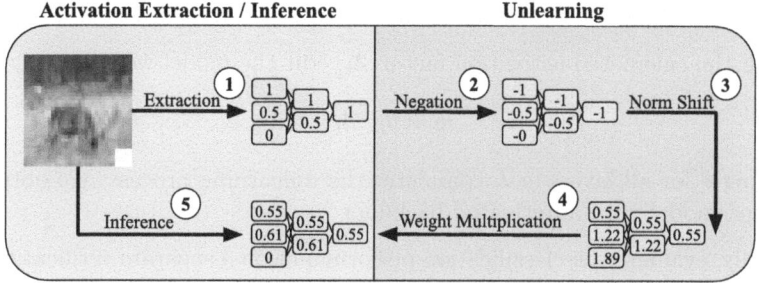

Fig. 3. Overview of proposed activation-guided model editing approach.

3.1 Assumption 1-BDK

As we assume we have backdoor information in this scenario, we poison D_U with the known backdoor trigger δ. In addition, for models with BN, we freeze Moving Average (MA) parameters during activation extraction because this was experimentally proven to be more effective for unlearning. Figure 3 shows an overview of the proposed model editing process, which works as follows.

(1) Extract model activation A for the whole dataset D_U. Therefore, iterate batch-wise over data X in D_U infected with trigger δ and forward it through model f_θ given $f_\theta(X+\delta)$. Capture and average the activations across all batches for each layer l in layer list L.

(2) Next, negate each value in the activation A as

$$A = -A. \tag{1}$$

Then, iterate over l in list L of the layers targeted for editing. For each layer l, check if the later multiplication of weights θ_l and activation A_l would result in a matrix-multiplication-error caused by shape mismatch. If this is the case, use adaptive average pooling to adjust the shape of A_l to match the shape of θ_l.

(3) Then, compute layer-wise mean and standard deviation statistics as

$$\mu_l = \frac{1}{m}\sum_{i=1}^{m} a_i, \quad \sigma_l^2 = \frac{1}{m}\sum_{i=1}^{m}(a_i - \mu_l)^2, \tag{2}$$

where $a_i \in A_l$. Then, use the calculated statistics to normalize activation A_l for each layer l as

$$A_l = \frac{A_l - \mu_l}{\sqrt{\sigma_l^2 + \epsilon}}, \tag{3}$$

where ϵ is a small value to avoid division by zero. Next, rescale A_l with scale and shift hyperparameters γ and λ as

$$A_l = \gamma \cdot A_l + \lambda. \tag{4}$$

④ With the calculated activation factor A_l, edit the model weights θ_l as

$$\theta_l = \theta_l \cdot A_l. \tag{5}$$

Modifying θ for all layers in L completes the unlearning process. We obtain an unlearned model with a mitigated backdoor.

⑤ Finally, evaluate the classification performance on a separate verification test set D_V to evaluate the unlearning process.

In Eq. 4, we scale and shift the activation with values 0.5 and 1.0 for γ and λ hyperparameters, which changes the activation mean to 1.0. With this, we aim to preserve the utility of the model during unlearning and mitigate the internal covariate shift, especially after significant distribution changes by Eq. 1. Choosing those values should lead to minimal change in the inherent mean of the weights when multiplying with the shifted activation. We experimentally confirmed the performance with those values in the supplementary material.

After model editing, we observe that the model tends to unlearn the whole class instead of unlearning backdoored samples only. To address this issue, we introduce an optional repair phase to restore some model utility if we have unlearned more than intended. Specifically, we fine-tune the model on the samples in D_U and on the same samples poisoned with δ for one epoch each, using the correct ground truth as the target label in both cases.

3.2   Assumption 2-¬BDK

In this scenario, we do not have any information about the backdoor trigger or algorithm (¬BDK). Therefore, we cannot poison the unlearning dataset D_U and use the clean D_U as it is in the unlearning process. For model editing with ¬BDK, we perform the process in Fig. 3 with two modifications: (1) backdoor trigger δ is zero as we do not have any information about it, and (2) we update MA parameters during activation extraction to aid unlearning of dataset D_U for models with BN layers, unlike the unlearning process with BDK. We perform the optional repairing by fine-tuning on the clean D_U set.

4   Experiments

In this section, we perform various experiments to show the effectiveness of our method in different settings and compare it with other existing backdoor unlearning methods. We conducted all experiments three times, and the averaged results are summarized as follows.

Table 1. Evaluation of proposed unlearning on five different datasets. Additional repairing was performed with learning rate value of 1e−2. We compare base state after unlearning with one after repairing. We used Top-5 accuracy for CIFAR100 and TinyImageNet. Best results are highlighted in bold.

Dataset (ASR/ACC)	BDK	State	Metric ASR (↓)	ACC (↑)	CTCA (↑)
MNIST (99.99/98.98)	✓	Base	0.0±0.0	86.04±1.2	0.0±0.0
		+repair	0.41±0.16	**97.77±0.37**	**96.09±2.51**
	✗	Base	66.81±5.87	77.37±2.84	59.19±11.56
		+repair	67.3±12.4	**97.31±1.27**	**98.13±0.31**
CIFAR10 (94.53/70.5)	✓	Base	0.0±0.0	**65.49±0.39**	0.0±0.0
		+repair	9.79±0.88	59.87±3.04	**42.07±7.72**
	✗	Base	**0.3±0.21**	59.18±2.63	0.14±0.19
		+repair	23.09±8.07	61.83±1.33	**55.02±4.41**
CIFAR100 (93.84/63.06)	✓	Base	0.92±0.33	50.68±1.36	3.91±2.8
		+repair	4.13±1.61	**60.42±0.96**	**41.27±10.27**
	✗	Base	**0.0±0.0**	37.88±2.19	0.0±0.0
		+repair	20.3±14.16	**60.83±0.81**	**58.78±12.86**
CINIC10 (96.28/57.35)	✓	Base	0.0±0.0	**53.08±0.49**	0.0±0.0
		+repair	0.0±0.0	10.0±0.0	0.0±0.0
	✗	Base	0.01±0.01	**41.89±0.83**	0.08±0.1
		+repair	0.0±0.0	10.0±0.0	0.0±0.0
TinyImageNet (95.25/45.21)	✓	Base	0.48±0.32	34.21±1.0	1.59±2.24
		+repair	0.86±0.33	**44.33±0.16**	4.14±3.75
	✗	Base	0.0±0.0	20.26±1.87	0.0±0.0
		+repair	0.72±0.6	**44.03±0.21**	**7.72±5.71**

4.1 Setup

Datasets. We explored our method on MNIST [26], CIFAR10, CIFAR100 [23], CINIC10 [6], and TinyImageNet [25].

Models. We used ResNet18 [18], VGG16 [40], EfficientNetV2-S [41], and small MobileNetV3 [20].

Backdoors. We considered eight different triggers applied with three state-of-the-art attack methods: six different patch triggers with BadNets [14], an invisible trigger with Steganography [27], and an optimized trigger with Narcissus [50]. Among the methods, Narcissus is the only one that solely infects target class samples, thus making it more stealthy without requiring a label change. Examples of the applied triggers are visualized in Fig. 2.

Baselines. We considered four suitable backdoor unlearning methods for comparison, two of which require BDK: (1) fine-tuning, which penalizes a high difference between activations of clean and poisoned data (actFT) [37], and (2)

Table 2. Evaluation of proposed unlearning on different models. Additional repairing was performed with learning rate value of $1e-2$. We compare base state after unlearning with one after repairing. Best results are highlighted in bold.

Model	BDK	State	Metric		
(ASR/ACC)			ASR ($\downarrow$)	ACC ($\uparrow$)	CTCA ($\uparrow$)
	✓	Base	**0.0±0.0**	**65.49±0.39**	0.0±0.0
ResNet18		+repair	9.79±0.88	59.87±3.04	**42.07±7.72**
(94.53/70.5)	✗	Base	**0.3±0.21**	59.18±2.63	0.14±0.19
		+repair	23.09±8.07	61.83±1.33	**55.02±4.41**
	✓	Base	0.0±0.0	71.23±0.38	0.0±0.0
VGG16		+repair	4.36±1.15	70.15±2.71	**65.5±4.69**
(96.25/78.39)	✗	Base	**0.0±0.0**	70.13±1.74	0.0±0.0
		+repair	92.17±1.17	72.23±0.32	**57.83±7.3**
	✓	Base	0.0±0.0	**47.49±2.79**	0.0±0.0
EfficientNetV2-S		+repair	6.82±2.96	31.51±5.45	**7.94±3.66**
(89.74/48.83)	✗	Base	0.0±0.0	11.07±1.5	0.0±0.0
		+repair	1.71±2.02	**31.41±6.19**	4.6±4.35
MobileNetV3	✓	Base	**0.0±0.0**	**55.8±1.84**	0.0±0.0
(small)		+repair	26.28±7.53	50.63±3.63	**49.18±11.63**
(95.67/62.3)	✗	Base	6.09±4.54	54.35±2.93	4.19±2.66
		+repair	84.69±6.09	51.05±3.68	**46.51±15.49**

BaEraser, which uses gradient ascent for unlearning [33]. The other two methods work with ¬BDK: (3) fine-tuning on clean D_U in the same way as the initial training (basicFT), and (4) Neural Attention Distillation (NAD), a knowledge distillation approach where the basicFT model, acting as the teacher model, only passes on its ability to clean data [28].

Evaluation Metrics. For evaluation, we used the Attack Success Rate (ASR), which is the ratio of a backdoor sample being misclassified as y_t, Clean Test Accuracy (ACC), and Clean Target Class Accuracy (CTCA), which is the ACC for samples of class y_t. We were interested in examining the change in CTCA because our unlearning method often leads to a drop in CTCA alongside ASR. Repairing can mitigate this side effect.

We introduce a two-part scoring function to estimate the forgetting and utility quality after unlearning combined in one value. The forgetting quality is estimated by subtracting the ASR ratio of the unlearned (U) and victim model (V) from 1. A higher drop in ASR after unlearning indicates a higher score for the forgetting part. The ACC ratio of the unlearned and the retrained model (R), which is trained on D_C from scratch, estimates the utility part. We strive to achieve the same or even higher ACC on the unlearned model compared to the retrained model that was never poisoned before. We use the retrained model for this ratio because, especially in cases with a high poisoning rate ρ, the poisoning

Table 3. Score (↑) comparison of proposed method with state-of-the-art unlearning methods on different infected backdoors. For backdoor verification, *alpha* value is multiplied by three up to maximum of 100%. Best results for each trigger are highlighted in bold.

Infected Trigger	poisoned (ASR/ACC)	ρ	alpha	BDK actFT [37]	BaEraser [33]	Ours
(a) White [14]	(94.53/70.5)	5%	1.0	**94.62±2.61**	59.5±4.62	93.02±0.66
(b) Mean	(88.49/68.71)	10%	1.0	61.81±6.09	70.85±7.33	**95.23±1.79**
(c) Apple	(99.23/69.63)	5%	1.0	85.21±11.5	62.87±11.29	**92.36±2.11**
(d) TEST1	(99.75/70.73)	5%	1.0	90.69±2.43	36.99±8.55	**92.19±0.87**
(e) TEST2	(99.99/69.66)	5%	0.15	88.3±1.19	48.57±19.28	**91.31±1.41**
(f) Gaussian Noise	(85.77/68.19)	5%	0.25	52.97±2.19	89.6±1.56	**93.57±0.35**
(g) Invisible [27]	(97.97/61.4)	50%	-	84.99±3.83	40.51±12.39	**92.24±2.38**
(h) Narcissus [50]	(99.32/63.72)	5%	0.2	4.09±1.61	51.35±14.04	**95.45±3.1**

Table 4. Score (↑) comparison of proposed method with state-of-the-art unlearning methods on different infected backdoors. For backdoor verification, *alpha* value is multiplied by three up to maximum of 100%. Best results for each trigger are highlighted in bold.

Infected Trigger	poisoned (ASR/ACC)	ρ	alpha	¬BDK basicFT	NAD [28]	Ours
(a) White [14]	(94.53/70.5)	5%	1.0	61.9±3.63	65.33±3.04	**83.73±3.29**
(b) Mean	(88.49/68.71)	10%	1.0	50.01±25.7	71.79±2.52	**89.95±1.73**
(c) Apple	(99.23/69.63)	5%	1.0	69.1±2.7	70.86±0.15	**80.41±3.78**
(d) TEST1	(99.75/70.73)	5%	1.0	31.44±23.82	62.26±3.38	**83.4±3.89**
(e) TEST2	(99.99/69.66)	5%	0.15	48.56±14.03	68.17±4.35	**82.47±0.65**
(f) Gaussian Noise	(85.77/68.19)	5%	0.25	41.92±22.62	31.19±13.93	**82.81±0.77**
(g) Invisible [27]	(97.97/61.4)	50%	-	58.66±14.52	73.7±1.1	**87.26±0.58**
(h) Narcissus [50]	(99.32/63.72)	5%	0.2	38.52±16.51	**75.75±1.38**	17.73±11.77

can negatively influence the ACC of the victim model, thus not representing a clean model performance. The final score value is calculated as

$$\text{Score} = (1 - \frac{\text{ASR}^U}{\text{ASR}^V}) \cdot \frac{\text{ACC}^U}{\text{ACC}^R}. \tag{6}$$

Base Configuration. We used specific base configurations if not stated otherwise for an experiment. Experiments were performed on the CIFAR10 dataset and ResNet18 as the victim model. The training dataset D_T was infected with a poisoning rate ρ of 5%, with the trigger displayed in Fig. 2a. The backdoor tar-

Table 5. Performance of proposed unlearning when using different numbers of samples for unlearning. Results represent the model state after unlearning without repairing. Best results are highlighted in bold.

Number of samples	BDK			¬BDK		
	ASR (↓)	ACC (↑)	CTCA (↑)	ASR (↓)	ACC (↑)	CTCA (↑)
2	0.0±0.0	63.92±1.91	0.0±0.0	24.6±17.84	61.1±1.94	**6.48±9.02**
4	0.0±0.0	65.18±0.91	0.0±0.0	**0.92±1.15**	60.77±2.25	0.0±0.0
8	0.0±0.0	65.14±0.65	0.0±0.0	5.76±6.26	60.19±2.96	0.0±0.0
16	0.0±0.0	65.17±0.68	0.0±0.0	4.82±6.75	60.77±2.91	0.0±0.0
32	0.0±0.0	65.47±0.44	0.0±0.0	9.26±12.93	**63.77±0.58**	0.07±0.1
64	0.0±0.0	64.7±1.45	0.0±0.0	7.63±8.32	61.9±2.22	1.94±2.74
128	0.0±0.0	65.53±0.45	0.0±0.0	4.38±5.58	61.67±1.22	0.07±0.1
256	0.0±0.0	65.51±0.41	0.0±0.0	**0.97±1.02**	58.86±2.42	0.2±0.16
512	0.0±0.0	65.49±0.39	0.0±0.0	**0.3±0.21**	59.18±2.63	0.14±0.19
5000	0.0±0.0	64.57±1.53	0.0±0.0	10.48±11.24	61.41±0.82	**6.75±9.54**

get class y_t was two, representing birds. Unlearn dataset D_U consisted of 5000 samples, but our method only used 512 by default.

4.2 Results

We examined the performance of our unlearning in different settings.

Different Datasets. In this experiment, we trained ResNet18 models poisoned with backdoor triggers on five datasets: MNIST, CIFAR10, CIFAR100, CINIC10, and TinyImageNet. Table 1 summarizes the evaluation of the proposed unlearning method with the different datasets in terms of ASR, ACC, and CTCA. The proposed method effectively reduced the ASR on every dataset, except MNIST (grayscale images), when we had ¬BDK. Repairing improved ACC for several datasets and restored CTCA while increasing ASR by a lesser extent. There was an exclusively negative influence on performance with CINIC10 repairing.

Different Models. In this experiment, we trained different models: ResNet18, VGG16, EfficientNetV2-S, and MobileNetV3 (small version). Table 2 presents the performance of the proposed unlearning method with the different models. The proposed method with BDK was effective on every tested model. After unlearning, we retained a good ACC on EfficientNetv2 with BDK, while the utility was lost with ¬BDK. However, repairing both models resulted in similar final performance, which benefited ¬BDK but decreased performance for BDK.

Comparison with State-of-the-Art Methods. In this experiment, we trained models and performed unlearning with different backdoor triggers. As described in Sect. 4.1, we considered four baseline methods: actFT and BaEraser under BDK, and basicFT and NAD under ¬BDK with eight poison triggers for comparison. The models were trained with different poisoning budgets ρ and *alpha* values of the RGBA-coded trigger. Figure 2 depicts the triggers.

Table 6. Efficiency of proposed unlearning compared with state-of-the-art methods. Experimented with 50%(5000), 5%(500), 0.5%(50), and 0.05%(5) of unseen CIFAR10 data for unlearning. Table displays only sample runs with highest and second-highest scores. Full table is displayed in supplementary material. Best results are highlighted in bold.

Method	Number of Samples	Metric Score (↑)	ASR (↓)	ACC (↑)	CTCA (↑)	Time (↓)
actFT [37]	5000	**92.95±4.72**	5.28±2.48	**69.64±0.92**	58.1±9.1	3.96±0.34
	500	7.0±2.09	87.17±2.86	69.41±1.02	61.47±7.99	2.81±0.04
BaEraser [33]	500	80.69±1.93	3.81±2.41	59.45±0.83	32.98±1.1	138.88±2.87
	5000	57.47±15.65	2.0±2.79	41.81±12.3	14.79±20.77	623.3±111.46
Ours(BDK)	50	**92.4±1.39**	**0.0±0.0**	65.29±0.58	0.0±0.0	**0.38±0.01**
	5	92.51±1.34	0.0±0.0	65.37±0.56	0.0±0.0	0.38±0.02
basicFT	5000	69.49±1.11	6.0±1.07	52.57±0.91	**33.86±4.07**	71.56±1.47
	5	14.17±0.11	0.0±0.0	10.01±0.0	0.0±0.0	71.49±0.03
NAD [28]	5000	71.23±3.2	4.54±1.48	52.95±1.41	32.05±1.3	114.9±1.45
	500	42.18±5.42	6.68±2.06	32.09±3.97	23.18±9.4	117.69±1.9
Ours(¬BDK)	50	**89.9±2.02**	**0.0±0.0**	**63.52±1.24**	0.0±0.0	**0.36±0.02**
	5	89.5±1.72	0.0±0.0	63.23±0.86	0.0±0.0	0.4±0.01
Retraining	47500	-	4.03±0.99	70.27± 0.66	60.55±4.3	423.56±73.96

Tables 3 and 4 summarize the performance of the proposed unlearning method with the different baseline methods in terms of score (see Sect. 4.1). The score metric measured the forgetting and utility quality after unlearning. For backdoor verification, the *alpha* value was multiplied by three up to a maximum of 100%. For actFT to be effective, we multiplied the *alpha* value for unlearning by the same magnitude. Our method outperformed the previous methods in terms of score with or without BDK for most triggers.

4.3   Analysis

We analyze the proposed unlearning method in terms of sample efficiency, time efficiency, and potential backdoor detection application.

Sample Efficiency. Table 5 shows the performance of the proposed unlearning method when using different numbers of samples for unlearning. With BDK, the performance did not depend on the sample count. With ¬BDK, the performance with different sample counts did not follow a clear pattern. Notably, the ASR with two samples was exceptionally high compared with others. Therefore, we recommend using a minimum of four samples for unlearning with ¬BDK.

Table 6 presents a performance comparison of the proposed unlearning and state-of-the-art methods in terms of several metrics, including the time required for unlearning. The baseline methods compared with ours depended more on a high sample count in D_U. For most of the baselines, more samples resulted in a higher score. An exception is BaEraser, which had the best performance with 500 samples.

Time Efficiency. In the particular scenario where training data is available and retraining is feasible, assessing the computational cost saved with unlearning compared with retraining is an important metric. When unlearning is not drastically more time efficient, retraining is the preferred choice to perfectly remove the influence of the data we want to forget. The unlearning time in our scenario with training data unavailability is not a deciding factor. Still, we have to consider the trade-off between unlearning performance and the cost of computing for the benefit of scalability.

As evident in Table 6, our method requires significantly less time and fewer samples for unlearning than other methods. Our method uses only a single forward pass to extract the activation, and the remaining operations are simple matrix operations. In comparison, all baseline methods require optimization with backpropagation, which generally is more computationally expensive, resulting in a higher unlearning time.

Target Class Detection. Our experiments show that the proposed unlearning method reduced the backdoor class accuracy (CTCA). To address this issue, we introduce a repair step to preserve utility. Before repairing, we can utilize significant decreases in target class accuracy with ¬BDK to detect a backdoor and the target class. We carried out a simple experiment on poisoned models on all ten classes of CIFAR10. We can usually observe an unusual decrease in accuracy for a single class. When we assumed the single class as the target class, we got a target class prediction accuracy of 80%. A formula sets the accuracy of the different classes into relation and returned a classification value. Comparing the value to a threshold value gives us a binary prediction for the existence of a backdoor. The backdoor detection accuracy was 67% when poisoned and 80% when having a clean model.

5   Discussion

We demonstrated a model-editing method that unlearns the backdoor trigger feature embedded in a backdoored model by utilizing the activation of clean or poisoned samples. Our method achieves consistent unlearning performance across various settings with different models, datasets, and backdoor triggers by state-of-the-art attacks. Apart from the unlearning performance, there are two key factors where our method exceeds current state-of-the-art methods by a significant margin. Our unlearning process is exceptionally fast to compute and, most of the time, requires only a handful of samples to unlearn the backdoor effectively. Additionally, we can use information gained after unlearning for backdoor presence and target class prediction.

We experimented with our algorithm and found specific activation-manipulating formulas that gave us the best unlearning performance for model editing. In Eq. 1, negating poisoned activation with BDK and clean activation with ¬BDK worked the best. With BDK, we negate the activation of the trigger-infected data we want to forget. Previously, Ilharco et al. [21] arrived at the same conclusion as we did, that moving in the negative direction of extracted information can lead to unlearning.

The most significant limitation of our method is that it disturbs the overall utility and unlearns the targeted class instead of only backdoor samples. Therefore, repairing is used to restore lost utility. The experimental scope was limited, and we covered only convolutional neural networks.

Hence, for future work, we shall explore the unlearning method with different architectures, such as vision transformers [8], mixers [43], *etc*. We shall investigate explainability methods to better understand the parts of the algorithm that are responsible for effective unlearning and ideally improve the unlearning performance without loss of utility. In addition, not limiting the method to backdoor unlearning, we shall expand the applications of unlearning, such as privacy-related unlearning applications. In this work, we analyzed backdoors in images, but for future work, we shall expand experiments to other data types, such as text or audio data.

6   Conclusion

Our method offers a new approach to tackling the security issue posed by backdoor attacks by mitigating the influence of attacks on a backdoor-infected model without requiring access to the original training data. Multiple experiments show the broad applicability of our method in various settings. It performs better than previous backdoor unlearning methods in most scenarios. Moreover, it executes faster and requires fewer samples for unlearning than the previous methods.

Acknowledgements. This work was partially supported by JSPS KAKENHI Grants JP21H04907, 23K19983, and JP24H00732, by JST CREST Grants JPMJCR18A6 and JPMJCR20D3 including AIP challenge program, by JST AIP Acceleration Grant JPMJCR24U3, by JST K Program Grant JPMJKP24C2 Japan, and by the project for the development and demonstration of countermeasures against disinformation and misinformation on the Internet with the Ministry of Internal Affairs and Communications of Japan.

References

1. Bagdasaryan, E., Veit, A., Hua, Y., Estrin, D., Shmatikov, V.: How to backdoor federated learning. In: International Conference on Artificial Intelligence and Statistics, pp. 2938–2948. PMLR (2020)
2. Baumhauer, T., Schöttle, P., Zeppelzauer, M.: Machine unlearning: Linear filtration for logit-based classifiers. Mach. Learn. **111**(9), 3203–3226 (2022)
3. Bourtoule, L., et al.: Machine unlearning (2020)
4. Cao, Y., Yang, J.: Towards making systems forget with machine unlearning. In: 2015 IEEE Symposium on Security and Privacy, pp. 463–480. IEEE (2015)
5. Chen, B., et al.: Detecting backdoor attacks on deep neural networks by activation clustering (2018)

6. Darlow, L.N., Crowley, E.J., Antoniou, A., Storkey, A.J.: CINIC-10 is not ImageNet or CIFAR-10. arXiv preprint arXiv:1810.03505 (2018)
7. Doan, B.G., Abbasnejad, E., Ranasinghe, D.C.: Februus: Input purification defense against trojan attacks on deep neural network systems. In: Proceedings of the 36th Annual Computer Security Applications Conference, pp. 897–912 (2020)
8. Dosovitskiy, A., et al.: An image is worth 16x16 words: transformers for image recognition at scale (2021)
9. Du, M., Jia, R., Song, D.: Robust anomaly detection and backdoor attack detection via differential privacy. arXiv preprint arXiv:1911.07116 (2019)
10. Gao, Y., Xu, C., Wang, D., Chen, S., Ranasinghe, D.C., Nepal, S.: STRIP: a defence against trojan attacks on deep neural networks (2020)
11. Golatkar, A., Achille, A., Soatto, S.: Eternal sunshine of the spotless net: selective forgetting in deep networks. In: Proceedings of the IEEE/CVF Conference on Computer Vision and Pattern Recognition, pp. 9304–9312 (2020)
12. Goodfellow, I.J., Bengio, Y., Courville, A.: Deep Learning. MIT Press, Cambridge, MA, USA (2016). http://www.deeplearningbook.org
13. Graves, L., Nagisetty, V., Ganesh, V.: Amnesiac machine learning (2020)
14. Gu, T., Dolan-Gavitt, B., Garg, S.: BadNets: identifying vulnerabilities in the machine learning model supply chain (2019)
15. Gu, T., Liu, K., Dolan-Gavitt, B., Garg, S.: BadNets: evaluating backdooring attacks on deep neural networks. IEEE Access 7, 47230–47244 (2019)
16. Guo, C., Goldstein, T., Hannun, A., Van Der Maaten, L.: Certified data removal from machine learning models. arXiv preprint arXiv:1911.03030 (2019)
17. Gupta, V., Jung, C., Neel, S., Roth, A., Sharifi-Malvajerdi, S., Waites, C.: Adaptive machine unlearning. Adv. Neural. Inf. Process. Syst. 34, 16319–16330 (2021)
18. He, K., Zhang, X., Ren, S., Sun, J.: Deep residual learning for image recognition (2015)
19. Hong, S., Chandrasekaran, V., Kaya, Y., Dumitraş, T., Papernot, N.: On the effectiveness of mitigating data poisoning attacks with gradient shaping. arXiv preprint arXiv:2002.11497 (2020)
20. Howard, A., et al.: Searching for MobileNetV3 (2019)
21. Ilharco, G., et al.: Editing models with task arithmetic (2023)
22. Kolouri, S., Saha, A., Pirsiavash, H., Hoffmann, H.: Universal litmus patterns: revealing backdoor attacks in CNNs (2020)
23. Krizhevsky, A., Hinton, G., et al.: Learning multiple layers of features from tiny images, Master's thesis, University of Tront (2009)
24. Kumar, R.S.S., et al.: Adversarial machine learning-industry perspectives. In: 2020 IEEE Security and Privacy Workshops (SPW), pp. 69–75. IEEE (2020)
25. Le, Y., Yang, X.: Tiny ImageNet visual recognition challenge. CS 231N 7(7), 3 (2015)
26. LeCun, Y., Bottou, L., Bengio, Y., Haffner, P.: Gradient-based learning applied to document recognition. Proc. IEEE 86(11), 2278–2324 (1998)
27. Li, S., Xue, M., Zhao, B.Z.H., Zhu, H., Zhang, X.: Invisible backdoor attacks on deep neural networks via steganography and regularization (2020)
28. Li, Y., Lyu, X., Koren, N., Lyu, L., Li, B., Ma, X.: Neural attention distillation: erasing backdoor triggers from deep neural networks (2021)
29. Li, Y., Jiang, Y., Li, Z., Xia, S.T.: Backdoor learning: a survey (2022)
30. Li, Y., Li, Y., Wu, B., Li, L., He, R., Lyu, S.: Invisible backdoor attack with sample-specific triggers. In: Proceedings of the IEEE/CVF International Conference on Computer Vision, pp. 16463–16472 (2021)

31. Lin, J., Xu, L., Liu, Y., Zhang, X.: Composite backdoor attack for deep neural network by mixing existing benign features. In: Proceedings of the 2020 ACM SIGSAC Conference on Computer and Communications Security, pp. 113–131 (2020)
32. Liu, K., Dolan-Gavitt, B., Garg, S.: Fine-pruning: defending against backdooring attacks on deep neural networks (2018)
33. Liu, Y., et al.: Backdoor defense with machine unlearning (2022)
34. Liu, Y., et al.: Trojaning attack on neural networks. In: Network and Distributed System Security Symposium (2018). https://api.semanticscholar.org/CorpusID: 31806516
35. Liu, Y., Xie, Y., Srivastava, A.: Neural trojans (2017)
36. Nguyen, T.A., Tran, A.: Input-aware dynamic backdoor attack. Adv. Neural. Inf. Process. Syst. **33**, 3454–3464 (2020)
37. Qiao, X., Yang, Y., Li, H.: Defending neural backdoors via generative distribution modeling (2019)
38. Schelter, S., Grafberger, S., Dunning, T.: HedgeCut: maintaining randomised trees for low-latency machine unlearning. In: Proceedings of the 2021 International Conference on Management of Data, pp. 1545–1557 (2021)
39. Shibata, T., Irie, G., Ikami, D., Mitsuzumi, Y.: Learning with selective forgetting. IJCAI **3**, 4 (2021)
40. Simonyan, K., Zisserman, A.: Very deep convolutional networks for large-scale image recognition (2015)
41. Tan, M., Le, Q.V.: EfficientNetV2: smaller models and faster training (2021)
42. Tarun, A.K., Chundawat, V.S., Mandal, M., Kankanhalli, M.: Fast yet effective machine unlearning. IEEE Trans. Neural Netw. Learn. Syst. (2023)
43. Tolstikhin, I., et al.: MLP-mixer: an all-MLP architecture for vision (2021)
44. Tran, B., Li, J., Madry, A.: Spectral signatures in backdoor attacks (2018)
45. Wang, B., et al.: Neural cleanse: identifying and mitigating backdoor attacks in neural networks. In: 2019 IEEE Symposium on Security and Privacy (SP), pp. 707–723. IEEE (2019)
46. Wang, J., Guo, S., Xie, X., Qi, H.: Federated unlearning via class-discriminative pruning. In: Proceedings of the ACM Web Conference 2022, pp. 622–632 (2022)
47. Wu, Y., Dobriban, E., Davidson, S.B.: DeltaGrad: rapid retraining of machine learning models (2020)
48. Xu, H., Zhu, T., Zhang, L., Zhou, W., Yu, P.S.: Machine unlearning: a survey (2023)
49. Xu, X., Wang, Q., Li, H., Borisov, N., Gunter, C.A., Li, B.: Detecting ai trojans using meta neural analysis (2020)
50. Zeng, Y., Pan, M., Just, H.A., Lyu, L., Qiu, M., Jia, R.: Narcissus: a practical clean-label backdoor attack with limited information. In: Proceedings of the 2023 ACM SIGSAC Conference on Computer and Communications Security, pp. 771–785 (2023)

Exploring Visual Multiple-Choice Question Answering with Pre-trained Vision-Language Models

Gia-Nghia Tran[1,4(✉)] ⓘ, Duc-Tuan Luu[1,2,3,4] ⓘ, and Dang-Van Thin[1,4] ⓘ

[1] University of Information Technology, VNU-HCM, Ho Chi Minh City, Vietnam
{nghiatg,tuanld,thindv}@uit.edu.vn
[2] University of Science, VNU-HCM, Ho Chi Minh City, Vietnam
[3] John von Neumann Institute, VNU-HCM, Ho Chi Minh City, Vietnam
[4] Vietnam National University, Ho Chi Minh City, Vietnam

Abstract. Visual question answering is a challenging task in computer vision and natural language processing that involves answering questions about an image using both visual and textual information. This task is more challenging when it comes to the Japanese language since there is a lack of research focus on Japanese compared to extensive studies for English and other languages. The ACCV Workshop on Large Vision - Language Model Learning and Applications (LAVA) has organized an interesting challenge that aims at benchmarking different systems on the multiple-choice visual question answering task across both Japanese and English. In this paper, we present a simple yet effective approach that competes in this LAVA Workshop Challenge. To provide a correct answer, our proposed framework needs to (1) Identify entities and understand the visual concepts and the underlying spatial relations in the image referred to in the question, (2) Align the multimedia representations of the visual content with the multiple-choice answers to determine the most accurate response. We believe that the size of the vision-language model affects the overall performance of the proposed system.

Keywords: LAVA Workshop Challenge · Vision-Language Model · Visual Question Answering

1 Introduction

Recent developments in the domains of artificial intelligence and machine learning have gained significant attention, specifically towards the advancement of large language models (LLMs) [1,2,17,36,42]. These sophisticated models have demonstrated exceptional capabilities in the processing and interpretation of extensive amounts of textual data, leading to impressive performance across a wide range of natural language processing (NLP) tasks. The success of these

M. Cho et al. (Eds.): ACCV 2024 Workshops, LNCS 15482, pp. 324–337, 2025.
https://doi.org/10.1007/978-981-96-2641-0_22

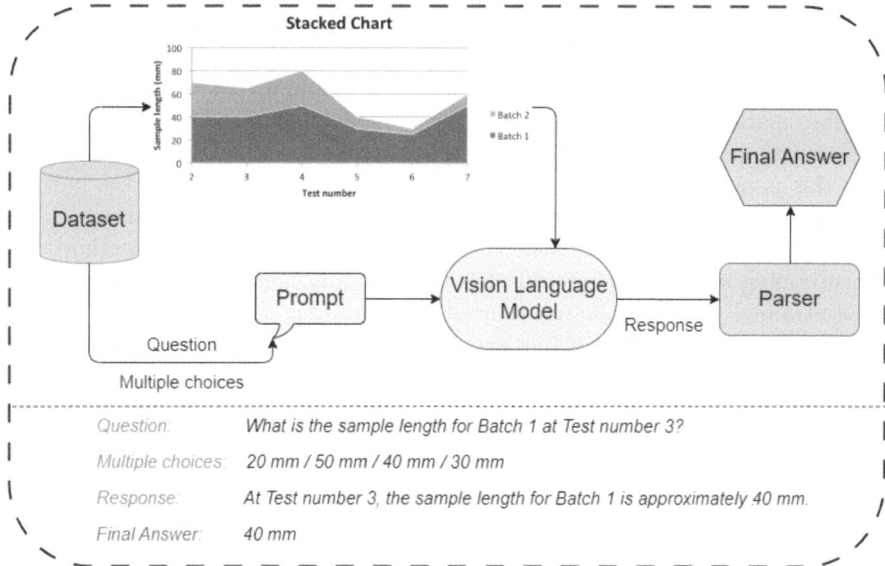

Fig. 1. Overview of our competition approach. For each sample, an image and the corresponding prompt are fed into a VLM. The model's response then goes through the parser as the post-processing step to obtain the final answer.

models has fostered a growing interest in extending their applicability beyond textual information to incorporate other modalities, including visual and auditory data, as well as the development of multi-modal inputs. This paradigm shift has led to the emergence of large vision-language models (VLMs), which focus on combining the strengths of both NLP and computer vision. Such integration aims to facilitate a more comprehensive and nuanced understanding of the world, bridging the gap between different forms of data.

A particular aspect in this field of research is visual question answering (VQA) [6,20,23]. In VQA tasks, models are requested to analyze images and subsequently answer questions regarding the visual content. This also consists of various formats, including multiple-choice VQA, in which the system is presented with a specific question alongside several potential answer options related to the visual input. The Large Vision - Language Model Learning and Applications (LAVA) workshop challenge is a newly organized competition that promotes comparative benchmarking of different multiple-choice VQA systems across complex visual data, such as diagrams, Gantt charts, building designs, and drawings.

In this challenge, participants are presented with multiple pairs of image-question. For each pair, each team must identify the most accurate answer from a set of four distinct natural language options. The dataset for this competition covers a wide range of world knowledge, featuring visual cues that present complex information (charts, diagrams, posters, drawings, etc.). Moreover, the competition incorporates both English and Japanese languages in its photos and question-answers set, adding an extra layer of complexity. The unique characteristics of the Japanese language, such as its complex writing system and rich cultural context, pose specific challenges for this multiple-choice VQA task. Therefore, participants are required to have not only sharp analytical skills but also linguistic proficiency and knowledge across multiple domains.

In this paper, we introduce our competition approach, which takes part in the LAVA 2024 challenge. To be specific, we evaluate various pre-trained VLMs and describe our strategies to achieve our competitive results. Our method can be easily integrated in a plug-and-play manner, allowing it to work with various pre-trained VLMs without the need for retraining or fine-tuning. Figure 1 demonstrates the workflow of our proposed approach. In general, each pair of image-prompt is passed through a VLM to obtain the response. It is filtered to achieve the most accurate multiple-choice answer.

2   Related Works

2.1   Multi-modal Pre-training

The integration of multiple modalities, such as vision and language, has become increasingly prominent in artificial intelligence research. Multi-modal pre-training aims to learn unified representations from diverse data types, enabling models to perform tasks that require understanding across different modalities. This approach leverages large-scale datasets to capture the complementary information inherent in each modality, leading to enhanced performance in downstream tasks like image captioning, visual question answering, and cross-modal retrieval.

Early works in multi-modal learning focused on task-specific architectures. For instance, VQA models [11,29] were designed to answer questions about images by combining vision encoder and text encoder. However, these models often suffered from limited generalization due to their reliance on task-specific data and architectures. The advent of transformer architectures [43] and the success of large-scale language models spurred the development of multi-modal transformers. ViLBERT [28] and LXMERT [39] extended the BERT model [19] to handle both visual and textual inputs by learning joint representations through co-attention mechanisms. CLIP [35] and ALIGN [16] employed contrastive learning techniques to align visual and textual representations in a shared embedding space. By training on large-scale image-text pairs collected from the internet, these models achieved remarkable zero-shot performance without explicit region-based features. Building upon this foundation, UNITER [7], OSCAR [24], and SimVLM [46] scale up pre-training data to better align visual and textual modalities. Additionally, CoCa [50] and Florence [52] integrated encoders with decoders, enabling both understanding and generation capabilities within a single, unified framework. More recently, significant progress has been made with large multi-modal models bridging vision and language at scale. GPT-4 [1], Gemini 1.5 [36,40], Llama 3 [10], demonstrated advanced cross-modal reasoning, enabling tasks like image analysis and visual question answering.

2.2 Visual Question Answering Dataset

Visual Question Answering is a complex yet essential task situated at the intersection of computer vision and natural language processing. It requires AI models to accurately answer questions based on visual input, combining the ability to interpret images with natural language comprehension. The introduction of the original VQA dataset [3] marked a significant milestone by providing the first standardized benchmark for evaluating such models. Building on its foundation, VQA 2.0 [12] addressed several limitations of the initial version, notably by enhancing the balance between questions and answers and minimizing biases, which contributed to a more robust model evaluation. Recent developments in VQA have expanded the scope and complexity of the task. New datasets now demand deeper reasoning abilities from models, moving beyond surface-level object recognition to require logical inferences, spatial reasoning, and contextual understanding [15,18,38,45,54]. Furthermore, VQA research has extended the task to different variations of visual input, including videos [47,51], scene text [5], and documents [34,41], broadening the applicability of the task across different domains. Similarly, specialized datasets have emerged for medical VQA [14,22,25], aimed at assisting in healthcare tasks, where questions are based on medical images such as MRIs or X-rays. Additionally, other datasets [32,33] focus on structured data in the form of plots, figures, and graphs, demanding models to interpret and generate insights from visualized data.

2.3 Visual Question Answering Approaches

The field of Visual Question Answering (VQA) has undergone significant transformation, with deep learning techniques now serving as the foundational framework for most contemporary methodologies. Early approaches [3,31,37] often involved separate encoders for visual and textual data to extract features from images and questions, respectively. These features were then fused using various strategies to merge the multi-modal information. The resulting combined representation was processed by either a classifier or a generator, depending on whether the answer generation was approached as a classification task or a generative problem. In recent years, there has been a notable shift towards Vision-Language Pre-training (VLP) [1,2,4,9,10,36,40], utilizing transformer architectures [43]. These models are pre-trained on extensive datasets comprising image-text pairs to learn generalized representations that span both modalities. By effectively capturing the complex relationships between visual and textual inputs, they can be fine-tuned for downstream tasks like VQA, leading to marked improvements in performance and generalization capabilities. This evolution towards transformer-based VLP models has not only enhanced the accuracy of VQA systems but also expanded their ability to tackle more intricate questions that demand deeper reasoning and contextual understanding. The transformer architecture enables models to dynamically attend to different parts of the input data, thereby improving interpretability and fostering more nuanced interactions between visual and textual information. Furthermore, this progression has opened up new avenues of research in VQA, including exploring zero-shot learning potentials [13,21,26], addressing inherent biases in datasets, and integrating commonsense knowledge to manage more sophisticated queries.

3   Our Proposed Approach

VLMs are designed to understand and generate text and images simultaneously. A crucial aspect of their performance lies in their ability to align both visual and textual feature representations. As mentioned earlier and shown in Fig. 1, in our proposed framework, each pair of image-prompt passes through a specific VLM and returns the related response, containing a detailed explanation. The response is further processed to obtain the final answers.

To be specific, we input the image I along with the guidance prompt containing the question Q and four multiple-choice options $o1, o2, o3, o4$. The VLMs then extract the image feature and process it simultaneously with the text prompt. The expected result consists of the number answer 1–4 and the corresponding explanation for the selected answer. Figure 2 illustrates our full guidance prompt used for all of our models. Regarding the VLM backbone, we explore different methods in order to find the best VLM that can perform well in both Japanese and English. Each method is presented separately for comprehensive understanding.

```
###Prompt: <Image><I></Image>
Imagine you are an expert in English and Japanese with good
knowledge. Please provide the answer with ONE number from 1-4. You
MUST give reason/explanation for your choice. The output MUST be
json format. Use the following JSON format:
  {
    "answer": "number",
    "explanation": "<text>"
  }
The following is the question and 4 choices:
<Question><Q></Question>
1 <Option><o1><Option>
2 <Option><o2><Option>
3 <Option><o3><Option>
4 <Option><o4><Option>
```

Fig. 2. Prompt guidance for all of our VLM approaches.

3.1   MiniCPM-V

MiniCPM-V [49] is a multi-modal large language model (MLLM) designed for efficient deployment on mobile devices, using approximately 8 billion parameters. The latest version of the MiniCPM-V series is utilized *(version 2.6)*, which is built on the SigLip-400M [53] and Qwen2-7B frameworks [48], achieving significant performance improvements over its predecessor, MiniCPM-Llama3-V 2.5 By being lightweight, this model is optimized explicitly for end-side deployment, meaning it can run efficiently on devices like mobile phones, tablets, and personal computers.

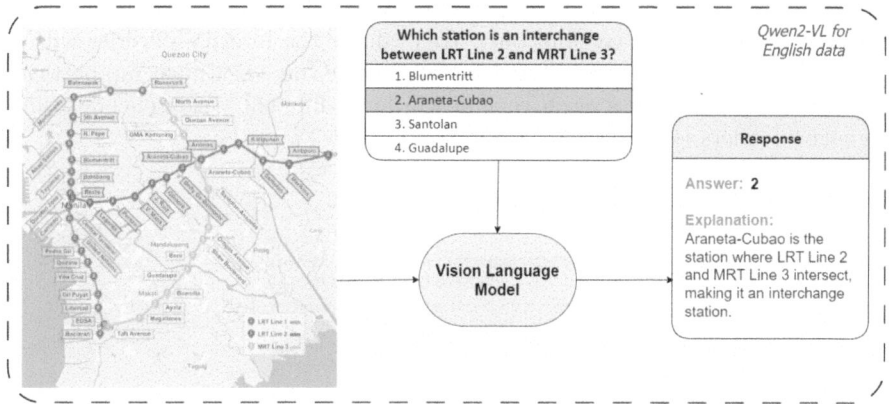

Fig. 3. An example of our best VLM approach for English data.

3.2   Qwen2-VL

Qwen2-VL [44] is the latest version of the VLM in Qwen model families [4,48] developed by Alibaba Cloud. It is the current state-of-the-art on several open-source visual understanding benchmarks, such as MathVista [30], DocVQA [34]and RealWorldQA[1]. Qwen2-VL has its text encoder as QwenV2 [48], including dense models and a mixture-of-experts model. These models are designed to perform well in multilingual settings, supporting over 29 languages, with the number of parameters ranging from 0.5 billion to 72 billion. Figure 3 and 4 demonstrate effective examples of both languages in the LAVA 2024 contest.

3.3   InternVL2

InternVL2 is one of the most powerful open-source MLLMs designed to handle complex multi-modal tasks involving text, images, and videos. It is the latest

---

[1] https://x.ai/blog/grok-1.5v.

version in the InternVL series [8,9], with various options for the model size, ranging from 1 billion to 108 billion parameters, making it highly versatile and scalable across various applications. It is particularly notable for its multi-modal input support, multitask output capabilities, and progressive alignment training strategy, which align its vision models natively with LLMs.

InternVL2 excels in a wide range of benchmarks, including tasks like visual question answering (VQA), OCR, and grounding. The models achieve performance levels comparable to commercial closed-source models like GPT-4V and Claude 3.5 Sonnet. For instance, InternVL2-8B and InternVL2-Pro have demonstrated superior performance in benchmarks such as MathVista [30], DocVQA [34], and ChartQA [32], showcasing its strengths in visual-linguistic tasks. Since the InternVL2-Pro is the paid commercial version, we decided not to deploy it, as we focus on open-source versions only. We evaluate the InternVL2 series with 4 different model sizes (8B, 26B, 40B and 76B). Since there is a huge gap between the number of parameters, each version will have different vision encoders and language encoders (see Table 1).

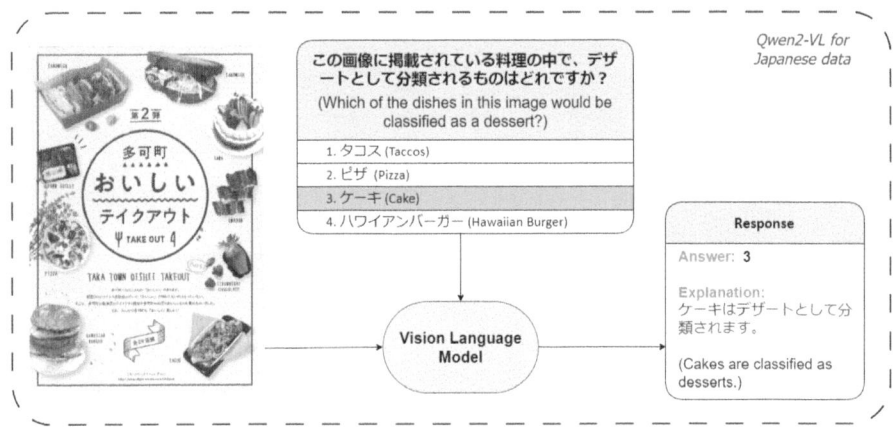

Fig. 4. An example of our best VLM approach for Japanese data. English translations are shown in parentheses.

3.4   GPT-4o Mini

GPT-4o mini[2] is a smaller and more cost-effective version of the larger GPT-4o model developed by OpenAI. It is designed to provide a balance between performance and cost-efficiency, making it suitable for a wide range of applications, especially for businesses and developers looking for powerful AI solutions at a lower price point. The model is created through a distillation process, where

---

[2] https://platform.openai.com/docs/models/gpt-4o-mini.

it learns to mimic the behavior and performance of the larger GPT-4o model, resulting in a smaller, more affordable version that retains much of the original's capabilities.

3.5   Gemini 1.5 Flash

Gemini 1.5 Flash [36] is a lightweight, fast, and cost-efficient AI model developed by Google DeepMind as part of the Gemini model family. It is optimized for high-speed and high-volume tasks, making it ideal for applications that require low latency and scalability. Gemini 1.5 Flash is designed to handle multi-modal reasoning, including text, images, and videos, and features a breakthrough long context window of up to one million tokens. This capability allows it to process large amounts of information, such as long documents, extensive codebases, and multimedia inputs.

3.6   Post-processing

As shown in Fig. 2, the response is expected to have the JSON format. However, in some cases, the VLM fails to produce the proper output in the first attempt. Since the model's output still contains the selected answer with its explanation, we do not want to regenerate the response, as the answer can differ from the first attempt. Instead, we prompt the VLM to fix the output in order to maintain our pre-designed JSON format. Finally, the final multiple-choice answers and their corresponding descriptions are extracted.

3.7   Ensemble

We also deploy an ensembling method to aggregate answers from multiple model results to produce a unified prediction for each data point. The primary objective of this ensembling approach was to combine outputs from various models and determine the most common prediction (or answer) for each sample across the different model outputs. By leveraging multiple model outputs and ensembling their predictions, we expect to reduce the variance of individual model predictions, thus enhancing the robustness and accuracy of the final result. The ensemble approach is considered as a major voting mechanism that ensures that the most common prediction among the models is selected, providing a reliable consensus output.

4   Experimental Results

All of our VLM approaches are implemented in a zero-shot setting, without any training or fine-tuning. We conducted our evaluations on two A100 80G GPUs and achieved comparable results. Table 1 presents an overview of our team approaches for each VLM separately with the related public scores. The closed-source GPT 4o-mini and Gemini-1.5 Flash share the same performance

Table 1. Performance comparison of various VLMs based on the LAVA challenge public dataset. Names of vision encoders and text encoders, along with #params and pixel image size, are also displayed. Model with the highest score is highlighted in red.

Model	Open Source	Vision Part	Language Part	Public Score
GPT-4o-mini [1]	✗	-	-	0.71
Gemini-1.5 Flash [36]	✗	-	-	0.71
MiniCPM-V-2.6-8B [49]	✓	SigLip-400M	Qwen2-7B	0.67
InternVL2-8B [9]	✓	InternViT-300M-448px	internlm2_5-7b-chat	0.70
InternVL2-26B [9]	✓	InternViT-6B-448px	internlm2-chat-20b	0.73
InternVL2-40B [9]	✓	InternViT-6B-448px	Nous-Hermes-2-Yi-34B	0.77
InternVL2-Llama3-76B [9]	✓	InternViT-6B-448px	Hermes-2-Theta-Llama-3-70B	0.76
Qwen2-VL-72B [44]	**✓**	**ViT - 675M - 224px**	**Qwen2-72B**	**0.83**

on the public dataset with a score of 0.71. Regarding the open-source models, the performance on the public leaderboard increases as we use the larger size of VLM. Qwen2-VL, with 76 billion parameters, performs the best, achieving a public score of 0.83. However, this only accounts for 30% of the competition's final score, with the remaining 70% is the score of the private dataset.

Table 2. Name of VLM chosen for the ensemble approach with the public scores.

Models	Public Score
InternVL2-(26B + 40B+ 76B)	0.77
InternVL2-(26B + 40B+ 76B) + Gemini-1.5-flash + Gpt-4o-mini	0.79
InternVL2-(26B + 40B+ 76B) + Qwen2-VL-72B	0.8
InternVL2-(26B + 40B+ 76B) + Gemini-1.5-flash + Gpt-4o-mini + Qwen2-VL-72B	0.83

Table 2 presents different ensemble configs of our attempt. In the first two settings (without Qwen2-VL), the ensemble approach improves the overall performance. However, when it comes to Qwen2-VL, the ensemble performance decreases compared to the score of the Qwen2-VL itself. This proves that Qwen2-VL outperforms other VLMs since the majority vote in the ensemble approach may neglect the correct multiple-choice answer from Qwen2-VL.

5   Discussion

5.1   About the Proposed Approach

Our method in the LAVA 2024 workshop challenge offers specific benefits: (1) **Robustness**: By exploring different large VLMs as well as implementing an ensemble mechanism to obtain the highest vote answers, we provide accurate answers with detailed explanations. However, explicitly verifying every explanation is time-consuming and requires much human force. Instead, we look for several samples in which the correct answers are effortlessly inferred by humans and use them as our validation set. (2) **Adaptability**: As presented earlier, our approach is capable of utilizing multiple pre-trained VLMs [8,27,36,44] without fine-tuning for retraining. This emphasizes that our method is a plug-and-play module and is capable of utilizing various multi-modal models.

Apart from the advantages, our approach still displays some limitations. Firstly, we do not control the language of the generated explanation. As a result, there are such cases where the language of the description differs from the language of the question. Secondly, our approach can be sensitive to the hallucination problem that exists in VLMs. Therefore, the explanation created by the model sometimes includes unnecessary details that are not related to the visual input. Resolving these limitations would necessitate improvement in the structure of the LLMs, which is not the scope of our competition approach.

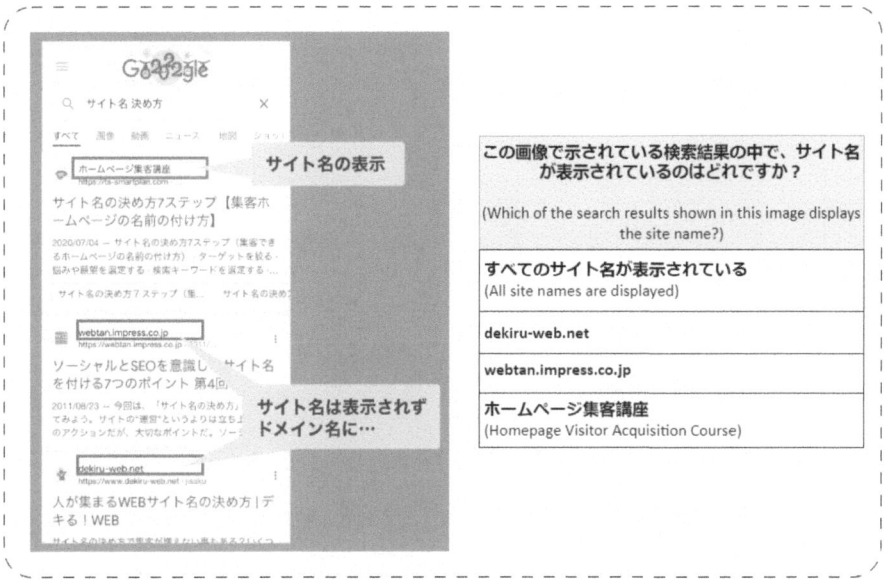

Fig. 5. Both Japanese and English appear in the content of the images as well as the multiple-choice answers, elevating the level of complexity. English translations are shown in parentheses.

5.2 About the Dataset

The LAVA 2024 workshop challenge is the first competition tackling the understanding of general knowledge for the task of multiple-choice VQA, especially where the Japanese language with a complex writing system is taken into account. The dataset covers a broad spectrum of domains (e.g., healthcare, education, or entertainment) by providing complex visual data (e.g., diagrams, charts, or drawings). One challenge in this competition is the mixed appearance in the visual content as well as the question-answers in a single sample. This happens in both public and private datasets, adding an extra layer of complexity. Figure 5 demonstrates a sample of mixed languages in the dataset, where the question is present in Japanese while the multiple-choice answers and the image content contain both English and Japanese sentences.

Additionally, we recognize an ambiguous case related to the dataset that can affect the performance in general. As shown in Fig. 6, there are 100 samples labeled ''ja'' instead of ''ja-JP'' or ''en-US''. In fact, these 100 samples are all written in the Japanese language. The problem of label inconsistency may hurt some automatic paradigms, where samples in both languages are executed in different pipelines.

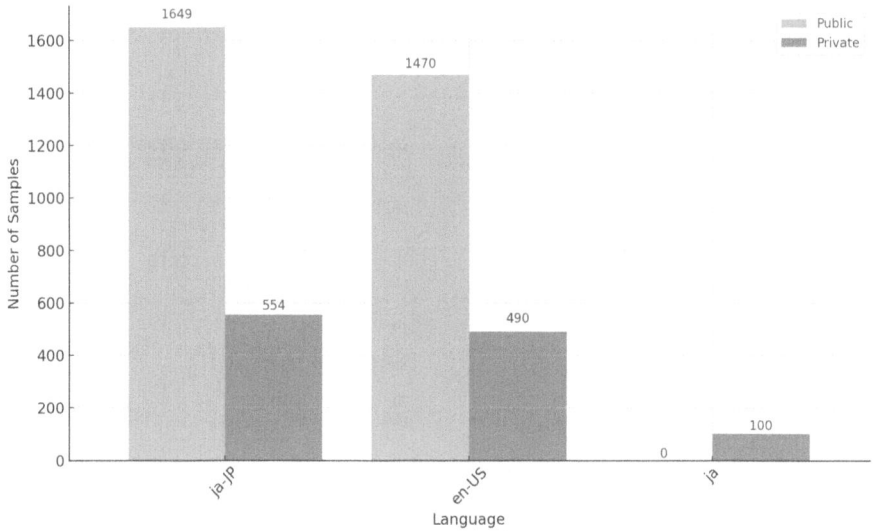

Fig. 6. Overview of the number of samples of each language in LAVA challenge.

6   Conclusion

In this paper, we present our multi-modal vision-language approach for the task of multiple-choice visual question answering. Our approach effectively utilizes large pre-trained VLMs and achieves competitive results in the LAVA 2024 workshop challenge. Through extensive experiments, we demonstrate that our method can handle the cases of Japanese, English, and even the mixture of both languages that appeared in the data samples. The limitations we identified suggest potential directions for future development to enhance the ability of multiple-choice visual question answering.

References

1. Achiam, J., et al.: GPT-4 technical report. arXiv preprint arXiv:2303.08774 (2023)
2. Anil, R., et al.: Palm 2 technical report. arXiv preprint arXiv:2305.10403 (2023)
3. Antol, S., Agrawal, A., Lu, J., Mitchell, M., Batra, D., Zitnick, C.L., Parikh, D.: VQA: visual question answering. In: Proceedings of the IEEE International Conference on Computer Vision, pp. 2425–2433 (2015)
4. Bai, J., et al.: QWEN-VL: a versatile vision-language model for understanding, localization, text reading, and beyond. arXiv preprint arXiv:2308.12966 (2023)
5. Biten, A.F., et al.: Scene text visual question answering. In: Proceedings of the IEEE/CVF International Conference on Computer Vision, pp. 4291–4301 (2019)
6. Chen, K., Wu, X.: VTQA: visual text question answering via entity alignment and cross-media reasoning. In: Proceedings of the IEEE/CVF Conference on Computer Vision and Pattern Recognition, pp. 27218–27227 (2024)
7. Chen, Y.-C., et al.: UNITER: UNiversal image-TExt representation learning. In: Vedaldi, A., Bischof, H., Brox, T., Frahm, J.-M. (eds.) ECCV 2020. LNCS, vol. 12375, pp. 104–120. Springer, Cham (2020). https://doi.org/10.1007/978-3-030-58577-8_7
8. Chen, Z.,et al.: How far are we to GPT-4V? Closing the gap to commercial multimodal models with open-source suites. arXiv preprint arXiv:2404.16821 (2024)
9. Chen, Z., et al.: InternVL: scaling up vision foundation models and aligning for generic visual-linguistic tasks. In: Proceedings of the IEEE/CVF Conference on Computer Vision and Pattern Recognition, pp. 24185–24198 (2024)
10. Dubey, A., et al.: The llama 3 herd of models. arXiv preprint arXiv:2407.21783 (2024)
11. Fukui, A., Park, D.H., Yang, D., Rohrbach, A., Darrell, T., Rohrbach, M.: Multimodal compact bilinear pooling for visual question answering and visual grounding. arXiv preprint arXiv:1606.01847 (2016)
12. Goyal, Y., Khot, T., Summers-Stay, D., Batra, D., Parikh, D.: Making the V in VQA matter: Elevating the role of image understanding in visual question answering. In: Proceedings of the IEEE Conference on Computer Vision and Pattern Recognition, pp. 6904–6913 (2017)
13. Guo, J., et al.: From images to textual prompts: zero-shot visual question answering with frozen large language models. In: Proceedings of the IEEE/CVF Conference on Computer Vision and Pattern Recognition, pp. 10867–10877 (2023)
14. He, X., Zhang, Y., Mou, L., Xing, E., Xie, P.: PathVQA: 30000+ questions for medical visual question answering. arXiv preprint arXiv:2003.10286 (2020)
15. Hudson, D.A., Manning, C.D.: GQA: a new dataset for real-world visual reasoning and compositional question answering. In: Proceedings of the IEEE/CVF Conference on Computer Vision and Pattern Recognition, pp. 6700–6709 (2019)
16. Jia, C., et al.: Scaling up visual and vision-language representation learning with noisy text supervision. In: International Conference on Machine Learning, pp. 4904–4916. PMLR (2021)
17. Jiang, A.Q., et al.: Mistral 7b. arXiv preprint arXiv:2310.06825 (2023)
18. Johnson, J., Hariharan, B., Van Der Maaten, L., Fei-Fei, L., Lawrence Zitnick, C., Girshick, R.: Clevr: A diagnostic dataset for compositional language and elementary visual reasoning. In: Proceedings of the IEEE Conference on Computer Vision and Pattern Recognition, pp. 2901–2910 (2017)
19. Kenton, J.D.M.W.C., Toutanova, L.K.: BERT: pre-training of deep bidirectional transformers for language understanding. In: Proceedings of NAACL-HLT. vol. 1, p. 2. Minneapolis, Minnesota (2019)

20. Khan, Z., Fu, Y.: Consistency and uncertainty: Identifying unreliable responses from black-box vision-language models for selective visual question answering. In: Proceedings of the IEEE/CVF Conference on Computer Vision and Pattern Recognition, pp. 10854–10863 (2024)

21. Lan, Y., Li, X., Liu, X., Li, Y., Qin, W., Qian, W.: Improving zero-shot visual question answering via large language models with reasoning question prompts. In: Proceedings of the 31st ACM International Conference on Multimedia, pp. 4389–4400 (2023)

22. Lau, J.J., Gayen, S., Ben Abacha, A., Demner-Fushman, D.: A dataset of clinically generated visual questions and answers about radiology images. Sci. Data **5**(1), 1–10 (2018)

23. Li, L., Peng, J., Chen, H., Gao, C., Yang, X.: How to configure good in-context sequence for visual question answering. In: Proceedings of the IEEE/CVF Conference on Computer Vision and Pattern Recognition, pp. 26710–26720 (2024)

24. Li, X., et al.: OSCAR: object-semantics aligned pre-training for vision-language tasks. In: Vedaldi, A., Bischof, H., Brox, T., Frahm, J.-M. (eds.) ECCV 2020. LNCS, vol. 12375, pp. 121–137. Springer, Cham (2020). https://doi.org/10.1007/978-3-030-58577-8_8

25. Liu, B., Zhan, L.M., Xu, L., Ma, L., Yang, Y., Wu, X.M.: Slake: A semantically-labeled knowledge-enhanced dataset for medical visual question answering. In: 2021 IEEE 18th International Symposium on Biomedical Imaging (ISBI), pp. 1650–1654. IEEE (2021)

26. Liu, C., Wang, C., Peng, Y., Li, Z.: ZVQAF: zero-shot visual question answering with feedback from large language models. Neurocomputing **580**, 127505 (2024)

27. Liu, Y., Liang, Z., Wang, Y., He, M., Li, J., Zhao, B.: Seeing clearly, answering incorrectly: a multimodal robustness benchmark for evaluating MLLMS on leading questions. arXiv preprint arXiv:2406.10638 (2024)

28. Lu, J., Batra, D., Parikh, D., Lee, S.: VilBERT: pretraining task-agnostic visiolinguistic representations for vision-and-language tasks. In: Advances in Neural Information Processing Systems, vol. 32 (2019)

29. Lu, J., Yang, J., Batra, D., Parikh, D.: Hierarchical question-image co-attention for visual question answering. In: Advances in Neural Information Processing Systems, vol. 29 (2016)

30. Lu, P., et al.: MathVista: evaluating mathematical reasoning of foundation models in visual contexts. arXiv preprint arXiv:2310.02255 (2023)

31. Malinowski, M., Rohrbach, M., Fritz, M.: Ask your neurons: a neural-based approach to answering questions about images. In: Proceedings of the IEEE International Conference on Computer Vision, pp. 1–9 (2015)

32. Masry, A., Long, D.X., Tan, J.Q., Joty, S., Hoque, E.: ChartQA: a benchmark for question answering about charts with visual and logical reasoning. arXiv preprint arXiv:2203.10244 (2022)

33. Mathew, M., Bagal, V., Tito, R., Karatzas, D., Valveny, E., Jawahar, C.: Infographicvqa. In: Proceedings of the IEEE/CVF Winter Conference on Applications of Computer Vision, pp. 1697–1706 (2022)

34. Mathew, M., Karatzas, D., Jawahar, C.: DocVQA: a dataset for VQA on document images. In: Proceedings of the IEEE/CVF Winter Conference on Applications of Computer Vision, pp. 2200–2209 (2021)

35. Radford, A., et al.: Learning transferable visual models from natural language supervision. In: International Conference on Machine Learning, pp. 8748–8763. PMLR (2021)

36. Reid, M., et al.: Gemini 1.5: unlocking multimodal understanding across millions of tokens of context. arXiv preprint arXiv:2403.05530 (2024)
37. Ren, M., Kiros, R., Zemel, R.: Exploring models and data for image question answering. In: Advances in Neural Information Processing Systems, vol. 28 (2015)
38. Schwenk, D., Khandelwal, A., Clark, C., Marino, K., Mottaghi, R.: A-okvqa: A benchmark for visual question answering using world knowledge. In: European Conference on Computer Vision, pp. 146–162. Springer (2022). https://doi.org/10.1007/978-3-031-20074-8_9
39. Tan, H., Bansal, M.: LXMERT: learning cross-modality encoder representations from transformers. arXiv preprint arXiv:1908.07490 (2019)
40. Team, G., et al.: Gemini: a family of highly capable multimodal models. arXiv preprint arXiv:2312.11805 (2023)
41. Tito, R., Karatzas, D., Valveny, E.: Document collection visual question answering. In: Lladós, J., Lopresti, D., Uchida, S. (eds.) ICDAR 2021. LNCS, vol. 12822, pp. 778–792. Springer, Cham (2021). https://doi.org/10.1007/978-3-030-86331-9_50
42. Touvron, H., et al.: LLAMA 2: open foundation and fine-tuned chat models. arXiv preprint arXiv:2307.09288 (2023)
43. Vaswani, A.: Attention is all you need. In: Advances in Neural Information Processing Systems (2017)
44. Wang, P.,et al.: Qwen2-VL: enhancing vision-language model's perception of the world at any resolution. arXiv preprint arXiv:2409.12191 (2024)
45. Wang, P., Wu, Q., Shen, C., Dick, A., Van Den Hengel, A.: FVQA: fact-based visual question answering. IEEE Trans. Pattern Anal. Mach. Intell. 40(10), 2413–2427 (2017)
46. Wang, Z., Yu, J., Yu, A.W., Dai, Z., Tsvetkov, Y., Cao, Y.: SimVLM: simple visual language model pretraining with weak supervision. arXiv preprint arXiv:2108.10904 (2021)
47. Xiao, J., Shang, X., Yao, A., Chua, T.S.: NExT-QA: next phase of question-answering to explaining temporal actions. In: Proceedings of the IEEE/CVF Conference on Computer Vision and Pattern Recognition, pp. 9777–9786 (2021)
48. Yang, A., et al.: Qwen2 technical report. arXiv preprint arXiv:2407.10671 (2024)
49. Yao, Y., et al.: MiniCPM-V: a GPT-4v level MLLM on your phone. arXiv preprint arXiv:2408.01800 (2024)
50. Yu, J., Wang, Z., Vasudevan, V., Yeung, L., Seyedhosseini, M., Wu, Y.: Coca: contrastive captioners are image-text foundation models. arXiv preprint arXiv:2205.01917 (2022)
51. Yu, Z., et al.: ActivityNet-QA: a dataset for understanding complex web videos via question answering. In: Proceedings of the AAAI Conference on Artificial Intelligence, vol. 33, pp. 9127–9134 (2019)
52. Yuan, L., et al.: Florence: a new foundation model for computer vision. arXiv preprint arXiv:2111.11432 (2021)
53. Zhai, X., Mustafa, B., Kolesnikov, A., Beyer, L.: Sigmoid loss for language image pre-training. In: Proceedings of the IEEE/CVF International Conference on Computer Vision, pp. 11975–11986 (2023)
54. Zhang, C., Gao, F., Jia, B., Zhu, Y., Zhu, S.C.: Raven: a dataset for relational and analogical visual reasoning. In: Proceedings of the IEEE/CVF Conference on Computer Vision and Pattern Recognition, pp. 5317–5327 (2019)

An Approach to Complex Visual Data Interpretation with Vision-Language Models

Thanh-Son Nguyen[1,2], Viet-Tham Huynh[1,2], Van-Loc Nguyen[1,2], and Minh-Triet Tran[1,2(✉)]

[1] Software Engineering Laboratory, University of Science, VNU-HCM, Ho Chi Minh City, Vietnam
{nthanhson,hvtham,nvloc}@selab.hcmus.edu.vn, tmtriet@fit.hcmus.edu.vn
[2] Vietnam National University, Ho Chi Minh City, Vietnam

Abstract. The LAVA Workshop 2024 challenge aimed to assess the capability of Large Vision-Language Models (VLMs) to interpret and understand complex visual data accurately. This includes intricate visual formats such as data flow diagrams, class diagrams, Gantt charts, and architectural blueprints. In response to this challenge, our research focuses on adapting the MMMU (Massive Multi-discipline Multimodal Understanding) benchmarks to better align with the requirements of visual data interpretation. We propose a comprehensive approach that leverages advanced prompt engineering techniques and incorporates a voting-based ensemble method for aggregating model predictions. This method improves the model's ability to generalize across different types of visual inputs. Our approach was rigorously evaluated within the context of the challenge, resulting in a total score of 0.85, ultimately securing the top position in the competition. This result demonstrates the effectiveness of combining prompt engineering with simple yet powerful ensemble strategies for enhancing the performance of VLMs on complex multimodal tasks.

Keywords: Visual language · MMMU · AGI

1 Introduction

Advancements in AI and machine learning, especially large language models (LLMs) like OpenAI's o1 or Meta's LLaMA, demonstrate great utilization in various applications in society. These language models have shown an exceptional ability to utilize large-scale text data to deliver outstanding results in various natural language understanding tasks. As a result, the research community has paid a lot of attention to expanding the capabilities of these language models to additional data modalities, such as images, videos, or audio, forming large vision-language models (LVLMs) like GPT-V or LLaVA.

T.-S. Nguyen, V.-T. Huynh, V.-L. Nguyen—Equal contributions.

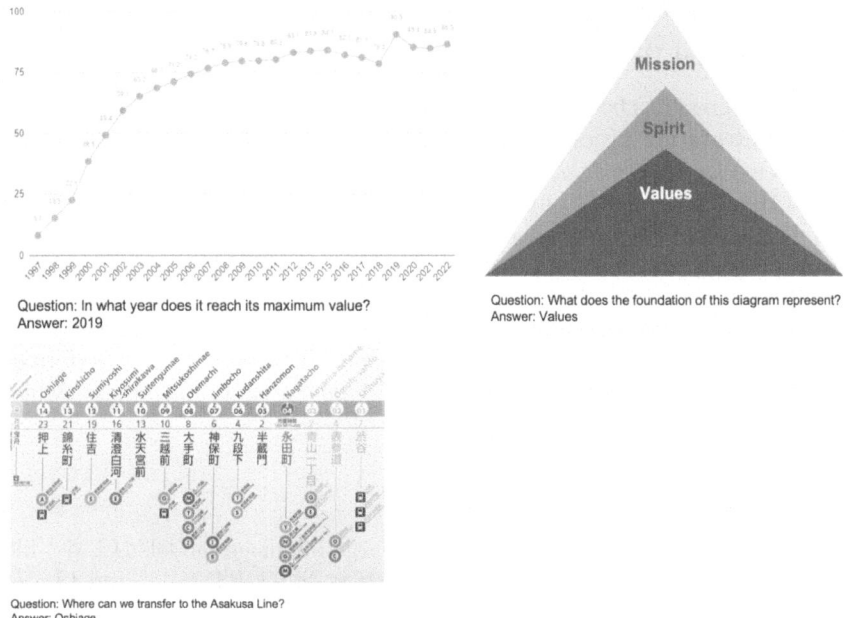

Question: In what year does it reach its maximum value?
Answer: 2019

Question: What does the foundation of this diagram represent?
Answer: Values

Question: Where can we transfer to the Asakusa Line?
Answer: Oshiage

Fig. 1. Examples of the inputs and outputs of LAVA workshop challenge

Large Vision-Language Model Learning and Applications Workshop (LAVA Workshop) is held to encourage researchers to unlock the full potential of research in large vision-language models by focusing on the integration of various modalities, such as text, images, and videos. In addition, the workshop serves as a forum to explore the usage of large vision-language models across various fields, such as healthcare, education, entertainment, transportation, finance, and more.

The LAVA workshop also contains the challenge, with the main objective to enhance the ability of large vision-language models to precisely interpret and comprehend complex visual data, including data flow diagrams (DFDs), class diagrams, Gantt charts, or building design drawings. Figure 1 illustrates examples of the inputs and outputs of the LAVA workshop challenge.

In this research, we propose investigating the capabilities of different large vision-language models in answering questions with complex visual data in various fields and the effect of changing prompts on the model's performance.

The next sections of this paper will be organized as follows. Section 2 briefly describes the related research to our work. Section 3 is our proposed method, including the information about the dataset, MMMU benchmark, model selection, the architectures of chosen models, and our prompt techniques. Section 4 is the experiments and results of our method, including our prompt and results on public and private datasets. Section 5 is our conclusion and discussion about future works.

2   Related Work

2.1   Multimodal Pre-training

Significant progress has been achieved recently in multimodal pre-training, which focuses on integrating vision and language into a unified model. Early works, including VinVL [45], Oscar [21], LXMERT [34], VilBERT [28], and UNITER [5], pioneered the development of universal models for vision-language tasks. These models often relied on pre-trained visual features like Faster RCNN to reduce training complexity. More recent approaches such as ALIGN [16], CoCa [41], CLIP [33], Fuyu [44] SimVLM [37], Flamingo [3], and BLIP-2 [19] have shifted toward training visual representations from scratch using Vision Transformers (ViT) [12] and vast datasets sourced from the web. These models have performed well in visual question answering (VQA) and image captioning tasks, which require less in-depth reasoning.

2.2   Multimodal Instruction Tuning

Building on the success of instruction-tuned large language models (LLMs) like Vicuna [8] and FLAN-T5 [9], new models such as MiniGPT-4 [11] and LLaVA [47] have been developed to enhance the instruction-following abilities of large multimodal models (LMMs). This has spurred advances in generating high-quality visual instruction data, with models like mPlug-OWL [39], LLaMA-Adapterv2 [13], LRV-Instruction [24], SVIT [26], and InstructBLIP [10] leading the charge. Another critical aspect of LMM research focuses on multimodal in-context learning, where models handle mixed examples of text and images. Notable models in this domain include M3IT [20], Sparkles [14], MetaVL [32], Otter [17], Flamingo [2], OpenFlamingo [4], and MMICL [46], which have contributed to improving multimodal training and instruction-following abilities.

2.3   LMM Benchmarks

The rapid advancement in multimodal pre-training and instruction tuning has outgrown traditional single-task benchmarks like MSCOCO [23], VQA [1], OK-VQA [30], and GQA [15]. These benchmarks are now insufficient for evaluating LMMs' broader abilities in perception and reasoning. Consequently, several comprehensive benchmarks have emerged, focusing on a range of LMM capabilities. These include optical character recognition (OCR) as explored in various studies, adversarial robustness, and hallucination issues, with benchmarks such as HaELM [35] and POPE [22] specifically addressing these challenges. Holistic evaluations are provided by LVLM-eHub [38], SEED [18,27], LAMM [40], and MM-Vet [42]. However, most benchmarks still emphasize basic perception tasks, which do not require deep domain knowledge or advanced reasoning. A recent benchmark, MathVista [29], evaluates models on visually complex questions in the mathematical domain. In contrast, MMMU [43] introduces more complex, expert-level tasks across 30 disciplines, demanding advanced perception and domain-specific knowledge for step-by-step reasoning. Concurrently,

GAIA [31] presents 466 questions testing models' abilities in reasoning, multi-modal comprehension, and tool usage.

3   Proposed Method

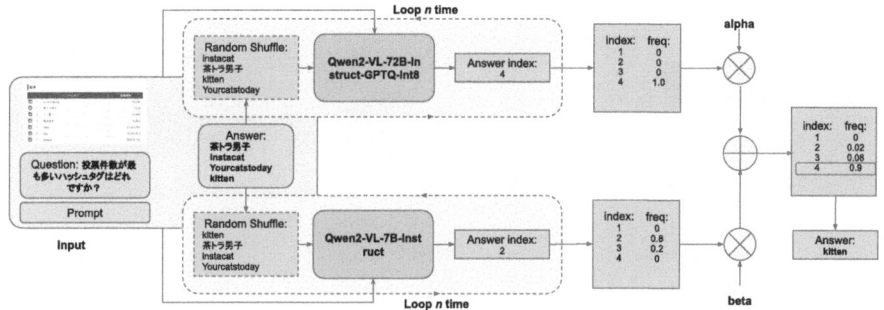

Fig. 2. Proposed method overview: our proposed method contains three steps. First, we use prompt engineering to produce input for pre-trained models. Second, the prompt is fed into the network to generate answers. Thirdly, we assemble all answers from multiple networks to decide the final result.

3.1   Overview of the Dataset

The dataset contains 2 subsets: public dataset and private dataset. The public dataset contains about 3000 data samples released by the organizing committee of the LAVA workshop. This subset is collected on the Internet. The private dataset, provided by the TASUKI team (SoftBank), consists of around 1100 samples.

Both subsets share the same structure: an image, a question, the four options to be chosen, and an indicator specifying the language of the question and options. There are 2 languages used in this dataset: English (*en-US*) and Japanese (*ja-JP*).

3.2   MMMU Benchmark

The Massive Multi-discipline Multimodal Understanding and Reasoning (MMMU) [43] benchmark, developed by Yue et al., is a novel benchmark designed to evaluate multimodal models on massive multi-discipline tasks requiring subject knowledge at a college level, and thoughtful reasoning. This benchmark is developed to measure 3 essential skills for multimodal models: perception, knowledge, and reasoning. There are 6 core disciplines covered by this dataset: Art & Design, Business, Science, Health & Medicine, Humanities & Social Science, and Technology & Engineering.

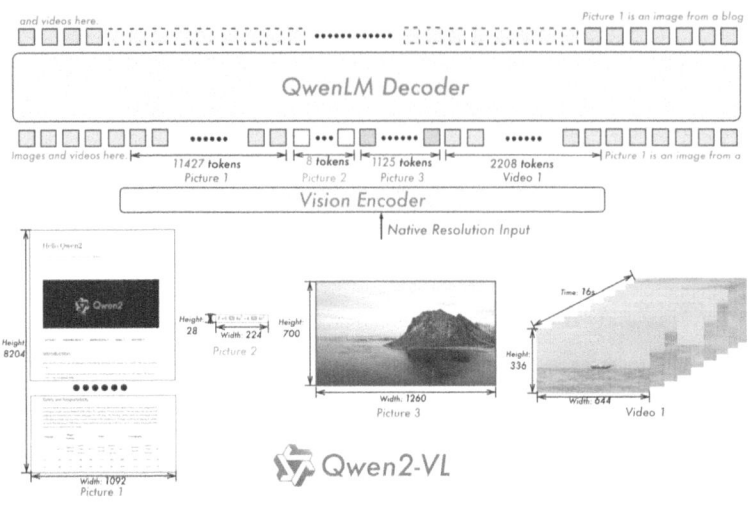

Fig. 3. Architecture of Qwen2-VL [36]

The results of submissions to the LAVA workshop challenge are evaluated by the MMMU metric, suggesting a relationship between the challenge's input-output structure and the MMMU benchmark. Therefore, we conducted our experiments on models with the highest MMMU benchmark results.

3.3 Model Selection

In the leaderboard of the MMMU benchmark, the model o1 from OpenAI achieves the highest result, which is even better than the Human Expert (Low) performance. The following positions contain several large language models and large vision-language models, such as GPT-4o also from OpenAI, Claude 3.5 Sonnet, Gemini 1.5 Pro, Qwen2-VL, InternVL2, LLaVA, etc.

In this paper, Qwen2-VL, InternVL2, and LLaVA are the main focus, especially Qwen2-VL [36][1] and InternVL2 [6,7][2], because these models are in top-ranking performance of the MMMU benchmark, as well as they are open-source and contain inference APIs on HuggingFace.

3.4 Model Architectures

Architecture of Qwen2-VL. The Qwen2-VL architecture builds upon the Qwen-VL framework, combining a Vision Transformer [12] (ViT) model with Qwen2 language models. Below are key elements of this architecture include:

---

[1] https://huggingface.co/Qwen/Qwen2-VL-2B-Instruct-GPTQ-Int4.
[2] https://huggingface.co/OpenGVLab/InternVL2-Llama3-76B.

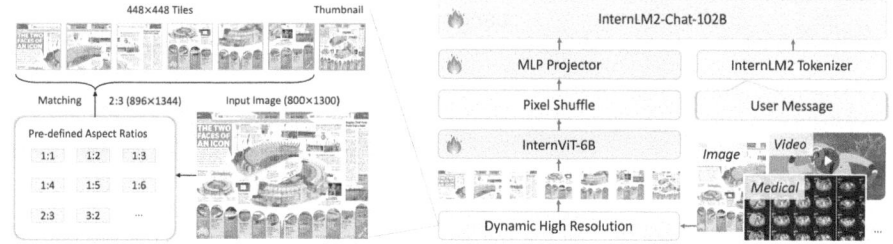

Fig. 4. Architecture of InternVL1.5 [6]

- **Vision Transformer (ViT) with 600M Parameters:** The model utilizes a ViT with approximately 600 million parameters to process both image and video inputs seamlessly.
- **Naive Dynamic Resolution Support:**
  - Qwen2-VL introduces Naive Dynamic Resolution, allowing the model to handle images of arbitrary resolutions.
  - This maps images into a variable number of visual tokens, ensuring that the input is consistent with the image's inherent information.
  - This mechanism mimics human visual perception, enabling the model to process images of different clarity and sizes.
- **Multimodal Rotary Position Embedding (M-ROPE):**
  - M-ROPE is an innovation that deconstructs the original rotary embedding into components representing temporal and spatial (height and width) dimensions.
  - This enables the model to capture and integrate positional information across 1D text, 2D images, and 3D videos, allowing simultaneous comprehension of various data types.

Architecture of InternVL2. Up until now, the architecture of InternVL2 has not been published yet. InternVL2 is an improvement of InternVL1.5, whose architecture is described in Fig. 4.

- **Strong Vision Encoder**: The majority of existing multimodal large language models use pre-trained ViTs [12]. However, these ViTs are popularly trained on image-text pairs scraped from the Internet with a low resolution, so their performance decreases when processing high-resolution images. To overcome this challenge, InternViT is introduced. The visual features learned by this model are broadly applicable, not only to specific large language models.
- **Dynamic High-Resolution** The authors adopt a dynamic high-resolution training approach for InternVL1.5 to adapt effectively to different input images' varying resolutions and aspect ratios.

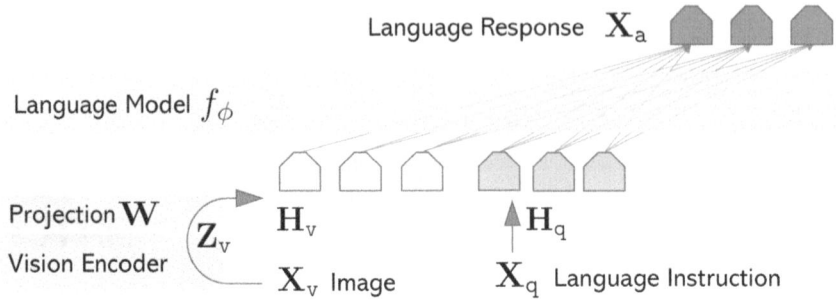

Fig. 5. Architecture of LLaVa 1.5 [25]

Architecture of LLaVa 1.5. LLaVA (Large Language and Vision Assistant) is an advanced AI model that handles both language and vision inputs. The architecture of LLaVA 1.5 is based on multimodal learning, combining both vision and language models to handle tasks like answering questions about images, image captioning, and generating detailed descriptions of visual content. Here's a high-level overview of its architecture:

– *Backbone Architecture*:
  • *Language Model (LM)*: LLaVA 1.5 integrates a pre-trained large language model like GPT or LLaMA as its language backbone. The language model handles text-based tasks, generates responses, and performs reasoning based on input queries. This LM is fine-tuned to handle the combination of visual and textual information.
  • *Vision Model*: For visual input, LLaVA employs a vision transformer (ViT) or a similar deep learning model trained on large-scale image datasets. This model processes image inputs, extracting visual features that are later combined with the language model's understanding. with linguistic tokens.
– *Multimodal Fusion Mechanism*: The key challenge in LLaVA's architecture is combining text and image features effectively. LLaVA 1.5 uses cross-attention layers to align the features from both the language model and vision model. The vision model encodes the image into a set of embeddings (image tokens), while the language model encodes the textual inputs (word tokens). These embeddings are fused together via a multimodal transformer layer, which learns how to relate visual features with linguistic tokens.

3.5   Prompt Techniques

Alongside selecting the appropriate models, this research focuses on prompt engineering for these models. For example, some improvement is made to the prompt fed into the models so that the model can perform at its best capability in understanding and answering questions with complex visual data. Moreover, when researching the MMMU benchmark, the inference on models is performed with the prompt of the benchmark itself.

Answer Shuffle. In addition, the shuffle on the order of answers is experimented to investigate the models' performance. To be more detailed, the regular order of the answers is: 1 for option A, 2 for option B, 3 for option C, and 4 for option D; after shuffling, one of the possible cases should be 3 for A, 1 for B, 4 for C, and 2 for D. The model will be asked several times, and the final option will be decided to be the answer with the most number of chosen times. See Fig. 6.

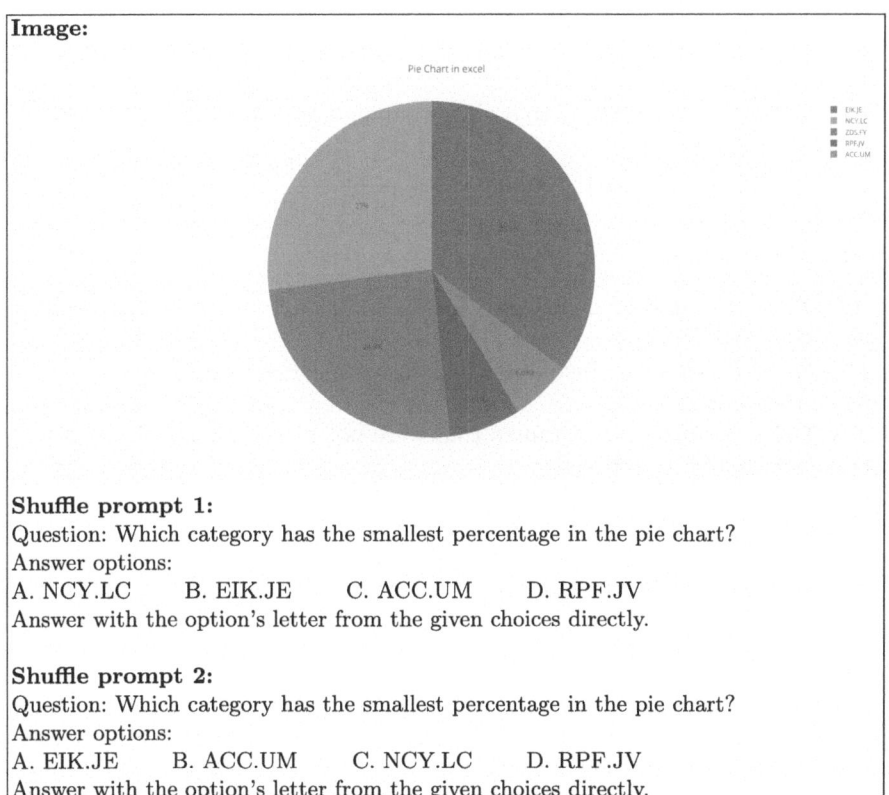

Shuffle prompt 1:
Question: Which category has the smallest percentage in the pie chart?
Answer options:
A. NCY.LC   B. EIK.JE   C. ACC.UM   D. RPF.JV
Answer with the option's letter from the given choices directly.

Shuffle prompt 2:
Question: Which category has the smallest percentage in the pie chart?
Answer options:
A. EIK.JE   B. ACC.UM   C. NCY.LC   D. RPF.JV
Answer with the option's letter from the given choices directly.

Fig. 6. An example of the random shuffle method: In the experiments, each sample will be shuffled 5 times, and the final answer will be the most frequent candidate.

Voting Ensemble. Additionally, the ensemble of several models is developed, combining the outputs from different sources to help improve the accuracy of the answer. For example, if answers of `Model1(image, prompt)` on 5 times shuffle is [1,2,2,3,4] and answers of `Model2(image, prompt)` on 5 times shuffle is [1,2,2,2,2], the final answer will be 2.

Post-processing. In the majority of experiments, the models can directly provide an answer from options A, B, C, or D. However, in rare instances where the model fails to generate a direct answer, we decide to automatically assign option A as the response, given the infrequency of such occurrences.

Additional Prompt. Another approach asks the models with given questions and whether it is true or false if the answer is A. The same things are implemented for B, C, and D. Based on the result of the model, the final answer to the question will be proposed.

4 Experiments and Results

4.1 Metric

LAVA Challenge used MMMU to evaluate submission results:

$$\text{Micro-averaged accuracy} = \frac{\sum (\text{TP})}{\sum ((\text{TP}) + (\text{FP}) + (\text{FN}))}$$

Which:

- **TP(True Positives):** Correctly predicted instances across all labels or classes.
- **FP(False Positives):** Instances incorrectly predicted as belonging to a class when they don't.
- **FN(False Negatives:** Instances that were not predicted as belonging to a class but should have been.

The total score will be calculated by:

$$\text{Total score} = 0.3(\text{Private score}) + 0.7(\text{Public score})$$

4.2 Settings

In experiments, about the shuffle of answer order, we asked the model 5 times, and the answer chosen the most number of times would be our final solution for the question. The ideal situation is that an option is selected in more than 2 times. However, there are some cases in which 2 options are equally responded by the model, for example, the model chooses option A 2 times, option B 2 times, and option C 1 time. In such cases, we will ask the model the same question again, and the answer now contains only the option selected most time, in the example, they are A and B. The model selection in this inference would be the final answer. The same approach is applied to the true-false strategy: if there is more than one true answer, the model will be asked again with only those answers.

In the actual examination, the Qwen2-VL model is noticed to produce better accuracy than the other selected models (LLaVA and InternVL2); therefore, most of our prompt improvement focuses on different versions of this model

(Qwen-VL-7B-Instruct-GPTQ-Int4, Qwen-VL-7B-Instruct-GPTQ-Int8, Qwen2-VL-7B-Instruct, etc.). Our answer order shuffle, true-false approach, and model assembly are applied to this model only.

For the LAVA Workshop challenge, the submission that receives the highest result on the public dataset is the assembly of Qwen2-VL-72B-Instruct-GPTQ-Int8 and Qwen2-VL-7B-Instruct models with 2 times vote results for Qwen2-VL-72B-Instruct-GPTQ-Int8 and 1 time for Qwen2-VL-7B-Instruct. This submission is conducted following the prompt (Table 1).

Table 1. Prompt 1

Question: <Question> Answer options:
A. <Option 1 (after being shuffled)>
B. <Option 2 (after being shuffled)>
C. <Option 3 (after being shuffled)>
D. <Option 4 (after being shuffled)>
Answer with the option's letter from the given choices directly.

The reason for having the last sentence (Answer with the option's letter from the given choices directly) is to force the model to answer the letter only, prevent it from answering a wall of text, and make the decision difficult.

There are different prompts used in our experiments, which are defined as follows (Tables 2, 3, 4, 5, 6, 7, 8 and 9):

Table 2. Prompt 2

Analyze the attached image and answer the following question with precision:
Question: <Question>
Options:
1. <Answer 1> 2. <Answer 2>
3. <Answer 3> 4. <Answer 4>
Please respond by selecting the option number that best fits the image's content. Be precise and choose only one option.

Table 3. Prompt 3

Look at the image and answer the following question:
Question: <Question>
Options:
1. <Answer 1> 2. <Answer 2>
3. <Answer 3> 4. <Answer 4>
Provide your answer by choosing the number corresponding to the correct option. For example, if the correct answer is the first option, respond with "1".

Table 4. Prompt 4

Based on the given image, answer the following question:

Question: <Question>

Options:

1. <Answer 1> 2. <Answer 2>

3. <Answer 3> 4. <Answer 4>

Your task is to analyze the image and choose the correct answer by providing the number of options you believe are correct. Only the number is required for your response.

Table 5. Prompt 5

Please answer the question: <Question>, by looking at the attached image and the following answers:

1. <Answer 1> 2. <Answer 2>

3. <Answer 3> 4. <Answer 4>

You must answer by number of the answer. For example, if you think the answer is option 1, please write 1.

Table 6. Prompt 6

Look at the image, based on the information in the image, answer the following question. Choose the correct answer among four options.

Question and answers are in <language>.

You should only choose the correct answer, without any further explanation.

Question: <Question>

Answers:

1. <Answer 1> 2. <Answer 2>

3. <Answer 3> 4. <Answer 4>

Table 7. Prompt 7

Of the 4 answers, based on the image and question, which is the correct answer (just write the correct answer number) to the following question:

Question: <Question>

Answers:

1. <Answer 1> 2. <Answer 2>

3. <Answer 3> 4. <Answer 4>

Table 8. Prompt 8

Qwen: You are an AI assistant specializing in multimodal understanding. Analyze the following question and multiple choice answers related to various topics. Your task is to select the most accurate response. Think through your reasoning carefully, but only output the letter corresponding to your final answer.

Question: <Question>

Answers:

A. <Answer A> B. <Answer B>

C. <Answer C> D. <Answer D>

Provide only your final answer as a single letter: A, B, C, or D.

Table 9. Prompt 9

The answer to the question: <question> is <option (option is four options A, B, C, D). True or False?

A. True

B. False

Answer with the option's letter from the given choices directly.

4.3 Results

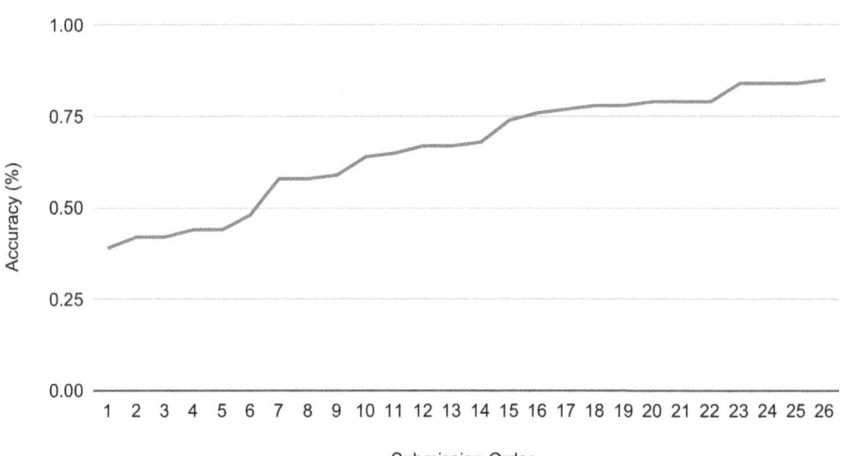

Fig. 7. Accuracy Trend by Submission Order

Table 10 illustrates our results on the public dataset of the LAVA workshop challenge, round by 2 decimal number. The highest result is 0.85, achieved by the assembly of Qwen2-VL-72B-Instruct-GPTQ-Int8 and Qwen2-VL-7B-Instruct, with the random shuffle of answer order, and prompt 1. From Table 10, we can have a brief summary of results based on the prompt:

– Prompt 1 and Prompt 8 consistently resulted in the highest scores, with top-performing models reaching up to 0.85.
– Prompts 2, 3, and 4 yielded lower results across models, with Prompt 3 showing the weakest performance.
– Prompt 7 and Prompt 9 also produced solid results, particularly for larger models like Qwen2-VL and InternV2.

Table 11 shows the results of the winning teams on the LAVA Workshop challenge. Our method scores the first prize on the challenge, with 0.85 on both public and private datasets.

Table 10. Results on public dataset of LAVA Workshop challenge

Model	Prompt	Answer order shuffle	Result
LLAVA-v1.5-7B	Prompt 2	✗	0.39
LLAVA-v1.6-34B	Prompt 2	✗	0.42
LLAVA-v1.6-34B	Prompt 3	✗	0.42
LLAVA-1.6-13B	Prompt 3	✗	0.44
LLAVA-v1.6-34B	Prompt 4	✗	0.44
Use the vote between 3 LLAVA models	Prompt 4	✗	0.48
LLAVA-v1.6-34B	Prompt 5	✗	0.58
InternVL2-Llama3-76B	Prompt 6	✗	0.58
Qwen2-VL-2B-Instruct	Prompt 7	✗	0.59
Qwen-VL-7B-Instruct-GPTQ-Int4	Prompt 7	✗	0.64
InternVL2-8B	Prompt 6	✗	0.65
Qwen-VL-7B-Instruct-GPTQ-Int8	Prompt 7	✗	0.67
InternVL2-26B	Prompt 6	✗	0.67
InternVL2-Llama3-76B	Prompt 7	✗	0.68
Qwen2-VL-7B-Instruct-GPTQ-Int4	Prompt 1	✗	0.74
InternVL2-40B	Prompt 1	✗	0.76
Qwen2-VL-7B-Instruct-GPTQ-Int4	Prompt 2	✗	0.77
Qwen2-VL-7B-Instruct-GPTQ-Int8	Prompt 2	✗	0.78
Qwen2-VL-7B-Instruct	Prompt 8	✗	0.78
Qwen2-VL-7B-Instruct-GPTQ-Int8	Prompt 1	✗	0.79
Qwen2-VL-7B-Instruct (True/False approach)	Prompt 9	✗	0.79
Qwen2-VL-7B-Instruct	Prompt 1	✓	0.79
Qwen2-VL-72B-Instruct-GPTQ-Int8	Prompt 1	✗	0.84
Qwen2-VL-72B-Instruct-GPTQ-Int8	Prompt 8	✗	0.84
Qwen2-VL-72B-Instruct-GPTQ-Int8	Prompt 1	✓	0.84
Qwen2-VL-72B-Instruct-GPTQ-Int8 *assembles with* **Qwen2-VL-7B-Instruct**	Prompt 1	✓	**0.85**

Table 11. Results on winning teams

	Public Score	Private Score	Total Score
WAS (Ours)	**0.85**	**0.85**	**0.85**
MMLAB-UIT	0.83	0.84	0.84
V1olet	0.82	0.82	0.82

5   Conclusion

Developing a Massive Multi-discipline Multimodal Understanding and Reasoning Benchmark for Expert AGI has become crucial in today's fast-paced and information-rich environment. This work is vital in supporting humans by enabling faster and more efficient access to knowledge across various disciplines. Understanding and reasoning through multimodal data is essential for handling complex, real-world problems that span multiple fields of expertise.

In conclusion, this research contributes to advancing the field by presenting a simple yet effective approach to solving the LAVA challenge. Using prompt engineering and an ensemble method based on voting, we demonstrate how multiple models can collaborate to improve performance on multimodal tasks. By leveraging the strengths of each model in the ensemble, the proposed method not only enhances accuracy but also showcases a scalable solution that can be applied to other similar tasks. This approach could be a foundation for future work in expert-level AGI systems, aiming to address increasingly complex multimodal problems with greater efficiency and precision.

Appendices

Table 12 shows the accuracy of the three models used in this research on the MMMU benchmark. Based on the results in this table, we have decided to proceed with Qwen2, which achieves the highest scores and has a public inference API, enabling readers to reproduce our results in most of our experiments easily.

Table 12. Accuracy of the Qwen2-VL, InternVL2, and LLaVA in MMMU benchmark

Model	MMMU-Pro	MMMU (Val)
Qwen2-VL-72B	46.2%	64.5%
InternVL2-Pro		62.0%
InternVL2-Llama3-76B	40.0%	58.3%
LLaVA-OneVision-72B	31.0%	56.8%
InternVL2-40B	34.2%	55.2%
InternVL2-8B	29.0%	51.2%

References

1. Agrawal, A., et al.: VQA: visual question answering (2016). https://arxiv.org/abs/1505.00468
2. Alayrac, J.B., et al.: Flamingo: a visual language model for few-shot learning (2022). https://arxiv.org/abs/2204.14198
3. Alayrac, J.B., et al.: Flamingo: a visual language model for few-shot learning. In: NeurIPS (2022)
4. Awadalla, A., et al.: OpenFlamingo: an open-source framework for training large autoregressive vision-language models (2023). https://arxiv.org/abs/2308.01390
5. Chen, Y.-C., et al.: UNITER: UNiversal image-TExt representation learning. In: Vedaldi, A., Bischof, H., Brox, T., Frahm, J.-M. (eds.) ECCV 2020. LNCS, vol. 12375, pp. 104–120. Springer, Cham (2020). https://doi.org/10.1007/978-3-030-58577-8_7
6. Chen, Z., et al.: How far are we to GPT-4V? Closing the gap to commercial multimodal models with open-source suites. arXiv preprint arXiv:2404.16821 (2024)
7. Chen, Z., et al: InternVL: scaling up vision foundation models and aligning for generic visual-linguistic tasks. arXiv preprint arXiv:2312.14238 (2023)
8. Chiang, W.L., et al.: VICUNA: an open-source chatbot impressing GPT-4 with 90%* ChatGPT quality (2023). https://lmsys.org/blog/2023-03-30-vicuna/
9. Chung, H.W., et al.: Scaling instruction-finetuned language models. arXiv:2210.11416 (2022)
10. Dai, W., et al.: InstructBLIP: towards general-purpose vision-language models with instruction tuning (2023). https://arxiv.org/abs/2305.06500
11. Dai, W., Li, X., Zhang, C., Hu, X., Li, J., Yin, X.: MiniGPT-4: enhancing ChatGPT with multimodal abilities. arXiv:2304.10592 (2023)
12. Dosovitskiy, A., et al.: An image is worth 16×16 words: transformers for image recognition at scale. In: ICLR (2021)
13. Gao, P., et al.: Llama-adapter V2: parameter-efficient visual instruction model (2023). https://arxiv.org/abs/2304.15010
14. Huang, Y., Meng, Z., Liu, F., Su, Y., Collier, N., Lu, Y.: Sparkles: Unlocking chats across multiple images for multimodal instruction-following models (2024). https://arxiv.org/abs/2308.16463
15. Hudson, D.A., Manning, C.D.: GQA: a new dataset for real-world visual reasoning and compositional question answering (2019). https://arxiv.org/abs/1902.09506
16. Jia, C., et al.: Align: scaling up visual and vision-language representation learning with noisy text supervision. In: ICML. PMLR (2021)
17. Li, B., Zhang, Y., Chen, L., Wang, J., Yang, J., Liu, Z.: Otter: A multi-modal model with in-context instruction tuning (2023). https://arxiv.org/abs/2305.03726
18. Li, B., Wang, R., Wang, G., Ge, Y., Ge, Y., Shan, Y.: Seed-bench: Benchmarking multimodal LLMS with generative comprehension (2023). https://arxiv.org/abs/2307.16125
19. Li, J., Hu, D., Zhao, H., Zhang, L., Li, X., Gao, J.: BLIP-2: bootstrapped language-image pre-training with frozen image encoders and large language models. arXiv:2301.12597 (2023)
20. Li, L., et al.: M^3it: a large-scale dataset towards multi-modal multilingual instruction tuning. arXiv preprint arXiv:2306.04387 (2023)
21. Li, X., et al.: OSCAR: object-semantics aligned pre-training for vision-language tasks. In: Vedaldi, A., Bischof, H., Brox, T., Frahm, J.-M. (eds.) ECCV 2020. LNCS, vol. 12375, pp. 121–137. Springer, Cham (2020). https://doi.org/10.1007/978-3-030-58577-8_8

22. Li, Y., Du, Y., Zhou, K., Wang, J., Zhao, W.X., Wen, J.R.: Evaluating object hallucination in large vision-language models (2023). https://arxiv.org/abs/2305.10355

23. Lin, T.Y., et al.: Microsoft coco: Common objects in context (2015). https://arxiv.org/abs/1405.0312

24. Liu, F., Lin, K., Li, L., Wang, J., Yacoob, Y., Wang, L.: Aligning large multi-modal model with robust instruction tuning. arXiv preprint arXiv:2306.14565 (2023)

25. Liu, H., Li, C., Wu, Q., Lee, Y.J.: Visual instruction tuning. In: NeurIPS (2023)

26. Liu, Y., Gehrig, M., Messikommer, N., Cannici, M., Scaramuzza, D.: Revisiting token pruning for object detection and instance segmentation. In: Proceedings of the IEEE/CVF Winter Conference on Applications of Computer Vision (WACV) (2024)

27. Liu, Y., et al.: MMBench: is your multi-modal model an all-around player? (2024). https://arxiv.org/abs/2307.06281

28. Lu, J., Batra, D., Parikh, D., Lee, S.: ViLBERT: pretraining task-agnostic visiolinguistic representations for vision-and-language tasks. In: NeurIPS (2019)

29. Lu, P., et al.: MathVista: evaluating mathematical reasoning of foundation models in visual contexts (2024). https://arxiv.org/abs/2310.02255

30. Marino, K., Rastegari, M., Farhadi, A., Mottaghi, R.: OK-VQA: a visual question answering benchmark requiring external knowledge (2019). https://arxiv.org/abs/1906.00067

31. Mialon, G., Fourrier, C., Swift, C., Wolf, T., LeCun, Y., Scialom, T.: Gaia: a benchmark for general AI assistants (2023). https://arxiv.org/abs/2311.12983

32. Monajatipoor, M., Li, L.H., Rouhsedaghat, M., Yang, L.F., Chang, K.W.: MetaVL: transferring in-context learning ability from language models to vision-language models (2023). https://arxiv.org/abs/2306.01311

33. Radford, A., et al.: Learning transferable visual models from natural language supervision. arXiv:2103.00020 (2021)

34. Tan, H., Bansal, M.: LXMERT: learning cross-modality encoder representations from transformers. In: EMNLP-IJCNLP. Association for Computational Linguistics (2019)

35. Wang, J., et al.: Evaluation and analysis of hallucination in large vision-language models. arXiv preprint arXiv:2308.15126 (2023)

36. Wang, P., et al.: Qwen2-VL: enhancing vision-language model's perception of the world at any resolution. arXiv preprint arXiv:2409.12191 (2024)

37. Wang, Y., et al.: SIMVLM: simple visual language model pretraining with weak supervision. arXiv:2108.10904 (2021)

38. Xu, P., et al.: LVLM-EHUB: a comprehensive evaluation benchmark for large vision-language models (2023). https://arxiv.org/abs/2306.09265

39. Ye, Q., et al.: MPLUG-OWL: modularization empowers large language models with multimodality (2024). https://arxiv.org/abs/2304.14178

40. Yin, Z.,et al.: LAMM: language-assisted multi-modal instruction-tuning dataset, framework, and benchmark. In: Advances in Neural Information Processing Systems, vol. 36 (2024)

41. Yu, J., et al.: COCA: contrastive captioners are image-text foundation models. arXiv:2205.01917 (2022)

42. Yu, W., et al.: MM-VET: evaluating large multimodal models for integrated capabilities (2023). https://arxiv.org/abs/2308.02490

43. Yue, X., et al.: MMMU: a massive multi-discipline multimodal understanding and reasoning benchmark for expert AGI. In: Proceedings of CVPR (2024)

44. Zhang, M., et al.: FUYU: fully unified vision-language models for multimodal tasks. arXiv:2305.05999 (2023)
45. Zhang, P., et al.: VINVL: revisiting visual representations in vision-language models. In: CVPR. IEEE (2021)
46. Zhao, H., et al.: MMICL: empowering vision-language model with multi-modal in-context learning (2024). https://arxiv.org/abs/2309.07915
47. Zhu, Y., et al.: LLAVA: large language and vision assistant. arXiv:2304.08485 (2023)

Author Index

The manufacturer's authorised representative in the EU is Springer
Nature Customer Service Centre GmbH, Europaplatz 3, 69115 Heidelberg,
Germany. If you have any concerns regarding our products, please
contact ProductSafety@springernature.com

Printed and bound by CPI Group (UK) Ltd, Croydon, CR0 4YY
27/04/2026
02097586-0012